IFIP Advances in Information and Communication Technology 701

IFIP Advances in Information and Communication Technology

The IFIP AICT series publishes state-of-the-art results in the sciences and technologies of information and communication. The scope of the series includes: foundations of computer science; software theory and practice; education; computer applications in technology; communication systems; systems modeling and optimization; information systems; ICT and society; computer systems technology; security and protection in information processing systems; artificial intelligence; and human-computer interaction.

Edited volumes and proceedings of refereed international conferences in computer science and interdisciplinary fields are featured. These results often precede journal publication and represent the most current research.

The principal aim of the IFIP AICT series is to encourage education and the dissemination and exchange of information about all aspects of computing.

More information about this series at https://link.springer.com/bookseries/6102

Christophe Danjou · Ramy Harik ·
Felix Nyffenegger · Louis Rivest ·
Abdelaziz Bouras
Editors

Product Lifecycle Management

Leveraging Digital Twins, Circular Economy, and Knowledge Management for Sustainable Innovation

20th IFIP WG 5.1 International Conference, PLM 2023
Montreal, QC, Canada, July 9–12, 2023
Revised Selected Papers, Part I

Editors
Christophe Danjou
Polytechnic Montreal
Montréal, QC, Canada

Ramy Harik
University of South Carolina
Columbia, SC, USA

Felix Nyffenegger
Ostschweizer Fachhochschule
Rapperswil, Switzerland

Louis Rivest
École de Technologie Supérieure
Montréal, QC, Canada

Abdelaziz Bouras
Qatar University
Doha, Qatar

ISSN 1868-4238 ISSN 1868-422X (electronic)
IFIP Advances in Information and Communication Technology
ISBN 978-3-031-62577-0 ISBN 978-3-031-62578-7 (eBook)
https://doi.org/10.1007/978-3-031-62578-7

This Springer imprint is published by the registered company Springer Nature Switzerland AG
The registered company address is: Gewerbestrasse 11, 6330 Cham, Switzerland

Preface

Since 2003 the International Conference on Product Lifecycle Management (PLM) has brought together researchers, developers and users of Product Lifecycle Management. It aims to integrate business approaches to the collaborative creation, management and dissemination of product and process data throughout the extended enterprises that create, manufacture and operate engineered products and systems. The conference aims to involve all stakeholders of the wide concept of PLM, hoping to shape the future of this field and advance the science and practice of enterprise development.

PLM 2023 was hosted by Polytechnique Montréal, Canada, from July 9th to July 12th, 2023. The conference started on Monday, July 10th, the industrial day was held on Tuesday, July 11th and the doctoral day was a success on Sunday, July 9th. The leading organizer was Prof. Christophe Danjou.

For its 20th anniversary, PLM 2023 registered 132 participants from 16 countries who had the chance to attend two Special Sessions (Digital engineering, and Artificial intelligence in support of digital transformation), an overall set of 20 scientific sessions plus one industrial day and an amazing social program to celebrate this anniversary. In addition, PLM 2023 proposed Academic Keynotes, a PLM Pioneer presentation, Industrial Keynotes and round tables.

4 academic keynotes:

- Catherine Beaudry, Polytechnique Montréal
- Felix Nyffenegger, Eastern Switzerland University of Applied Sciences
- Louis Rivest, École de Technologie Supérieure
- Hany Moustapha, École de Technologie Supérieure

PLM Pioneer speaker:

- Francis Bernard, 1st PLM pioneer award recipient

Industrial Keynotes:

- Mohamed-Ali El Hani (Impararia) & Frédéric Gal (Bouygues Construction) – Smart modular home construction
- Wissem Maazoun (BusPas) – The Bus Stop, Reimagined
- Fethi Chebli (VPorts) – Shaping the future of advanced air mobility
- Nick Giannias (CAE) – DOME: The Digital Operations and Maintenance Environment for the offshore renewable energy lifecycle
- Jonathan Brodeur (MILA) – Adopting applied AI in manufacturing SMEs: A project lifecycle approach

Round table on "The future of PLM with AI and Digital twins":

- Moderator: Benoît Eynard, Université de Technologie de Compiègne
- Panelists: Éric Boutin (Siemens), Kevin Chagnon Beaulieu (Techso), Arnaud Guitton (Impararia), François Lamy (PTC), Florent Salako (Dassault Systems)

All submitted papers were double-blind peer-reviewed by at least two reviewers. In total, 116 contributions were submitted, of which 62 were accepted. Program chairs Christophe Danjou and Ramy Harik proposed an amazing scientific conference with presentations grouped in 20 thematic sections.

We would like to thank everyone who directly or indirectly contributed to making the PLM 2023 conference a success, particularly by participating in rich scientific discussions and meeting all of us in Montreal, Canada.

Christophe Danjou
Ramy Harik
Felix Nyffenegger
Louis Rivest
Abdelaziz Bouras

Organization

Conference Chair

Hany Moustapha	École de Technologie Supérieure, Canada

Program Committee Chairs

Christophe Danjou	Polytechnique Montréal, Canada
Ramy Harik	University of South Carolina, USA

Local Organizing Committee

Enzo Domingos	Polytechnique Montréal, Canada
Claudia François Eudoxie	Polytechnique Montréal, Canada
Lucas Jacquet	Polytechnique Montréal, Canada
Syrine Njah	Polytechnique Montréal, Canada
Loïc Parrenin	Polytechnique Montréal, Canada
Suzanne Pirié	Polytechnique Montréal, Canada
Julien Pons	Polytechnique Montréal, Canada
Zeina Tamaz	Polytechnique Montréal, Canada

Steering Committee

Chairs

Felix Nyffenegger	Eastern Switzerland University of Applied Sciences (OST), Switzerland
Louis Rivest	École de Technologie Supérieure, Canada

Members

Abdelaziz Bouras	Qatar University, Qatar
Anderson Luis Szejka	Pontifical Catholic University of Paraná, Brazil
Balan Gurumoorthy	Indian Institute of Science, India
Benoit Eynard	University of Technology of Compiègne, France
Christophe Danjou	Polytechnique Montréal, Canada
Clement Fortin	Skolkovo Institute of Science and Technology, Russia
Darli Rodrigues Vieira	Université du Québec à Trois-Rivières, Canada
Debashish Dutta	University of Michigan, USA
Frédéric Noël	Grenoble Institute of Technology, France
José Rios	Polytechnic University of Madrid, Spain
Klaus-Dieter Thoben	University of Bremen, Germany
Néjib Moalla	University of Lyon, France

Osiris Canciglieri Junior	Pontifical Catholic University of Paraná, Brazil
Paolo Chiabert	Polytechnic of Turin, Italy
Ramy Harik	University of South Carolina, USA
Romeo Bandinelli	University of Florence, Italy
Sebti Foufou	University of Sharjah, UAE
Sergio Terzi	Politecnico di Milano, Italy

Doctoral Workshop Committee

Monica Rossi	Politecnico di Milano, Italy
Yacine Ouzrout	University of Lyon, France

Scientific Committee

Abdelhak Belhi	Qatar University, Qatar
Aicha Sekhari	University of Lyon, France
Alain Bernard	Central School of Nantes, France
Alexander Smirnov	Russian Academy of Sciences, Russia
Alison McKay	University of Leeds, UK
Amaratou Saley	University of Lyon, France
Ambre Dupuis	Polytechnique Montréal, Canada
Améziane Aoussat	Arts et Métiers ParisTech, France
Anderson Luis Szejka	Pontifical Catholic University of Paraná, Brazil
Anneli Silventoinen	LUT University, Finland
Balan Gurumoorthy	Indian Institute of Science, India
Benoit Eynard	University of Technology of Compiègne, France
Brendan Sullivan	Polytechnic of Milan, Italy
Bruno Agard	Polytechnique Montréal, Canada
Camelia Dadouchi	Polytechnique Montréal, Canada
Chen Zheng	Northwestern Polytechnical University, China
Chris Mc Mahon	Technical University of Denmark, Denmark
Christophe Danjou	Polytechnique Montréal, Canada
Claudio Sassanelli	Politecnico di Bari, Italy
Clement Fortin	Skolkovo Institute of Science and Technology, Russia
Daniel Schmid	Zurich University of Applied Sciences, Switzerland
Darli Rodrigues Vieira	Université du Québec à Trois-Rivières, Canada
Debashish Dutta	University of Michigan, USA
Detlef Gerhard	Ruhr University Bochum, Germany
Dimitris Kiritsis	Federal Institute of Technology in Lausanne, Switzerland
Eduardo Zancul	University of São Paulo, Brazil
Felix Nyffenegger	Eastern Switzerland University of Applied Sciences, Switzerland
Fernando Mas	University of Seville, Spain
Frédéric Demoly	University of Technology of Belfort-Montbéliard, France
Frédéric Noël	Grenoble Institute of Technology, France

Frédéric Segonds	Arts et Métiers ParisTech, France
George Huang	University of Hong Kong, China
Gianluca D'Antonio	Polytechnic of Turin, Italy
Guilherme Brittes Benitez	Pontifical Catholic University of Paraná, Brazil
Hamidreza Pourzarei	University of Quebec, Canada
Hannele Lampela	University of Oulu, Finland
Hannu Kärkkäinen	Tampere University of Technology, Finland
Hervé Panetto	University of Lorraine, France
Ilkka Donoghue	LUT University, Finland
Johan Malmqvist	Chalmers University of Technology, Sweden
Jong Gyun Lim	Samsung Advanced Institute of Technology, South Korea
José Rios	Polytechnic University of Madrid, Spain
Julie Jupp	University of Technology Sydney, Australia
Julien Le Duigou	University of Technology of Compiègne, France
Klaus-Dieter Thoben	University of Bremen, Germany
Lionel Roucoules	Arts et Métiers ParisTech, France
Louis Rivest	University of Quebec, Canada
Marcelo Rudek	Pontifical Catholic University of Paraná, Brazil
Margherita Peruzzini	University of Modena and Reggio Emilia, Italy
Mariangela Lazoi	University of Salento, Italy
Matthieu Bricogne	University of Technology of Compiègne, France
Michael Schabacker	University of Magdeburg, Germany
Michele Marcos de Oliveira	Pontifical Catholic University of Paraná, Brazil
Mickael Gardoni	University of Quebec, Canada
Monica Rossi	Polytechnic of Milan, Italy
Néjib Moalla	University of Lyon, France
Nickolas S. Sapidis	University of Western Macedonia, Greece
Nicolas Maranzana	Arts et Métiers ParisTech, France
Nikolaos Bilalis Technical	University of Crete, Greece
Nikolay Shilov	Innopolis University, Russia
Nopasit Chakpitak	Chiang Mai University, Thailand
Osiris Canciglieri Junior	Pontifical Catholic University of Paraná, Brazil
Paolo Chiabert	Polytechnic of Turin, Italy
Paul Hong	University of Toledo, USA
Peter Hehenberger	University of Applied Sciences Upper Austria, Austria
Ramy Harik	University of South Carolina, USA
Rebeca Arista	Airbus, France
Ricardo Jardim-Goncalves	New University of Lisbon, Portugal
Rob Vingerhoeds	ISAE-SUPAERO, France
Romain Pinquié	Grenoble Institute of Technology, France
Romeo Bandinelli	University of Florence, Italy
Samira Keivanpour	Polytechnique Montréal, Canada
Sebti Foufou	University of Sharjah, UAE
Sehyun Myung	Youngsan University, South Korea
Sergio Terzi	Politecnico di Milano, Italy

Contents – Part I

Organisation: Knowledge Management, Change Management, Frameworks for Project and Service Development

Modelisation: CAD and Collaboration, Model-Based System Engineering and Building Information Modeling

Contents – Part II

Learning and Training: From AI to a Human-Centric Approach

Smart Processes: Prediction, Optimization and Digital Thread

Technology Implementation: Augmented Reality, CPS and Digital Twin

Benefits of Digital Twin Applications Used to Study Product Design and Development Processes

Milad Attari Shendi[1(✉)], Vincent Thomson[1], Haoqi Wang[2], and Gaopeng Lou[2]

[1] McGill University, 845 Sherbrooke St W, Montreal, QC H3A 0G4, Canada
milad.attarishendi@mail.mcgill.ca
[2] Zhengzhou University of Light Industry, Zhongyuan District 450001, Henan, China

Abstract. Fast-paced technological advancements and ever-growing demand for customized products trigger the need for a transition from traditional manufacturing to intelligent manufacturing. This review paper focuses on identifying the benefits of Digital Twin applications for the theme of product development processes. Benefits are analyzed according to the 10 knowledge areas defined by the Project Management Institute's project management body of knowledge (PMBOK). Classification by knowledge area helps to understand where benefits can best be found according to desirable product characteristics and corporate goals. This paper analyzes which PMBOK knowledge area has proven, practical applications and sound theoretical underpinning, thus indicating the likelihood of a successful implementation of Digital Twin.

Keywords: Product Development Process · Digital Twin · Digital Twin Benefits

1 Introduction

Transition to intelligent manufacturing has become a priority for various enterprises due to the advancement of technologies such as Digital Twin (DT), artificial intelligence, cloud computing, and the internet of things [1]. The advent and development of DT technology as a tool to facilitate the shift to intelligent manufacturing has gained remarkable attention in recent years [2]. Moreover, there is irrefutable growth in the application of DT in the manufacturing industry [3]. However, the use of DT brings many challenges such as IT infrastructure requirements, quality data requirements, privacy and security, trust, lack of a standardized modeling approach, and more importantly, financial feasibility studies of DT applications [4]. Therefore, the current paper focuses on the benefits of DT applications for decision-making.

The rest of this paper is organized as follows. The background on DT applications in product development processes (PDPs) is given in Sect. 2. Section 3 has the method for article selection. Section 4 discusses findings and the relationship between DT application benefits and the Project Management Institute's project management body of knowledge (PMBOK). Finally, Sect. 5 presents conclusions. The main contribution of

Published by Springer Nature Switzerland AG 2024
C. Danjou et al. (Eds.): PLM 2023, IFIP AICT 701, pp. 3–13, 2024.
https://doi.org/10.1007/978-3-031-62578-7_1

this paper is to identify the benefits of DT applications in the manufacturing and mapping the benefits to different knowledge areas to assist managers in dealing with challenges in shifting to smart manufacturing.

2 Background

In Sub-Sect. 2.1, a generic summary of the PDP is presented. As the major objective is to analyze the benefits of DT applications in PDPs, a quick introduction to the concept of a DT is provided in Sect. 2.2. Following is the literature review of DT applications in PDPs in Sect. 2.3, and the introduction of PMBOK knowledge domains as the classification method of the literature review is offered in Sect. 2.4.

2.1 Product Development Processes

PDPs are a sequence of activities to create, design, and market a product. The six phases are planning, concept development, system-level design, detailed design, testing and refinement, and production ramp-up. We analyze the use of a DT model where the benefits can be from the PDP lifecycle or a single phase.

2.2 DT Concept

A physical product, a virtual product, and their connections make up the three primary components of the DT concept [5]. It was introduced by Grieves as a "Conceptual Ideal for Product Lifecycle Management (PLM)" demonstrating the connection between real and virtual spaces [6]. Liu et al. described DT as a digital entity that replicates a physical entity's behavior rules and is updated during a real entity's lifecycle [3]. The application of DTs, especially in manufacturing, has grown significantly. DT has various applications such as assisting decision-making as well as simulating, monitoring, and elevating design performance [6].

2.3 Review of DT Applications Based on Different Classification Methods

DT applications in product design and development processes have experienced significant growth in recent years [6]. Several authors have written review articles. The current paper presents an analysis according to the classification methods.

Lo et al. presented a general classification and stage-based categorization by research area including manufacturing, automotive, computing, energy, aerospace, etc. [6]. Stages of the product lifecycle (product design, manufacturing, logistics, use, and end of life) as well as stages of new product development (idea generation, market analysis, product design, testing, and commercialization) were introduced as two main approaches to investigate DT applications. Liu et al. classified DTs by product lifecycle stages, the content of literature including concepts, key technologies, paradigms and frameworks, and applications [3]. Kritzinger et al. carried out a categorical literature review by research type (definition, review, concept, and case study) [7]. Moreover, focus area (layout planning, product lifecycle, process design, manufacturing, and maintenance) as well as key technologies were also applied as categorization criteria. Phanden et al. analyzed simulation-based DT and DT-based models in aerospace, manufacturing, and robotics fields [8]. Fuller et al. presented smart cities, manufacturing, and healthcare applications [4]. They classified papers based on enabling technologies, such as the internet of things, augmented reality, artificial intelligence, etc. Zhang et al. reviewed integrated methods associated with DTs such as TRIZ, lean and green design to create a DT-driven, smart product design framework [9]. Hoiler et al. presented classifications based on industry (aerospace, automotive, manufacturing, …), literature domain (product development, manufacturing, lifecycle management, modeling and simulation,…), and geographic distribution [10].

As shown in Table 1, the majority of review papers classified articles based on application domain or industry. Furthermore, categorizing papers based on PLM stages as well as enabling technologies were the most common criteria. However, an important factor has been neglected: identifying the benefits obtained from DT application in manufacturing and mapping the benefits to different knowledge areas that exist in each PDP. The current paper does this.

The PMBOK knowledge areas cover all stages of a PDP, all required tools, techniques, and outcomes; therefore, it is a more comprehensive manner to classify benefits. Moreover, classification based on knowledge areas provides a clearer way for decision-makers to find the proper solution for challenges they are facing when shifting to smart manufacturing.

Table 1. Literature review paper classification criteria

Authors	Year	Classification criteria								
		Application domain	Product life cycle stages	NPD stages	Enabling technologies	Frameworks	Article content type	Simulation modeling	Geographic distribution	Integrated methods
Lo et al. [6]	2021	●	●	●						
Liu et al. [3]	2021	●			●	●				
Kritzinger et al. [7]	2018	●	●		●		●			
Phanden et al. [8]	2021	●	●					●		
Fuller et al. [4]	2020	●			●					
Zhang et al. [9]	2020									●
Hoiler et al. [10]	2016	●	●					●	●	

2.4 PMBOK Project Management Knowledge Areas

The Project Management Institute (PMI), the premier management organization in the world, uses 10 major knowledge areas of project management in their most influential guideline, PMBOK Guide. Each PMBOK knowledge area involves several project management processes and covers different parts of each process: requirements, tools, techniques, and outcomes. The 10 PMBOK project management knowledge areas are: integration, project scope, time, cost, quality, human resources, communications, risk, procurement, and stakeholder, which the authors used to map DT benefits [11].

3 Methodology in Research Selection

A systematic search was conducted to identify articles. The search was carried out in the Scopus, Google Scholar, and Science Direct databases. A comprehensive search was also conducted by our team in China's national knowledge internet and Engineering Village databases to obtain papers in the Chinese language. Collected articles were screened and refined through a three-step approach.

First, articles were selected via keywords and titles from 2017 to 2022. The search was focused on quality publications in journals, conference papers, and book chapters. Since our paper is centered on DT applications in manufacturing, different combinations of keywords that relate DT to PDPs were examined. In addition to DT, terms such as "virtual twin" and "cyber twin" were used. Selected papers were re-examined to ascertain that they included DT applications in PDPs. For the second step, a secondary source for articles used cited references from the articles selected in step 1. In step 3, from the papers collected in steps 1 and 2, articles were selected that specifically focused on DT applications and gave benefits for PDPs. Finally, 26 papers in English and 19 papers in Chinese were included in the classification.

4 Research Findings

4.1 Mapping DT Application Benefits to PMBOK Knowledge Areas

Given the methodology described in Sect. 3 articles are reviewed for the benefits of DT applications in product design and development. Benefits are mapped to knowledge areas based on the PMBOK definitions as shown in Table 2.

4.2 In-Depth Analysis

A framework based on DT for smart development processes was created using 45 articles of which 20 articles built an actual DT model and 9 presented simulation results of a real-world DT implementation and compared results to statistically support benefits. Furthermore, articles that built an actual DT model provided more benefits in multiple knowledge areas. Benefits are categorized qualitatively, e.g., increased product understanding, and quantitatively, e.g., machining reduced by 26%. Figure 1 gives the number of articles that cover each knowledge area using data from Table 2. Figure 2 gives the number of benefits for each knowledge area based on Table 2, where an article can have more than one benefit and benefits can be in more than one knowledge area.

From our point of view, time, quality, and risk are not only the most used knowledge areas, but also the most important. Time has the most articles (19) with proven benefits (Table 2) and all such articles are quantitative. Conversely, procurement, communication,

Table 2. DT application benefits as related to knowledge areas.

PMBOK Knowledge area	DT application benefit	Articles
Integration	Multidisciplinary collaborative design	[12–14]
	Analyzing the composition and integration of mechanical, electrical, and other subsystems	[15]
	Description of a DT's technical architecture	[16, 17]
	Increasing product understanding	[18]
	Mechanisms and integration of assembly processes	[19]
	Physical and virtual cutting tool lab test platform	[20]
	DTs with integrated production knowledge	[21]
	Engineering product family design and optimization	[22]
Scope	DT applications in all stages of PLM	[17, 20, 23, 24]
	Reducing the number of manufacturing operations	[25]
	DT applications for metalized film capacitors	[16]
	DT driven data flow for the cutting tool lifecycle	[20]
	DT applications for product design	[9]
Time (Schedule)	Minimizing debugging time	[12]
	Shortening the development lifecycle	[14, 18, 26–30] 28–30]
	Reducing blade machining time by 26%	[31]
	Reducing design and operation time	[15]
	Optimizing design schedule	[16, 24, 32–34]
	Shortening assembly time to 46%	[19]
	Improving combat command efficiency	[35]
	Shortening personnel training time	[36, 37]
Cost	Minimizing total cost	[12, 16, 18, 38]
	Reducing physical validation cost	[27]
	Reducing maintenance cost	[33]
	Developing a cost estimation model	[39]
	Cost effectiveness validation	[40]
	Cost effective solutions using DT technology	[38]

(*continued*)

Table 2. (*continued*)

PMBOK Knowledge area	DT application benefit	Articles
	Logistics precise distribution	[24]
Quality	Improving product quality	[26, 38, 41]
	Increasing machining precision by 23%	[31]
	Improving inspection	[25]
	Decreasing the defective rate of products	[25]
	Monitoring equipment online	[24]
	Intelligent maintenance	[29]
	Optimizing the quality of an entire system	[42]
	Improving model accuracy	[35]
	Tracking vibration and surface quality of a fixture	[12]
	Solving quality specific problems	[43]
	Improving DT accuracy	[44]
	Reference model for factory construction quality	[30]
Human Resource	Involving process personnel in developing a real DT	[31]
	Describing human resources role in DT development	[31]
	Reducing the number of workers	[25]
	Effect of human resource knowledge	[28]
	Effect of personnel position	[36]
	Effect of macro-supervision and precise supervision	[45]
Communications	Communication connection	[19]
	Intelligent decision-making through real-time output	[35]
	Communications for interaction between physical and virtual worlds	[46]
	Communication operating modes	[47]
Risk	Finding types of component failures	[48]
	Identifying high failure rate subsystems	[49]
	Reducing the risks for new conceptual products	[50]

(*continued*)

Table 2. (*continued*)

PMBOK Knowledge area	DT application benefit	Articles
	Condition monitoring and providing early warnings	[16]
	Addressing future challenges of DT applications	[21]
	Improving reliability	[33, 51]
	Decreasing maintenance frequency	[33]
	Predictive maintenance	[40]
	Improving maintenance and repair processes	[38]
	Future predictions	[38]
	Improving the success rate of innovative design	[34]
	Predicting stability	[43]
	Identifying problems in the design to make process	[37]
Stakeholder	Logistics precise distribution	[24]
	Increasing customer satisfaction	[32]

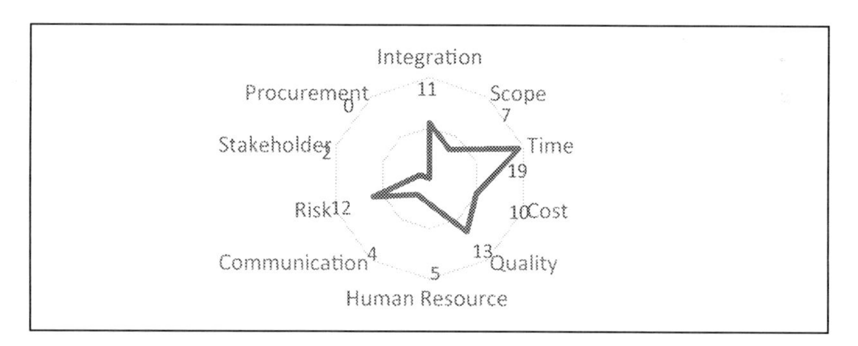

Fig. 1. Number of articles in each PMBOK knowledge area

and stakeholders areas seem to be neglected by researchers. Nevertheless, these areas seem useful for a decision-making team to decide whether DT could assist them in their challenges.

Fig. 2. Qualitative and quantitative measures of DT application benefits

5 Conclusion and Future Directions

The main goal of the current paper is to facilitate the process of shifting to smart manufacturing by decision-makers. To facilitate this goal, first, this paper made an endeavor to find the articles with benefits for the application of DT to PDPs. Second, the paper mapped the obtained benefits to PMBOK knowledge areas. The PMBOK knowledge areas cover all aspects of a PDP; therefore, decision-makers that deal with challenges in the transition to smart manufacturing can find examples in their area of interest and understand how the benefits were achieved. This should help decision-makers to know the likelihood of being successful in shifting to smart manufacturing based on the knowledge area in which the company faces challenges. The knowledge areas with the most research were time, quality, and risk. There were few DT studies in the procurement and stakeholder knowledge areas, which highlights areas where more research could be done. Moreover, one of limitations of the study is that the frameworks, techniques, and tools used to achieve benefits are not discussed in detail.

Acknowledgement. This research was funded by the Natural Science and Engineering Research Council for the Canadian authors and by the National Natural Science Foundation of China (No. 52275277) for the Chinese authors.

References

1. Sassanelli, C., Rossi, M., Terzi, S.: Evaluating the smart maturity of manufacturing companies along the product development process to set a PLM project roadmap. Int. J. Prod. Lifecycle Manag. **12**(3), 185–209 (2020)
2. Ríos, J., Staudter, G., Weber, M., Anderl, R.: Enabling the digital twin: a review of the modelling of measurement uncertainty on data transfer standards and its relationship with data from tests. Int. J. Prod. Lifecycle Manag. **12**(3), 250–268 (2020)

3. Liu, M., Fang, S., Dong, H., Xu, C.: Review of digital twin about concepts, technologies, and industrial applications. J. Manuf. Syst. **58**, 346–361 (2021)
4. Fuller, A., Fan, Z., Day, C., Barlow, C.: Digital twin: enabling technologies, challenges and open research. IEEE Access **8**, 108952–108971 (2020)
5. Ríos, J., Staudter, G., Weber, M., Anderl, R., Bernard, A.: Uncertainty of data and the digital twin: a review. Int. J. Prod. Lifecycle Manag. **12**(4), 329–358 (2020)
6. Lo, C., Chen, C., Zhong, R.Y.: A review of digital twin in product design and development. Adv. Eng. Inform. **48**, 101297 (2021)
7. Kritzinger, W., Karner, M., Traar, G., Henjes, J., Sihn, W.: Digital Twin in manufacturing: a categorical literature review and classification. Ifac-PapersOnline **51**(11), 1016–1022 (2018)
8. Phanden, R.K., Sharma, P., Dubey, A.: A review on simulation in digital twin for aerospace, manufacturing and robotics. Mater. Today Proc. **38**, 174–178 (2021)
9. Zhang, M., Sui, F., Liu, A., Tao, F., Nee, A.: Digital twin driven smart product design framework. In: Digital Twin Driven Smart Design, pp. 3–32. Elsevier (2020)
10. Holler, M., Uebernickel, F., Brenner, W.: Digital twin concepts in manufacturing industries-a literature review and avenues for further research. In: Proceedings of the 18th International Conference on Industrial Engineering (IJIE), Seoul, Korea, pp. 10–12 (2016)
11. Rose, K.H., Indelicato, G.: Book Review: A Guide to the Project Management Body of Knowledge (PMBOK® Guide). SAGE Publications, Los Angeles (2009)
12. Wu, Y., Zhou, L., Zheng, P., Sun, Y., Zhang, K.: A digital twin-based multidisciplinary collaborative design approach for complex engineering product development. Adv. Eng. Inform. **52**, 101635 (2022)
13. Lin, T.Y., et al.: Evolutionary digital twin: a new approach for intelligent industrial product development. Adv. Eng. Inform. **47**, 101209 (2021)
14. Li, L., Li, H., Gu, F., Ding, N., Gu, X., Luo, G.: Multidisciplinary collaborative design modeling technologies for complex mechanical products based on digital twin. Comput. Integr. Manuf. Syst. **25**(6), 1307–1319 (2019)
15. Hu, T., Kong, T., Ye, Y., Tao, F., Nee, A.: Digital twin based computerized numerical control machine tool virtual prototype design. In: Digital Twin Driven Smart Design, pp. 237–263. Elsevier (2020)
16. Zhang, Y.-X., et al.: Digital twin accelerating development of metallized film capacitor: Key issues, framework design and prospects. Energy Rep. **7**, 7704–7715 (2021)
17. Tao, F., et al.: Digital twin-driven product design framework. Int. J. Prod. Res. **57**(12), 3935–3953 (2019)
18. Jones, D.E., Snider, C., Kent, L., Hicks, B.: Early stage digital twins for early stage engineering design. In: Proceedings of the Design Society: International Conference on Engineering Design, vol. 1, no. 1, pp. 2557–2566. Cambridge University Press (2019)
19. Yi, Y., Yan, Y., Liu, X., Ni, Z., Feng, J., Liu, J.: Digital twin-based smart assembly process design and application framework for complex products and its case study. J. Manuf. Syst. **58**, 94–107 (2021)
20. Xie, Y., Lian, K., Liu, Q., Zhang, C., Liu, H.: Digital twin for cutting tool: modeling, application and service strategy. J. Manuf. Syst. **58**, 305–312 (2021)
21. Wagner, R., Schleich, B., Haefner, B., Kuhnle, A., Wartzack, S., Lanza, G.: Challenges and potentials of digital twins and industry 4.0 in product design and production for high performance products. Procedia CIRP **84**, 88–93 (2019)
22. Lim, K.Y.H., Zheng, P., Chen, C.-H., Huang, L.: A digital twin-enhanced system for engineering product family design and optimization. J. Manuf. Syst. **57**, 82–93 (2020)
23. Tao, F., Cheng, J., Qi, Q., Zhang, M., Zhang, H., Sui, F.: Digital twin-driven product design, manufacturing and service with big data. Int. J. Adv. Manuf. Technol. **94**, 3563–3576 (2018)

24. Hu, C., Gao, W., Xu, C., Ben, K.: Study on the application of digital twin technology in complex electronic equipment. In: Duan, B., Umeda, K., Hwang, W. (eds.) Proceedings of the Seventh Asia International Symposium on Mechatronics. LNEE, vol. 589, pp. 123–137. Springer, Singapore (2020). https://doi.org/10.1007/978-981-32-9441-7_14
25. Ma, J., et al.: A digital twin-driven production management system for production workshop. Int. J. Adv. Manuf. Technol. **110**, 1385–1397 (2020)
26. Wang, C., Li, Y.: Digital-twin-aided product design framework for IoT platforms. IEEE Internet Things J. **9**(12), 9290–9300 (2021)
27. Huang, S., Wang, G., Lei, D., Yan, Y.: Toward digital validation for rapid product development based on digital twin: a framework. Int. J. Adv. Manuf. Technol., 1–15 (2022)
28. Cheng, Z., Tong, S., Tong, Z., Zhang, Q.: Review of digital design and digital twin of industrial boiler. J. ZheJiang Univ. (Eng. Sci.) **55**(8), 1518–1528 (2021)
29. Hao, L., Fei, T., Haoqi, W.: Integration framework and key technologies of complex product design-manufacturing based on digital twin. Comput. Integr. Manuf. Syst. **25**(6), 1320–1336 (2019)
30. Zhao, H., Zhao, N., Zhang, S.: Factory design approach based on value stream mapping and digital twin. Comput. Integr. Manuf. Syst. **25**(06), 1481–1490 (2019)
31. Zhang, X., Zhu, W.: Application framework of digital twin-driven product smart manufacturing system: a case study of aeroengine blade manufacturing. Int. J. Adv. Rob. Syst. **16**(5), 1729881419880663 (2019)
32. Mourtzis, D., Angelopoulos, J., Panopoulos, N.: Equipment design optimization based on digital twin under the framework of zero-defect manufacturing. Procedia Comput. Sci. **180**, 525–533 (2021)
33. Wang, L.: Application and development prospect of digital twin technology in aerospace. IFAC-PapersOnLine **53**(5), 732–737 (2020)
34. Li, X., Hou, X., Yang, M., Wang, L., Wang, Y.: Construction and application of CMF design service model for industrial products driven by digital twins. Comput. Integr. Manuf. Syst. **27**(02), 307–327 (2021)
35. Guang, J., Jianguo, H., Zhenwei, Z.: Design scheme of virtual twin system for UAV combat. Acta Armamentarii **43**(8), 1902 (2022)
36. Kaiyu, L., Qi, C., Xinxin, C., Yuqing, C., Liu, P., Yifang, Z.: Application of digital twin technology in the design phase of floating nuclear power plants. 核动力工程 **43**(1), 197–201 (2022)
37. Zhang, P., Feng, H., Yang, T., Zhao, B., Sun, J., Tan, R.: Innovative design process model of TRIZ and digital twin integration iterative evolution based on parameter deduction. Comput. Integr. Manuf. Syst. **25**(6), 1361–1370 (2019)
38. Mendi, A.F., Erol, T., Doğan, D.: Digital twin in the military field. IEEE Internet Comput. **26**(5), 33–40 (2021)
39. Farsi, M., Ariansyah, D., Erkoyuncu, J.A., Harrison, A.: A digital twin architecture for effective product lifecycle cost estimation. Procedia CIRP **100**, 506–511 (2021)
40. Zheng, P., Lim, K.Y.H.: Product family design and optimization: a digital twin-enhanced approach. Procedia CIRP **93**, 246–250 (2020)
41. de Oliveira Hansen, J.P., da Silva, E.R., Bilberg, A., Bro, C.: Design and development of automation equipment based on digital twins and virtual commissioning. Procedia CIRP **104**, 1167–1172 (2021)
42. Zhuang, C., Liu, J., Xiong, H., Ding, X., Liu, S., Weng, G.: Connotation, architecture and trends of product digital twin. Comput. Integr. Manuf. Syst. **23**(4), 753–768 (2017)
43. Songhua, M., Kaixin, H., Tianliang, H.: Digital twins of fixtures supporting rapid design and performance tracking. Comput. Integr. Manuf. Syst. **28**(9), 2718 (2022)
44. Yan, Y.: Digital twin based optimization design method for aerospace electric thruster. J. Astronaut. **43**(4), 518 (2022)

45. Jinjiang, W., Haotian, Y., Fengli, Z., Laibin, Z.: Design and development of intelligent oil and gas stations based on digital twin
46. Tao, F., et al.: Digital twin and its potential application exploration. Comput. Integr. Manuf. Syst. **24**(1), 1–18 (2018)
47. Li, H., et al.: Concept, system structure and operating mode of industrial digital twin system. Comput. Integr. Manuf. Syst. **27**(12), 3373–3390 (2021)
48. Cai, H., Zhu, J., Zhang, W.: Quality deviation control for aircraft using digital twin. J. Comput. Inf. Sci. Eng. **21**(3) (2021)
49. Li, S., Wang, J., Rong, J., Wei, W.: A digital twin framework for product to-be-designed analysis based on operation data. Procedia CIRP **109**, 179–184 (2022)
50. Kim, J.-W., Kim, S.-A.: Prototyping-based design process integrated with digital-twin: a fundamental study. J. KIBIM **9**(4), 51–61 (2019)
51. Xie, J., Wang, X., Yang, Z.: Design and operation mode of production system of fully mechanized coal mining face based on digital twin theory. Comput. Integr. Manuf. Syst. **25**(6), 1381–1391 (2019)

A Digital Twin Framework for Industry 4.0/5.0 Technologies

Mansur Asranov[1,2], Khurshid Aliev[1,2], Paolo Chiabert[1,2(✉)], and Jamshid Inoyatkhodjaev[2]

[1] Politecnico di Torino, Corso Duca degli Abruzzi 24, 10129 Torino, Italy
{mansur.asranov,khurshid.aliev,paolo.chiabert}@polito.it
[2] Turin Polytechnic University in Tashkent, Kichik Halqa Yuli 17, 100095 Tashkent, Uzbekistan
jamshid.inoyatkhodjaev@polito.uz

Abstract. Industry 4.0 Paradigm unleashed tremendous opportunities to boost economical and societal transitions for the business and improved living standards of society, while Industry 5.0 is extending previous technological breakthrough in steering transformation, stimulating industry and society to be human-centric, sustainable, resilient, and green. Digital Twins, in turn, play a key role as enabler in such transformation. Heterogeneity of manufacturing technologies imposes specific challenges in developing and adopting Digital Twins. The main goal of this paper is to propose, assess and justify a robust and open framework of Digital Twins for manufacturing technologies, such as collaborative and mobile robots, as well as subtractive manufacturing machines. A framework eventually can be used by SME, while also in Educational and Research Institutions.

Keywords: Industry 4.0/5.0 · Digital Twins · Industrial Internet of Things · Robotics

1 Introduction

A Digital Twin (DT) framework is an emerging concept that is becoming increasingly popular in the context of Industry 4.0 and Industry 5.0. The idea is to create a digital replica of a physical asset or system, such as a factory or a product, which can be used to simulate and optimize its performance [1]. This approach can help organizations to better understand the behavior of their assets and make more informed decisions about how to manage and optimize them.

In the context of Industry 4.0, a Digital Twin framework can be used to improve the efficiency and productivity of manufacturing processes. By creating a digital replica of a factory or production line, organizations can simulate different scenarios and identify opportunities to optimize processes and reduce waste. This can help to increase throughput, reduce costs, and improve quality [2].

In the context of Industry 5.0, a Digital Twin framework can be used to improve the design and performance of products. By creating a digital replica of a product, organizations can simulate its behavior under different conditions and identify opportunities to

Published by Springer Nature Switzerland AG 2024
C. Danjou et al. (Eds.): PLM 2023, IFIP AICT 701, pp. 14–24, 2024.
https://doi.org/10.1007/978-3-031-62578-7_2

improve its design and functionality. This can help to create more innovative products that better meet the needs of customers [3].

A Digital Twin framework typically involves the integration of various technologies, including the Internet of Things (IoT), artificial intelligence (AI), advanced robotics, and big data analytics. By leveraging these technologies, organizations can collect data from sensors and other sources, analyze it in real-time, and use it to improve the performance of their assets and systems [4].

A Digital Twin framework has some benefits that could be included for Industry 4.0/5.0 technologies [5]:

- Data Acquisition: The first step in creating a Digital Twin is to acquire the necessary data from physical systems. This includes sensor data, machine logs, and other relevant data sources. The framework should provide guidelines on how to collect and manage this data.
- Modeling: Once the data has been collected, it must be used to create a virtual model of the physical system. The framework should define best practices for creating accurate and reliable models.
- Integration: The Digital Twin should be integrated with other systems and platforms, such as Enterprise Resource Planning (ERP) systems, Manufacturing Execution Systems (MES), and other Industry 4.0/5.0 technologies. The framework should provide guidelines for how to accomplish this integration effectively.
- Analytics: The Digital Twin generates vast amounts of data, which can be analyzed to gain insights into the performance of the physical system. The framework should define best practices for analyzing this data to identify patterns, trends, and opportunities for optimization.
- Visualization: The Digital Twin can be used to create visual representations of the physical system, such as 3D models or Augmented Reality (AR) displays. The framework should provide guidelines for how to create effective visualizations that support decision-making.
- Maintenance: The Digital Twin should be maintained over time to ensure that it remains an accurate representation of the physical system. The framework should provide guidelines for how to update and maintain the Digital Twin effectively.

The objective of this article is to introduce a Digital Twin framework concept that can be utilized in manufacturing scenarios, concentrating on collaborative and mobile robots. Furthermore, the article outlines the process of Digital Twin implementation and how communication and services take place between physical, simulation, and digital assets. The results section provides information on the framework's application layer, which is intended to be beneficial to small and medium-sized businesses, as well as educational and research establishments.

2 Related Works

The concept of Digital Twins has gained significant attention in recent years, becoming a key aspect of Industry 4.0/5.0 and enabling companies to optimize their operations, reduce costs, and improve performance across a wide range of industries and applications. This literature review explores the origins and evolution of the Digital Twin concept, its key components and applications, and the current state of research and development in this field.

An overview of Digital Twins and how they can be used to improve manufacturing systems which includes a discussion of key elements of a Digital Twin framework is provided by authors [6]. Another research paper discusses the challenges and opportunities of creating Digital Twins for Industry 4.0. It provides a framework for developing Digital Twins and describes how they can be used to support various industrial processes [7]. Another study demonstrated the potential of digital twin technology by simulating a UR3 collaborative robot in the V-REP environment and utilizing the Modbus communication protocol [8]. A book by Alasdair Gilchrist provides a comprehensive introduction to Industry 4.0 and its key technologies, including the Internet of Things, cyber-physical systems, and cloud computing [9]. A systematic approach of implementing a digital twin, driven by information from various lifecycle stages is proposed by researchers. This approach uses a digital thread as linker between the different type of information to the digital twin applications [10].

Other literatures provide a comprehensive overview of Digital Twins and their application in Industry 4.0 that includes examples of Digital Twin implementations in various industries and a discussion of key applications, such as predictive maintenance, quality control, and supply chain optimization [11, 12].

A different article investigates the development of Industry 5.0 and its fundamental traits, including a focus on humans, sustainability, and resilience. It suggests a three-dimensional framework for implementing Industry 5.0, analyzes significant drivers, potential applications, and challenges. The paper concludes by emphasizing the importance of a comprehensive Industry 5.0 system and by suggesting future research directions. The primary aim of this paper is to encourage discussion and cooperation in the Industry 5.0 sector [13].

Nahavandi et al. provides an overview of Industry 5.0 and its key features, including the integration of human skills and the use of advanced technologies like AI and robotics. It also discusses the role of Digital Twins in supporting Industry 5.0 [14].

Above mentioned literatures provide concepts of the Digital Twin but does not present a solution with a case study applications. Therefore, this paper aims to provide steps to develop a Digital Twin from the Physical assets using proposed framework concept.

3 Digital Twin Framework Concept

The proposed framework supports conceptual overview of technology stack used for building a DT. Higher level abstraction of the framework incorporates functional planes and interfaces. The Fig. 1 illustrates the structure of the framework.

Fig. 1. The Digital Twin framework structure abstraction.

As can be seen from the Fig. 1, the framework consists of the following planes and interfaces:

- *Physical Plane.* Physical assets that include shop floor technologies, such as, collaborative and industrial robotic systems, additive and subtractive manufacturing systems, and mobile robotic systems reside at this plane.
- *Southbound Interfaces.* This plane represents interfaces that support communication and plays a key role in providing necessary communication services for monitoring and control data exchange between Physical and the other functional planes.
- *Data Plane.* This plane is responsible for supporting data acquisition, processing, storage, and management services for other functional planes of the framework.
- *Control plane.* The plane represents services for enforcement of control functions on physical plane assets, as well as their digital replicas in the Visualization plane.
- *Northbound interfaces.* These interfaces support communication services between the digital assets of Visualization plane and Data, Control planes of the framework.
- *Visualization plane.* Digital counterparts of the physical assets are concentrated at this plane. Moreover, this plane provides access to HMI interfaces of the physical and digital assets, therefore giving opportunity of extended control and visualization for end-users.

4 Digital Twin Implementation

In the case study, implementation of the Digital Twin relies on the framework abstraction described in above section. In the following sections, items of the technology stack are described in more detail starting from Physical plane up to Digital Twins, finally being concluded with process flow between physical assets and their digital counterparts. The implementation relies on a physical laboratory for additive and subtractive manufacturing processes integrating robots, cobots and mobots.

4.1 Physical Twin Description

The physical twins of the framework are represented by heterogenous manufacturing technologies available in the lab and can be subdivided into two categories:

- *Fixed Robotic Systems.* This subcategory represents collaborative and industrial robotic systems, which are statically fixed in a cell and implement manipulation function in a manufacturing process. In the case study, collaborative robots from Universal Robots: UR10e, UR3e, as well as ABB IRB 4400 industrial robot, are used as the Fixed Robotic Systems.
- *Mobile Robotic Systems.* Mobile robots, typically performing transportation services for supporting internal logistics of the manufacturing process, are represented in this category. The case study focuses on a mobile robot with extended manipulation capabilities. Specifically, mobile robot-manipulator is based on UR3 manipulator, fixed on the frame on top of Mobile Industrial Robot MiR100. This robot supports execution of mixed tasks, such as pick-and-place and transportation of the workpieces in different manufacturing scenarios.

4.2 Communication and Services Infrastructure

The communication and services infrastructure described in this section, allows for building a robust bridge between physical and digital world.

The infrastructure that supports the communication and provision of the necessary services for the workflows of DT is described using extended version of the previously introduced conceptual framework. Specifically, each functional plane and interface include corresponding enabling technology that provides necessary services for DT. The following part of this section describes and discusses each enabling technology in detail, focusing on their specific role in completing overall workflow.

The Fig. 2 depicts a detailed technology stack of the DT framework.

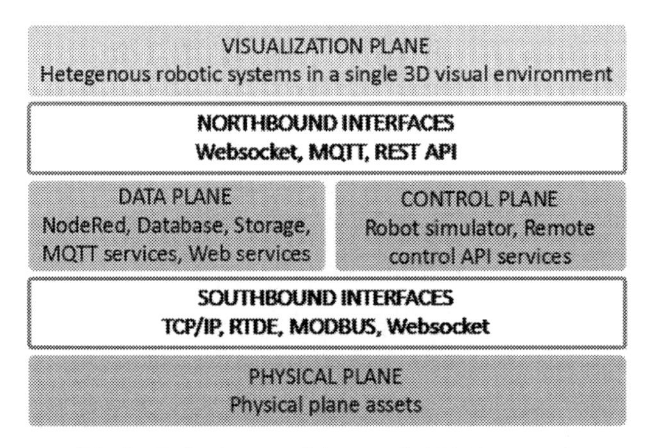

Fig. 2. Technology stack in extended DT framework

As can be seen from the Fig. 2, each plane and interface contain specific enabling technologies that support the services:

- *Visualization Plane.* Digital replicas of heterogenous Robotic Systems, previously introduced in Physical twin description section, reside at this plane. The visual behavior of digital replicas relies on the data received from Data and Control planes, through the appropriate Northbound Interfaces.
- *Northbound Interfaces.* Heterogeneity imposes specific challenges in integration of robotic systems from different vendors. Therefore, careful adoption of different interfaces and Application Programming Interfaces (API), provided by vendors, are required. For this reason, Websocket, MQTT and REST API are included in the framework as the Northbound Interfaces for the case study. As mentioned before, these interfaces provide communication services between Data, Control and Visualization planes.
- *Data Plane.* The Data Plane comprehends a set of services, focused on horizontal and vertical data flow support, which is necessary to drive overall DT framework. This plane represents NodeRed service for data acquisition and management, while Database service records time series data related to hardware health status of the

physical twins. Additionally, MQTT service facilitates data transmission procedures, while the Storage is responsible for provision of file system service, necessary for storing digital assets, configurations, robotic programs, and related data. Finally, Web services provide engine for running the unified Web Applications Environment discussed in Sect. 4.4.

- *Control Plane.* This plane provides access to the simulators and remote-control API services, therefore, allowing for extended control of the Robotic Systems in both Physical and Visualization planes. The Control plane relies on the Data plane in terms of exchange of control information, and relies on Southbound Interfaces for enforcing control functions
- *Southbound Interfaces.* A boundary between physical and digital parts of the framework are represented by these interfaces. Similarly, as Northbound interfaces, realizations of communication at this plane relies on accurate decisions about selecting appropriate interfaces. In the case study, TCP/IP, RTDE, MODBUS and Websocket interfaces are selected. These interfaces are used for acquisition of raw hardware status data in upstream direction, while in downstream direction, they support communication for the control functions enforcement, originated by the Control Plane.
- *Physical Plane* is presented in this section to complete the logical sequence of discussion, but it has been already described in detail in the Sect. 4.1.

The assets and services in Data, Control and Visualization Planes are organized as microservices (Fig. 3).

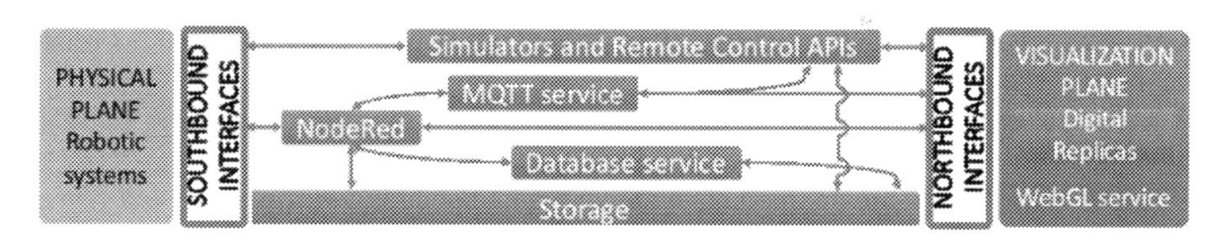

Fig. 3. Microservice architecture for supporting DT framework.

Arrows between microservices represent horizontal and vertical data exchange during interaction between the enablers from various planes of the DT framework, as discussed previously. Each microservice is enclosed in a container (green boxes) and performs the desired functions. The containers for microservices are developed using Docker software, and they include necessary runtime environment (middleware libraries and binaries) to run the services.

4.3 Digital Twin Description

The Visualization Plane, hosting the digital replicas of Robotic Systems, gives a unified access for end-users to monitor, control and manage the DT. In a web application terminology, it can be considered as front-end system, which stays near to end-users, while the other microservices reside in a back-end, supporting overall workflow of DT.

In the case study, this front-end system is accessed by the end-users using web browsers, and incorporates all the necessary digital assets and tools to support interaction between end-users and DT:

- Digital replicas – 3D models of Robotic Systems: ABB IRB 4400 industrial robot equipped with high-speed spindle, Universal Robots UR10e collaborative robot, mobile robotic manipulator system based on Universal Robots UR3 collaborative robot and Mobile Industrial Robot MiR100.
- Dashboard services – these services give the access to NodeRed dashboard user interface, which contains performance real time and historical data. Such data are based on the readings from different sensors of Robotics System and managed using NodeRed flow design capabilities.
- Control services – this capability is supported by virtual instances of Universal Robots Simulation software (URSim), running in containers, as well as control interfaces on the physical counterparts. To provide seamless end-user access to these control services in the same browser, open-source RPORT software is utilized. The RPORT software is also packed into a container and added to the set of containerized microservices.

Obviously, the 3D models are designed in CAD and 3D modelling software. The workflow that integrates digital replicas of different robotic systems, monitoring and control tools, kinematic models and allows for their visualization in a browser, is represented in Fig. 4:

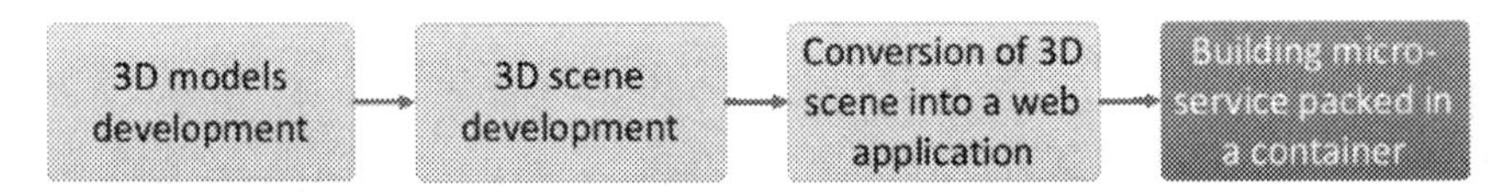

Fig. 4. Digital replicas web application microservice development workflow.

The workflow illustrated in Fig. 4 in composed of the following four phases:

- *3D models development.* 3D models are developed using CAD/3D modelling software (SolidWorks). Important decisions are made regarding solid components. The modeling of robots' joints has a relevant impact on the replication of robot movements in the digital environment with respect to the true motion the robot carries out in the physical environment. Finally, 3D models are converted into appropriate file format and passed to the next phase of 3D scene development.
- *3D scene development.* This phase requires complex development and configuration process, where 3D models and their kinematics are synchronized with their physical counterparts, along with integration of monitoring and control tools in a single 3D environment. Therefore, 3D game development software is used (Unity3D), which provides comprehensive toolbox for 3D scene design. 3D models are imported into the 3D scene, necessary kinematics behavior of robots is configured, and important monitoring and control tools are integrated using C# and Python APIs of the Unity3D software. As discussed before, Robots' behavior, monitoring and control tools rely

on the communication provided by Northbound and Southbound interfaces, as well as on appropriate data provision supported by NodeRed and MQTT services.

- *Conversion of 3D scene into a web application.* Developed and properly configured 3D scene is converted into a web application using built-in tool of Unity3D software. Unity3D software generates a web application files package, which is then passed to the next phase. Unity3D software utilizes WebGL framework for converting 3D applications into a web applications (Fig. 3).
- *Building microservice packed in a container.* Finally, the web application files package, the output from previous phase, is packed into a container, along with necessary runtime environment (software libraries and binaries), and added into the microservice architecture, as described in Fig. 3.

4.4 Process Flow Between Digital and Physical Systems

The process flow that supports interaction between the Digital and Physical systems, relies on the unified Web Application Environment. This environment aggregates different activity modes, assets and microservices in a single workspace (a web browser). The Fig. 5 illustrates that process flow, which incorporates previously described elements of the DT framework along with an end user, another actor of the process.

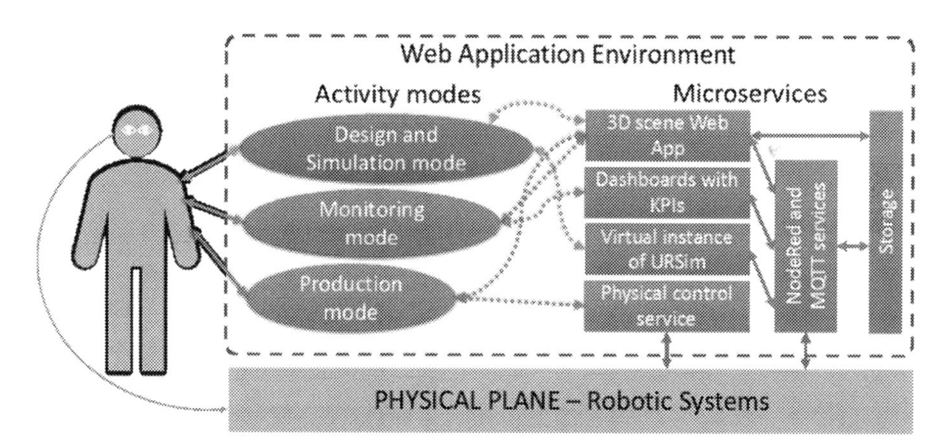

Fig. 5. Process flow of the DT framework.

In Fig. 5, the end-user interacts with the Web application environment that engages digital resources and communicates with the physical Robotics Systems.

Digital resources are grouped into two categories:

- *Activity modes* – elements of this category, represent capabilities that can be leveraged by utilizing each corresponding mode;
- *Microservices* – repository of previously introduced microservices and assets, which support capabilities, provided by the Activity modes.

Activity modes includes the following:

- *Design and simulation mode* – a web page, which allows for designing and simulating the manufacturing processes (blue solid arrow). Corresponding microservices are retrieved (blue dashed arrows) from the Microservices repository to support the capability. Manufacturing process can be designed and simulated using virtual instance of URSim software. The activities are immediately visualized in the 3D scene Web application.
- *Monitoring mode* – a web page, which supports observation of KPIs of Physical Robotics Systems and overall manufacturing process (orange solid arrow). This activity mode relies on 3D scene Web application along with Dashboards and KPIs services (orange dashed arrows).
- *Production mode* – at this web page, the end-user is provided with capability to interact with the Robotic Systems of Physical Plane (green solid arrow). The robotic programs, simulated and verified in the Design and Simulation mode, can now be tested in physical Robotic Systems, using Physical control service. This service gives direct access to the control interfaces of the Robotics systems in the same browser, the end-user utilizing, and the results of such interaction can be visually inspected in the 3D scene web application (green dashed arrows), as well as in a physical workspace (grey solid arrow).

Finally, the services, supporting the Activity modes, exchange real time monitoring and control information with NodeRed and MQTT broker services. They also store and retrieve the time series hardware health, configuration, and 3D assets data, provided by the Storage service (green solid arrows).

Additionally, Physical control, NodeRed and MQTT services can communicate with physical robotics systems to acquire real time status data, while also enforce control actions on the physical Robotics Systems (red solid arrows).

The Fig. 6 shows the implementation of the proposed DT framework, using the available assets in the lab of the department. As can be seen from the figure DT counterparts of physical robotic systems are combined in a single 3D environment. Figure 6.A depicts the Digital Twins of the robotic systems, while Fig. 6.B and Fig. 6.C show their physical counterparts. Interaction of the assets is supported by the proposed framework. The work on building manufacturing scenario using the assets and supporting DT framework is undergoing.

Fig. 6. Implementation of DT framework

5 Results, Discussion and Conclusions

Nowadays, web applications, being popular in a digital services domain, relies on existence of appropriate browser and web server to deliver a digital service. This, in turn, attracts end-users since it requires minimum efforts to consume the service. Therefore, in the case study, in order ensure the best user experience, Web application approach has been selected for supporting interaction between end-users and DT framework. However, such approach is still needing an assessment in terms of cognitive load, which may impact the overall user experience and quality of service.

Microservices approach provides flexibility, scalability, portability, and robustness of the DT framework. This approach is well adopted by IT professionals. Thank to such adoption, the proposed DT framework can be easily deployed and maintained in both industrial companies and research communities. Moreover, the containerization allows for deployment of DT framework in any container orchestration platform: starting from Docker Desktop on a standalone PCs, to advanced and highly scalable enterprise orchestration engines, such as Kubernetes, Mesos, Docker Swarm, OpenShift etc.

The configuration of containers in the current state of the DT framework is performed manually and requires relevant effort for adding new manufacturing technologies into the Physical plane. Therefore, there is a room for improvement to add automation of container configurations procedures, to provide flexibility of DT framework to different changes of manufacturing scenarios.

References

1. Tao, F., et al.: Digital twin-driven product design framework. Int. J. Prod. Res. **57**(12), 3935–3953 (2019)
2. Tao, F., Zhang, M., Nee, A.Y.C.: Digital Twin Driven Smart Manufacturing. Academic Press, London (2019)
3. Huang, S., Wang, B., Li, X., Zheng, P., Mourtzis, D., Wang, L.: Industry 5.0 and Society 5.0—Comparison, complementation and co-evolution. J. Manuf. Syst. **64**, 424–428 (2022)
4. Fuller, A., Fan, Z., Day, C., Barlow, C.: Digital twin: enabling technologies, challenges and open research. IEEE Access **8**, 108952–108971 (2020)
5. Qi, Q., et al.: Enabling technologies and tools for digital twin. J. Manuf. Syst. **58**, 3–21 (2021)
6. Semeraro, C., Lezoche, M., Panetto, H., Dassisti, M.: Digital twin paradigm: a systematic literature review. Comput. Ind. **130**, 103469 (2021)
7. Stavropoulos, P., Mourtzis, D.: Digital twins in industry 4.0. In: Design and Operation of Production Networks for Mass Personalization in the Era of Cloud Technology, pp. 277–316. Elsevier (2022)
8. Pires, F., Cachada, A., Barbosa, J., Moreira, A.P., Leitão, P.: Digital twin in industry 4.0: technologies, applications and challenges. In: 17th International Conference on Industrial Informatics (INDIN), vol. 1, pp. 721–726. IEEE (2019)
9. Gilchrist, A.: Industry 4.0: The Industrial Internet of Things. Apress, Berkeley (2016)
10. Monnier, L.V., Shao, G., Foufou, S.: A methodology for digital twins of product lifecycle supported by digital thread. In: Proceedings of the ASME 2022 International Mechanical Engineering Congress and Exposition, Volume 2B: Advanced Manufacturing, Columbus, Ohio, USA, October 30–November 3 2022. V02BT02A023
11. Grieves, M.: Digital twin: manufacturing excellence through virtual factory replication. White Paper (2014)

12. Mihai, S., et al.: A digital twin framework for predictive maintenance in industry 4.0. (2021)
13. Leng, J., et al.: Industry 5.0: prospect and retrospect. J. Manuf. Syst. **65**, 279–295 (2022)
14. Nahavandi, S.: Industry 5.0 – a human-centric solution. Sustainability **11**(16), 4371 (2019)

A Data Structure for Developing Data-Driven Digital Twins

Oghenemarho Orukele[1](✉), Arnaud Polette[1], Aldo Gonzalez Lorenzo[2], Jean-Luc Mari[2], and Jean-Philippe Pernot[1]

[1] École nationale supérieure d'arts et métiers, Aix-en-Provance, France
oghenemarho.orukele@ensam.eu
[2] Aix-Marseille University, Marseille, France

Abstract. Digital twins have the potential to revolutionize the way we design, build and maintain complex systems. They are high-fidelity representations of physical assets in the digital space and thus allow advanced simulations to further optimize the behaviour of the physical twin in the real world. This topic has received a lot of attention in recent years. However, there is still a lack of a well-defined and sufficiently generic data structure for representing data-driven digital twins in the digital space. Indeed, the development of digital twins is often limited to particular use cases. This research proposes a data structure for developing modular digital twins that maintain the coherence between the digital and physical twins. The data structure is based on a hierarchical representation of the digital twin and its components; the proposed data structure uses concepts from distributed systems and object-oriented programming to enable the integration of data from multiple sources. This enables the development of a digital twin instance of the system and facilitates maintaining the coherence between the digital twin and the physical twin. We demonstrate the effectiveness of our approach through a case study involving the digital twin of an industrial robot arm. Our results show that the proposed data structure enables the efficient development of modular digital twins that maintain a high degree of coherence with the physical system.

Keywords: Digital and physical twin · Data structure · Digital coherence

1 Introduction

In recent years, the industrial sector has been undergoing a revolution that seeks to exploit the increase in computational power, advances in artificial intelligence, machine learning, connected devices and availability of data to improve the manufacturing or industrial processes. This revolution, known as industry 4.0, aims

This work has been supported by the French ANR PRC grant COHERENCE4D (ANR-20-CE10-0002).

Published by Springer Nature Switzerland AG 2024
C. Danjou et al. (Eds.): PLM 2023, IFIP AICT 701, pp. 25–35, 2024.
https://doi.org/10.1007/978-3-031-62578-7_3

to improve the efficiency of production which reduces the cost and increases the profit margin for the producers, while also benefiting the customers by allowing for a better user experience with customization, selections and improved availability of products for the customers based on their behaviours. Some technologies have been developed to achieve the goal of industry 4.0, a notable example is the internet of things (IoT) or the industrial internet of things (IIoT), which have allowed devices to communicate with each other, use the data generated from its operation to optimize them and recognize patterns that can be used for predictive maintenance, for instance.

One of the concepts that have been developed, which is considered as a key aspect of the industry 4.0, is the Digital Twin (DT). This is a technological concept that seeks to create a high-fidelity digital representation of a physical system, and then connect this digital representation to the physical system [1,2]. This connection allows for communication between the digital representation and the physical system, remote monitoring of the physical system with the digital representation and conducting optimization in the virtual space using data from the physical system.

This paper presents a DT data structure (DTDS) for representing physical entities in the virtual space. This data structure is built with the aim of being modular and generic in order to facilitate fast developments of DTs, and also expanding existing DTs built with the proposed data structure when there is a modification to the composition of the physical system. The paper is organized as follows: Sect. 2 reviews literature on industrial DTs and existing DTDS, Sect. 3 proposes a DTDS, Sect. 4 explains the implementation of the DTDS, while Sect. 5 discusses the implementation, provides conclusions and future outlook.

2 Literature Review

There is a consensus that the concept of a DT was introduced by M. Grieves in 2003, as explained in the white paper [1], however, there is no generally accepted definition of a DT. Grieves and Vickers defined a DT in [2] as "a set of virtual information constructs that fully describes a potential or actual physical manufactured product from micro atomic level to the macro geometrical level". Another definition argues that a DT should be understood as a virtual entity that is linked to a real-world (physical) entity, which describes a planned or actual real-world object with the best available accuracy [3]. While [4] argues that a DT is "a comprehensive digital representation of a physical asset, comprising the system design and configuration parameters, the system state and system behaviour".

Despite the different definitions, there are distinct features that are present in all of these definitions, which are (1) the physical entity, (2) the virtual entity, and (3) the connection between physical and virtual entities. These features can be considered the core components of the DT. The physical entity refers to the physical asset that is to be twinned, while the virtual entity on the other hand is the digital representation of the physical entity. For the physical entity to communicate with the virtual entity there needs to be a connection between both of them. This connection between the entities is a bidirectional automatic data connection which allows the physical entity to send data or its state to the virtual entity, and for the virtual entity to send commands or control the physical entity.

2.1 Digital Twins in Industry

The concept of DTs is used in different sectors. This work focuses on the application of the DT in the industrial sector. The function of DTs in the industrial sector can be classified into three categories: (1) process monitoring, (2) process control and (3) process optimization, simulation or planning [6].

Process monitoring is the most common or in some cases the default application of DTs in the industrial sector. As the name implies, this application involves the use of the DT to track the real-time status of specific parameters or attributes of the physical twin. In [5] a DT was capable of obtaining the real time information of the manufacturing system using the internal sensors of a CNC machine as well as external sensors used to instrument other parts of the system, and relaying the information to the end user of the DT. Fang et al. [7] developed a DT for a mobile phone manufacturing system, which is used to monitor the real-time status of the production using CAD models for 3D visualization.

The process control application involves the use of the DT to control the physical twin. This application seeks to replicate any change made to parameters of the DT in the physical twin. Fonseca et al. changed the position and direction of a model naval vessel using the DT of the vessel [10], and Fen Yepeng et al. used a DT to control selected production parameters of an automotive camshaft production line [5].

In the process simulation, optimization or planning application, the data obtained from the physical twin is used for computation in the DT to simulate the behaviour of the physical twin when certain parameters change [8]. For example, a DT was developed for a CNC milling machine to calculate the residual lifetime of the ball screw actuator for determining when preventive maintenance should be carried out [9].

2.2 Digital Twin Data Structure

Data structures are an organization of data in a logical or mathematical model that is simple enough so that one can effectively process the data when necessary. In the context of DTs, the digital twin data structure (DTDS) refers to the organization of the data of the physical twin in the virtual space.

In the literature surveyed, there is a noticeable lack of DTDS in the different implementations of DTs. In the DTs' implementations which have a DTDS, the DTDS acts as a central data aggregator for the multi-source data from the physical entities for DT services to utilize, and to maintain the coherence between the physical twins and the DT services as shown in Fig. 1. The DTDS can be implemented with XML [9,14], python [12] or OWL [15]. The data structure in some cases can be developed using an object-oriented approach where different classes are created for components of the physical twin [12,15].

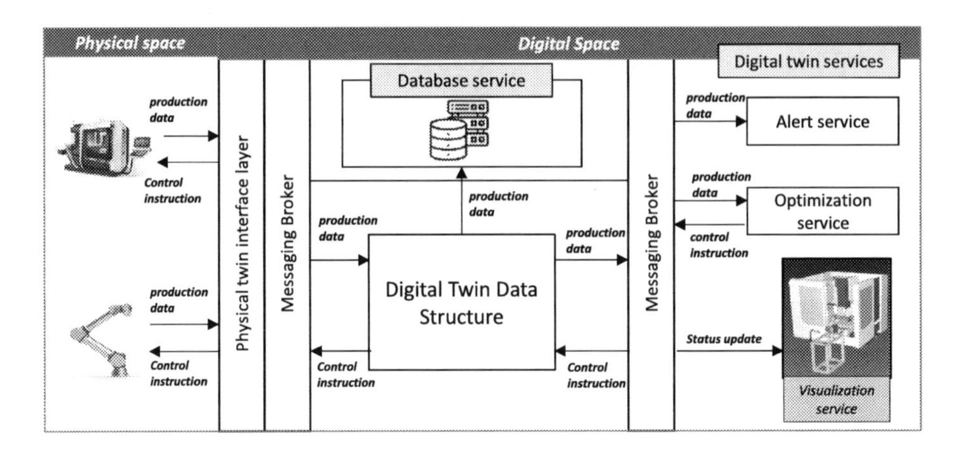

Fig. 1. Digital twin data structure in the DT concept.

Weichao Luo et al. developed a data structure for the DT of CNC machine tools [9]. This data structure allowed for representation of the different components of the CNC milling machine in a hierarchical manner as well as representing the relationships between the components. In another implementation, discrete event system (DES) modelling theory was used to create a data structure for the DT of manufacturing shop floor [11]. This data structure was able to represent the flow of a manufacturing system, but lead to a complex relational structure because of the multiple components. Similarly, a DT data structure was used in the development of a DT for a mobile log crane [13], this data structure represented the different components in the system and the data state in a detailed manner. However, it had a high level of complexity due to the numerous types of classes.

The DTDS allow for the development of digital twin instances (DTI) [16], which allow for faster development of DTs of complex systems that are made up of multiple instances of the same physical entity. DTIs enable the creation of isolated instances of the DT which can be used to conduct experiments based on a test data without affecting the principal DT instance. Another benefit of the DTDS is that it allows for scaling of the DT services without having to carry out a one-to-one connection with the components in the physical entity.

Building upon the layer axis of the Reference Architectural Model Industrie 4.0 (RAMI 4.0) framework [17], the DTDS proposes an implementation for the integration and communication layers. Within the integration layer, the DTDS facilitates the digital availability of the physical twin, allowing the data exchange by various digital services that reply on them. Concerning the communication layer of RAMI 4.0, the DTDS establishes a standardized interface that enables manufacturers, customers, or third parties to communicate with the physical twin in a manner that is independent of the underlying communication protocol [18].

The DTDS in the literature surveyed lacked sufficient genericity which makes them mostly suitable for only developing DTs for the specific use case under study or the family of systems under study in the literature. This limits the application of the DTDS in the industrial section; this paper introduces a generic DTDS that allows for modular development of DTs for various industrial systems.

3 Proposed Digital Twin Data Structure

For the development of a modular and sufficiently generic DT, it is important to develop a data structure for the virtual entity that reflect these properties as well. With this in mind, we propose a DTDS that is hierarchical, allows for multiple data connections, has standardized operations through an application programming interface (API), and implements abstraction and encapsulation to ensure data safety.

The digital twin data structure consists of four hierarchical classes: *attribute*, *component*, *system*, and *environment*. Each of these classes is discussed in detail below. This digital twin data structure idea is developed from a combination of systems programming concepts and Object Oriented Programming (OOP) principles, this approach allows for the reuse of pre-configured classes by instantiating a new instance of the class. These hierarchical classes can be used to build more complex systems, and the relationships between these classes are illustrated in Fig. 2.

The *attribute* class represents physical properties in the virtual world and functions as an abstract class. This ensures a standardized and generic design in all classes that implement it, despite any unique implementation details. The attribute class can be defined as a set $A = \{I, N, V, R, T\}$ where I is a unique identifier, N is the name of the attribute, V is the value of the attribute, R is the list of relationship of the attribute and T is the last update time of the attribute. The relationship R is a set of Ids I of the different attributes, components or systems associated with a particular class; it can be defined as $R = \{I_1, .., I_n\}$.

The *component* class consists of a collection of attribute classes that define a component in the physical twin. Standardization enables the creation of reusable component classes that can be used repeatedly. The class includes functions for accessing other classes within it. The component class consist of the following properties $C = \{I, N, F, A_c, C_s, R\}$ where I is the unique identifier of the component, N is the name of the component, F is the path to the CAD file of the component, A_c is the component attributes where $A_c = \{A_{c1}, .., A_{cn}\}$, C_s is the list of sub-components in the component with $C_s = \{C_{s1}, .., C_{sn}\}$, and R is a list of relationships of the component.

The *system* class is composed of component and attribute classes, enabling the creation of systems that consist of multiple components and attributes. It

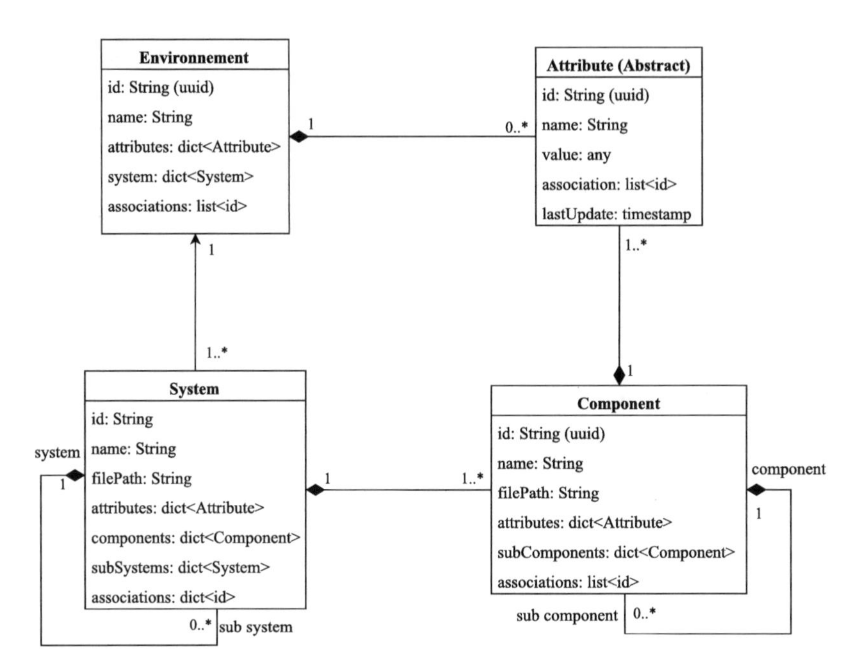

Fig. 2. Digital twin data structure classes and relationships.

features a collection of sub-systems, supporting a recursive design approach. The system class is described as $S = \{I, N, F, A_s, C, S_s, R\}$ where I is the unique identifier of the system, N is the name of the system, F is the path to the CAD file of the system, A_s is the system attributes where $A_s = \{A_{s1}, .., A_{sn}\}$, C is the system components where $C = \{C_1, .., C_n\}$, S_s is the sub-systems of the system with $S_s = \{S_{s1}, .., S_{sn}\}$, and R is the relationships of the system.

Lastly, the *environment* class serves as a wrapper for all other classes in the system, providing the starting point for accessing the digital twin data structure. It is implemented as a singleton class. Similarly, the environment class is defined as $S = \{I, N, A_e, S, R\}$ where I is the unique identifier of the environment, N is the name of the environment, A_e is the set of environment attributes with $A_e = \{A_{e1}, .., A_{en}\}$, S is the set of systems in the environment with $S = \{S_1, .., S_n\}$, and R is the relationships of the environment.

The standardized nature of the DTDS allows for algorithms to be developed for accessing the data in the data structure and sending commands to the physical twin using the DTDS. Figure 3 presents the flow chart of an algorithm for reading the data of a specified attribute in the digital twin data structure; the algorithm consists of multiple search operations that try to find and return an attribute with the specified name in the DTDS. Similarly, other algorithms can be developed to control the physical twin using the DTDS.

These four classes can be used to build digital twins of complex systems if the systems have been decomposed into attributes, components and systems of

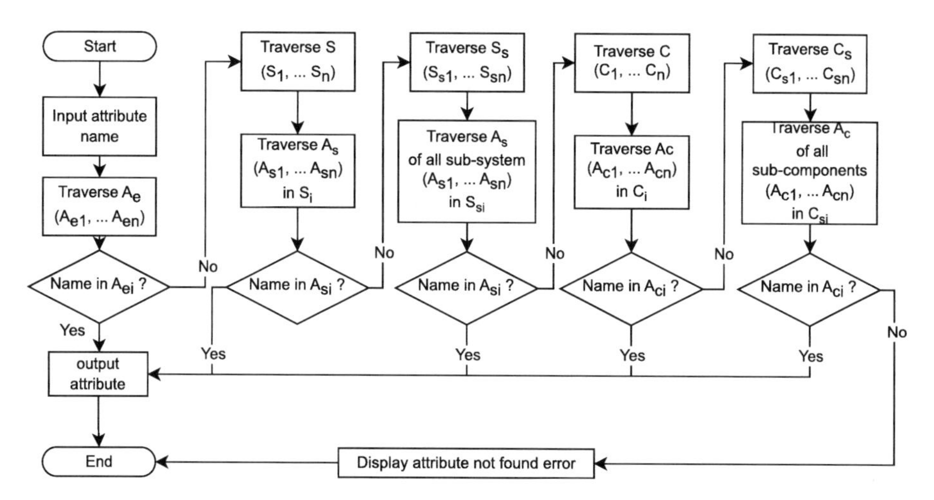

Fig. 3. Flow diagram for reading data of attribute in the DTDS.

interest. This approach allows for modelling a heterogeneous industrial environment made up of different systems while allowing the digital twin services access to the same parameters regardless of the implementation details of the system. This allows for a generic approach in the development of DTs.

4 Implementation of the Digital Twin Data Structure

This section presents an implementation of the DTDS in the development of a DT for an industrial robot arm. The industrial robot is a UR5e 6-axis collaborative robot that has integrated sensors which can be used to obtain data from the robot.

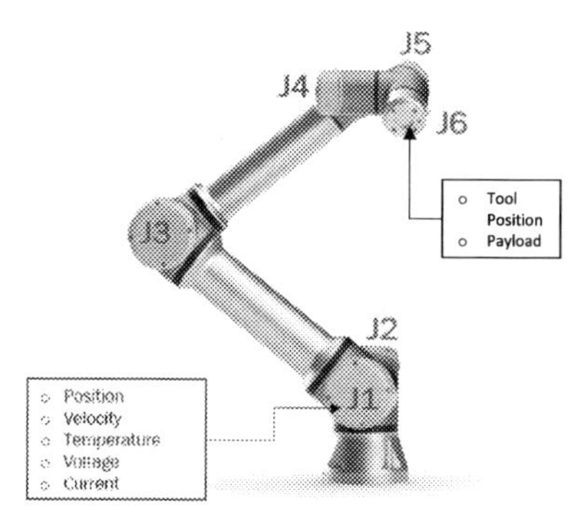

Fig. 4. UR5e robot with joints and joint attributes.

To create a DT, it is important to first identify the aspects of physical entity that are to be replicated in the virtual space. In this case, the aspects to be replicated are the joints (Jn) and the tool centre point (TCP) of the robot shown in Fig. 4. The joint attributes of interest include position, velocity, temperature, voltage, and current while the TCP attributes of interest include tool position and payload. These attributes were selected based on authors' preferences, additional attributes can be added as needed.

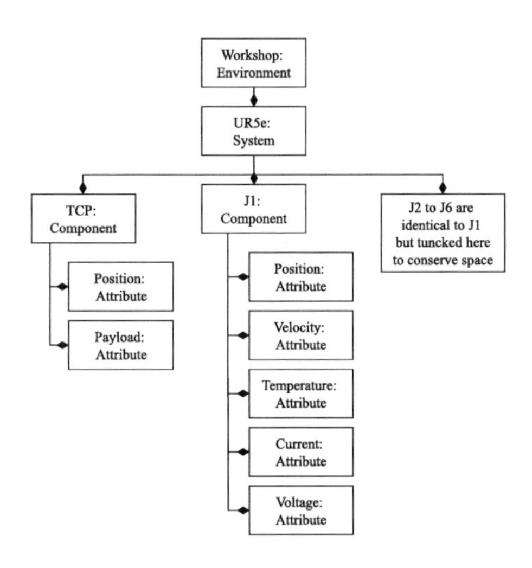

Fig. 5. Representation of the UR5e robot with the DTDS.

In representing the UR5e robot using the DTDS, a bottom-up approach is adopted i.e. building the DTDS from the attribute to environment class. The different attribute of interest are modelled with the attribute class. This allows for the implementation details for the data exchange of the different attributes to be implemented independently. As an example, the joint temperature is a scalar quantity while the TCP position is a vector quantity, despite the different quantity types, they can be represented using the attribute class and integrate seamlessly into the DTDS. Within the DTDS, the joint component is represented with a component class which comprises the specific attributes associated with the joint. This facilitates the creation of five similar components for the remaining joints of the UR5e robot, with distinct names and ids assigned to each.

Likewise, the DTDS employs a component class to model the TCP of the UR5e. Subsequently, the associated attributes are incorporated into this class, effectively capturing the essential characteristics of the TCP within the DTDS framework. Subsequently, the UR5e robot is modelled within the DTDS using the system class, comprising of the six joints and the TCP components. Finally, an environment class is instantiated, serving as a container for the UR5e system within the DTDS framework. The resulting representation of the UR5e robot with the DTDS is illustrated in Fig. 5.

The process of modelling the UR5e robot with the DTDS shows how it can be applied to a system by decomposing it into attributes and components. It also illustrates the modularity of the DTDS, as seen in the modelling of a joint component with its attributes and the subsequent reusability in the instantiation of the other joints in the UR5e robot.

5 Discussion and Conclusion

5.1 Discussion

The DTDS representation of the UR5e robot facilitates the consolidation of the implementation details of data exchange between the physical and digital twins in the respective attribute classes of the DTDS, which minimizes the need for redundant implementation effort. Consequently, various DT services can seamlessly interact with the physical twin by leveraging the DTDS's API, ensuring a cohesive and standardized approach. Thus the DTDS serves as a central hub in the digital space for DT services to interact with the data from the physical twin.

Additionally, the DTDS enables the creation of multiple DTIs that can be utilized for various purposes, such as simulating the system's behaviour using test datasets, as depicted in Fig. 6. This also ensures the efficient scaling and modularity of the DT in the event that a new instance of the physical robot or an entirely new system is introduced.

As of the time of writing this paper, a digital shadow has been developed utilizing the DTDS framework to acquire real-time data from two physical entities namely the UR5e robot and a 5-axis CNC machine. The data stored within the DTDS is seamlessly integrated with a data visualization dashboard developed with Node-RED [19], alongside a 3D visualization platform facilitated by RoboDK [20].

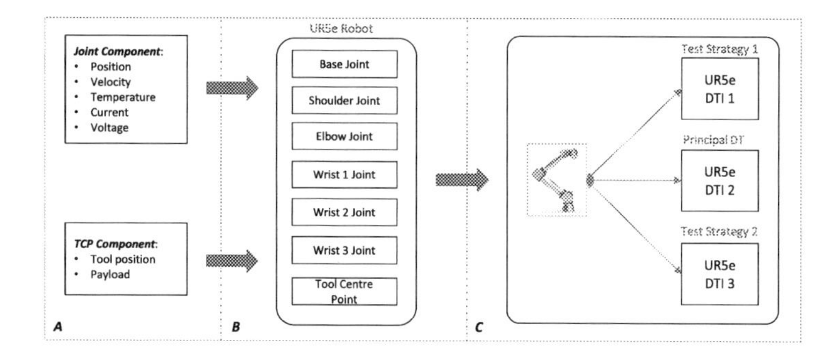

Fig. 6. Construction of digital twin instance for UR5e robot - (A) robot components (B) robot instance (C) digital twin instances of robots.

5.2 Conclusion and Future Work

This article presents a DTDS that is capable of representing a physical entity in the virtual space. The DTDS is modular, generic and aims to fill the research gap by serving as a central point for DT services to interact with the physical twin, and a generic data structure for developing digital twins of complex system. This allow for the efficient scaling of the DT services, creation of DT instances and the easy addition or removal of systems/components/attributes in the DTDS. This also allows for the development of DT for other systems in an efficient and modular manner.

The next stage of this work involves the integration of the DTDS with a decision-making service and the subsequent implementation of an interface for controlling the physical twin through the DTDS. By incorporating the decision-making service, the aim is to enable optimal and secure control of the physical twin from the virtual domain. As an integral part of validating the proposed DTDS, it will be extended to modelling the digital twins of additional industrial systems.

References

1. Grieves, M.: Digital twin: manufacturing excellence through virtual factory replication, March 2015
2. Grieves, M., Vickers, J.: Digital twin: mitigating unpredictable, undesirable emergent behavior in complex systems. In: Kahlen, F.-J., Flumerfelt, S., Alves, A. (eds.) Transdisciplinary Perspectives on Complex Systems, pp. 85–113. Springer, Cham (2017). https://doi.org/10.1007/978-3-319-38756-7_4
3. Autiosalo, J., et al.: A feature-based framework for structuring industrial digital twins. IEEE Access **8**, 1193–1208 (2020)
4. Hribernik, K., et al.: Autonomous, context-aware, adaptive Digital Twins State of the art and roadmap. Comput. Ind. **133**, 103508 (2021)
5. Fan, Y., et al.: A digital-twin visualized architecture for Flexible Manufacturing System. J. Manuf. Syst. **60**, 176–201 (2021)
6. Yi, L., et al.: Process monitoring of economic and environmental performance of a material extrusion printer using an augmented reality-based digital twin. Addit. Manuf. **48**, 102388 (2021)
7. Fang, L., Liu, Q., Zhang, D.: A digital twin-oriented lightweight approach for 3d assemblies. Machines **9**, 231 (2021)
8. Gopalakrishnan, S., et al.: Integrating materials model-based definitions into design, manufacturing, and sustainment: a digital twin demonstration of incorporating residual stresses in the lifecycle analysis of a turbine disk. J. Comput. Inf. Sci. Eng. **21**(2), 021006 (2020)
9. Luo, W., et al.: Digital twin for CNC machine tool: modeling and using strategy. J. Ambient. Intell. Humaniz. Comput. **10**, 1129–1140 (2019)
10. Fonseca, I.A., et al.: A standards-based digital twin of an experiment with a scale model ship. Comput. Aided Des. **145**, 103191 (2022)
11. Jiang, H., et al.: How to model and implement connections between physical and virtual models for digital twin application. J. Manuf. Syst. **58**, 36–51 (2020)

12. Zhidchenko, V., Startcev, E., Handroos, H.: Reference architecture for running computationally intensive physics-based digital twins of heavy equipment in a heterogeneous execution environment. IEEE Access **10**, 54164–5418 (2022)
13. Cai, Y., et al.: Using augmented reality to build digital twin for reconfigurable additive manufacturing system. J. Manuf. Syst. **56**, 598–604 (2020)
14. Schroeder, G., et al.: Digital twin data modeling with AutomationML and a communication methodology for data exchange. IFACPapersOnLine **49**, 12–17 (2016)
15. Liu, C., et al.: Web-based digital twin modelling and remote control of cyber-physical production systems. Robot. Comput.-Integr. Manuf. **64**, 101956 (2020)
16. Jones, D., et al.: Characterising the digital twin: a systematic literature review. CIRP J. Manuf. Sci. Technol. **29**, 36–52 (2020)
17. Baptista, L.F., Barata, J.: Piloting Industry 4.0 in SMEs with RAMI 4.0: an enterprise architecture approach. Procedia Comput. Sci. **192**, 2826–2835 (2021)
18. Abdel-Aty, T.A., et al.: Asset administration shell in manufactuing: applications and relationship with digital twins. IFAC PapersIbkube **55–10**, 2533–2538 (2022)
19. Node-Red Homepage. https://nodered.org/. Accessed 31 Jan 2023
20. RoboDK Homepage. https://robodk.com/. Accessed 31 Jan 2023

A Quality-Oriented Decision Support Framework: Cyber-Physical Systems and Model-Based Design to Develop Design for Additive Manufacturing Features

Claudio Sassanelli[1](✉), Giovanni Paolo Borzi[2], Walter Quadrini[3], Giuseppe De Marco[4], Giorgio Mossa[1], and Sergio Terzi[3]

[1] Department of Mechanics, Mathematics and Management, Politecnico di Bari, 70125 Bari, Italy
claudio.sassanelli@poliba.it

[2] Industry 4.0 Business Unit, EnginSoft Spa, 35129 Padova, Italy

[3] Department of Management, Economics and Industrial Engineering, Politecnico di Milano, 20135 Milano, Italy

[4] Living Space for Additive Technologies, Kilometro Rosso Spa, 24126 Bergamo, Italy

Abstract. Metal Additive manufacturing (AM) is a complex operation, which requires the fine-tuning of hundreds of processes parameters to obtain repeatability and a good quality design at dimensional, geometric, structural levels. Therefore, to be used as final product, metal AM parts must go through advanced quality control processes. This implies large capital equipment investment in measurement systems (i.e., tomography and lengthy inspection operations that adversely impact costs and lead times). A large amount of data can be collected in metal AM processes, as most industrial AM systems are equipped with sensors providing log signals, images and videos. This paper develops and proposes an innovative quality-oriented decision support framework, composed by a Model-based Design tool providing Design for Additive Manufacturing features, and a Cyber-Physical System created by integrating an AM asset with a real-time smart monitoring software application. Such framework caters to process engineers and quality managers needs to improve a set of quality and economic KPIs.

Keywords: metal additive manufacturing · model-based design · design guideline · cyber-physical systems · digital innovation hub

1 Introduction

Metal Additive Manufacturing (AM) is a complex operation, which requires the fine-tuning of hundreds of processes parameters to obtain repeatability and a good quality design at dimensional, geometric, structural levels (Frazier, 2014). To be used as final product, metal AM parts must go through advanced quality control processes. This implies large capital equipment investment in measurement systems (i.e., tomography

Published by Springer Nature Switzerland AG 2024
C. Danjou et al. (Eds.): PLM 2023, IFIP AICT 701, pp. 36–46, 2024.
https://doi.org/10.1007/978-3-031-62578-7_4

and lengthy inspection operations that adversely impact costs and lead times) (Wang et al., 2021). A large amount of data can be collected in metal AM processes, as most industrial AM systems are equipped with sensors providing log signals, images, and videos (Slotwinski et al., 2014). In particular, small and medium enterprises (SMEs) are increasingly applying data analytics to extract useful information (Lamperti et al., 2023) (e.g., by creating quality-oriented models and integrating them within their production processes (Baijens et al., 2022)). In this context, Digital Innovation Hubs (DIHs) and innovation districts act as one-stop-shop to support SMEs in benefitting from advanced digital technologies and industrial processes (Sassanelli and Terzi, 2022). The development of novel decision support frameworks help raising the benefits to SMEs, by clarifying the information flows, the development sequences, and the models and tools' target objectives and benefits. Accordingly, the objective of this research is to develop and propose a quality-oriented decision support framework, composed by a MBD tool providing Design for Additive Manufacturing (DfAM) features and a CPS.

The structure of the paper is the following. Section 2 presents the research context. Section 3 shows the research method adopted. Section 4 presents the results and Sect. 5 discusses them. Finally, Sect. 6 concludes the paper.

2 Research Context and Main Gaps

2.1 Metal Additive Manufacturing

Latest metal AM assets are equipped with several sensors capable to inspect operations. For example, Laser Powder Bed Fusion (LBPF) (known also as Direct Metal Laser Sintering (DMLS) or Selective Laser Melting (SLM)) machines (Aboulkhair et al., 2019), not only builds the desired part by melting thin layers of metal powder on which a laser beam acts (selectively melting the powder only in the areas of interest and functional to the construction of the component) but also relies on advanced sensors suited for detailed process monitoring (Slotwinski et al., 2014). However, data, due to the long process timeframe and their heterogeneity (e.g., process chamber % Oxygen, % Inert Gas, Pressure and Temperature, Laminar Gas Flow Speed, Laser Power, Platform Temperature), are stored but not analysed in real time to monitor the production quality. While the role of the human process specialist is key, a solution is needed to support the quality acceptance decision by means of smart, real-time process monitoring. In fact, there are no consolidated solutions in the industrial practice capable to analyse this data in real-time for quality control and to provide useful information to improve the part and process design (Sames et al., 2016). The LBPF can be equipped with an Optical Tomography (OT) system that can be flanked by software platform allowing to manage all sensors inside the system, the related process data and the current powder bed.

2.2 Process Simulation for AM

AM process simulation tools are a relatively new set of computer-aided engineering tools, introduced into the market over the past few years. These AM process simulation tools can predict residual stress, distortion, microstructure, porosity, and other characteristics of an AM part prior to its creation (Song et al., 2020). By simulating the effects of an AM process on a specific geometry, build failures can be reduced, and shape changes that occur in the part can be compensated before production, so that the part can be produced to a higher tolerance and with a higher probability of success. A detailed simulation of the whole process is typically infeasible (scanning length ranges from 10 to 1000 km for build volumes in the order of $10^4 \div 10^6$ mm^3). Numerical models can be classified on the basis of the dimensional scale of simulated phenomena in powder-scale modelling ($10^{-9} \div 10^{-6}$ m), layer-scale modelling ($10^{-6} \div 10^{-3}$ m) and part-scale modelling ($10^{-3} \div 1$ m). Several finite element analysis tools have been developed using average assumptions to form a solution for large, complex geometries. Multi-scale simulations alone are still too slow to enable a complete part simulation. New computational approaches for AM are still in development to reduce the solution time for large-scale AM problems (McMillan et al., 2017). Process simulation at part-scale can be used before part production to evaluate the additively part itself, since, if a failed build is predicted, it is possible to adapt the print by adding/strengthening support structure or by printing the part oversized and machining with specific design requirements.

3 Research Methodology

The research has been carried out in two overlapping steps (Fig. 1 and Table 1). The first step was aimed at creating and tuning the quality-oriented MBD/CPS framework. In it, the MBD approach contributes to make an abstraction of the technical requirements and to create the project file containing all the necessary instructions for the CPS, so to print the designed parts. Therefore, a quality check with CT-Scan, OT and metallography cut-up, has been conducted to obtain quality information. Finally, quality-oriented predictive models were created and integrated into the CPS. The second step was aimed at validating the framework in operation. To support the MBD approach, the CPS supplies the information useful to improve the project file and prints the part by maximizing the part quality based on the design specifications. Decision support is enabled by the process data analysis and through advanced data analytics, to provide information of the

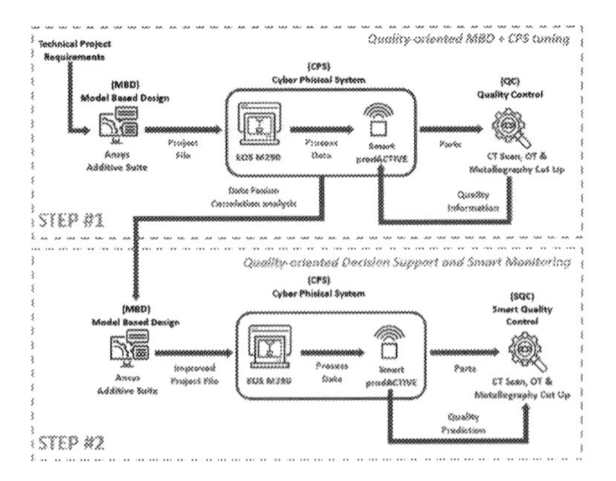

Fig. 1. Project implementation steps, and information flow

expected quality outcomes of a specific setup. Finally, AM production has been carried out, monitored and validated (to check the improved design).

Table 1. Research steps: Activities and details

Step	Activity	Details
1	Preparation of a catalogue	of the features for aluminum parts printed with a LBPF, and of defects, to capture the available design know-how about this technology limits
1	Definition of the targeted defects, and defects-parameters correlations a-priori identification	the MBD tool defines part geometry, machine, and process parameters (to identify defect risks), and generates a project file containing the machine instructions (e.g., AM part oriented geometry). The part geometry includes simple parts (e.g., a cylinder, to allow an efficient inspection process with the available technologies and accurate measures of the defect grade), and complex parts to induce feature-related defects (e.g., rugosity, deformations)
1	CPS creation and setup	connecting the smart monitoring solution to the AM system, to acquire machine and process data
1	Definition of a production plan	for the defined parts (including manufacturing and inspection activities). The actual AM process is carried out according to the Design of Experiment (DOE) (machine and process data are collected by the CPS at real time)
1	Quality information collection	traceability information by the OT and CT-Scan allows to link the quality information to the relevant job and part number. Collected quality information need to be consistent with the targeted KPIs
1	Advanced data analytics	a data fusion approach is evaluated (e.g., by integrating hot spots data and defects data). Quality-oriented predictive models are created, suitable to be executed in real-time. Data fusion consists of a multiple data elaboration to obtain a quality index (e.g., the volumetric energy supplied to the metal powder bed to be correlated to the material densification)
1	Models and CPS integration	to enable a smart process monitoring
2	Part design improvement	the MBD tool are applied to production parts representative of common AM challenges (e.g., a structural component) utilizing the information provided by the CPS to support decisions and improve the design
2	Improved parts monitoring and inspection	though the application of the quality-oriented models, to validate the improved design and the smart monitoring features
2	Approach contribution evaluation	to check business objectives achievement and plan future development activities accordingly

3.1 The Case and Its Partners

The partners of this research are three SMEs: EnginSoft (the technology provider), and KM Rosso and Pres-X (as technology adopters). EnginSoft role included the research coordination, MBD tool application, development of the CPS and framework integration. Kilometro Rosso role included performing DfAM activities, providing the AM asset and carrying out the actual prototyping production, and act as a testbed for the framework. Pres-X dealt with the AM processes, and in particular quality control & post processing processes, and materials characterization activities.

4 Results

The main result of this paper is a quality-oriented decision support framework, composed by a MBD tool providing DfAM features (i.e., ANSYS additive suite), and a CPS (created by connecting the EOS M290 machine with the Smart ProdACTIVE smart monitoring solution) (Fig. 2), also applied and validated in a pilot application case.

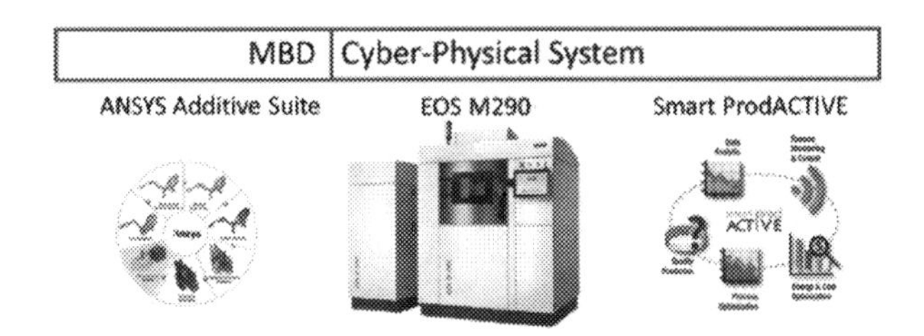

Fig. 2. Quality-oriented decision support framework: main elements

The Ansys Additive Suite delivers the critical insights required for the development and analysis of an AM product to avoid build failure and create parts that accurately conform to design specifications. This comprehensive solution spans the entire AM workflow (from DfAM through validation, print design, process simulation and exploration of materials). Related to the MBD, it offers dedicated tools to support the design of a product suitable to be realized with metal AM (in particular, LPBF).

The EOS M290 machine uses a LPBF process, combined with a software (EOS connect), that relies on advanced sensors suited for detailed process monitoring.

Smart ProdACTIVE is an integrated smart monitoring solution that connects production processes data sources (machines, sensors) with software modules offering smart features. It provides a centralized production monitoring via a network of data acquisition agents, a flexible database to persist and manage production processes data, a web-based application for remote monitoring of the process stability, and KPIs such as efficiency and costs, smart monitoring features enabling production optimization, active real-time application of predictive models (e.g., quality-oriented models). Data analysis approach have been performed on the data to create quality-oriented predictive models: such models have been integrated into the Smart ProdACTIVE system to integrate real time smart monitoring features, in turn enabling a better decision making.

4.1 Features and Defects Catalogue for Metal AM

A catalogue of the features for aluminum parts that can be printed with a LBPF, and of typical defects, has been created. It aims to capture the available design know-how regarding the advantages and limits of the metal AM technology. A specific focus has been given to LBPF process, and to the material used for project activities (AlSi10Mg). The catalogue is subdivided into 3 main sections: (i) process description, workflow and advantages; (ii) material details (with information on chemical and mechanical properties); (iii) aspects related to DfAM (about what may be the features to be controlled and verified in the components). These guidelines are useful to reduce the errors that could bring to the failure of the manufacturing process. Following these instructions, suggestions, and strategies, it is possible to achieve the best results in AM with fewer iterations. In this study, a qualitative comparison between MBD simulation output (e.g., stress concentration, final deformation, errors in post processing) and the actual process data and results (as evaluated by means of CT scan and metallographic inspections datasets) has been performed. The catalogue substantiates the link between the MBD and the CPS and can be generalised and applied to different AM processes and materials.

4.2 Application to the Pilot Application Case

The framework developed has been applied to the pilot case shown in Sub-Sect. 3.1 going through a series of steps, as follows.

Definition of the Targeted Defects and A-Priori Identification of the Potential Correlations. First, the targeted defects have been defined, focusing on porosity (as expressed by DMQ) and on shape variation from the nominal one. The a-priori identification of the potential correlations between such defects and parameters has been carried out by discussing process experience, reviewing nominal process parameters for the targeted alloy and the information available in literature, and by MBD DOEs execution.

Starting from the standard process parameters for the selected alloy, and qualitative considerations, multiple DOEs have been defined and executed using the MBD tool (ANSYS Additive Suite), to identify potential areas of interest for the subsequent physical DOEs. MBD simulation allowed to explore areas of the parameter space that are relatively distant from the standard parameters suggested for a given process-machine-alloy combination. The solver can predict lack-of-fusion porosity via the powder-solid state tracking but does not predict balling or keyhole phenomena. The simulations helped to identify areas of the parameters space where lack-of-fusion porosity would happen, providing relevant information about meltpool size and the resulting density.

Initial CPS Setup. The CPS has been created (installing Smart ProdACTIVE on a virtual machine connected to the EOS M290 system through the EOS CONNECT Core OPC UA interface module) and configured to acquire all available endpoints (49 tags).

Physical DOE Setup. Starting from the virtual DOEs results, the physical DOEs have been defined for machine and process constraints (e.g., the maximum laser power available for the EOS M290 of 370 W). Cylindrical specimens have been selected for the physical DOEs, being the most suitable shape to conduct inspection activities and to determine the quality output (i.e., DMQ). The first physical DOE has been defined by varying laser power and laser scan speed while keeping the other parameters constant.

CPS Adaptation and Improvement. The initial CPS tests clarified the need to install, configure and connect additional process monitoring technologies to expand the scope of the monitored process data.The a-priori identification of the potential correlations, confirmed by the MBD simulations, provided evidence of the need to identify suitable process data related to the volumetric energy density not provided by the EOS CONNECT system. Therefore, to grasp these data, the EOSTATE Monitoring Suite advanced technologies has been added. It is composed of EOSTATE Exposure OT, providing OT, and EOSTATE MeltPool Monitoring, measuring the quantity of energy transferred by the laser locally. Both technologies provided grey values, representing the intensity of the light emitted by the bed fusion or meltpool, offering measures of energy deposition. To acquire the grey values, as EOSTATE data is not provided by any protocol, a Smart ProdACTIVE client has been configured to poll the CSV file during production and ingest the grey value data, together with job, part and layer information, as they are available.

Finalization of the Catalogue of Features and Defects for Metal AM. Process defects guidelines have been augmented with qualitative considerations regarding the usage and outputs of process simulation tools. In this research, the study refers to the ANSYS product family, but can be easily expanded to include other simulation products.

Part Design Improvement by MBD Tool Application. The structural part selected for this activity is presented in Fig. 3. This complex design presents a set of geometrical features common for metal AM parts, therefore it can be considered representative of a typical DfAM challenge. Its complexity may induce feature-related defects (e.g., rugosity, deformations). The MBD tool (ANSYS Additive Suite) has been applied to generate different project files containing the machine instructions for the selected structural component. To determine the impact of the framework, two different project files have been targeted: a baseline project file, designed without considering the catalogue of features and defects; and an improved project file, designed considering the best practice described in the catalogue. Simulation results unveil that, compared to the nominal shape, the improved project shows a relevant reduction of the residual stresses (50 MPa reduction), and a 200% reduction of the maximum deformation. A visual comparison between the nominal and the resulting shapes confirms that the baseline job may generate problematic shape variations in critical features (e.g., support holes).

Shape variations were measured on the improved design by applying OT inspection. Results shown that the variations are limited and inferior to the values predicted by the process simulation. Table 2 describes the features catalogue applied to improve the baseline job design towards KPI4 (Reduce the complexity of design by abstraction).

 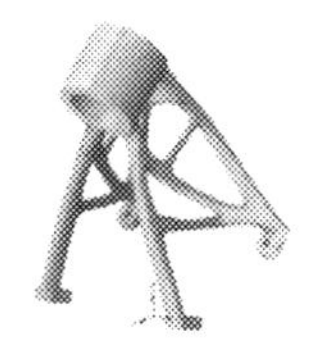 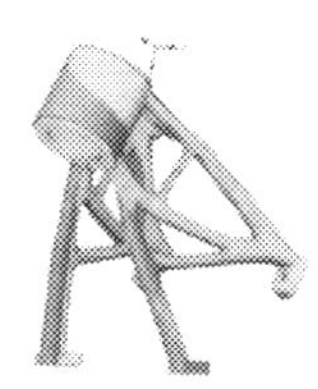

Fig. 3. Left: Structural component selected for the framework application. Center (improved job) and right (baseline job): Nominal (grey) vs resulting (green) shape comparison.

Table 2. Features catalogue applied to improve the baseline job design towards KPI4

Category	Feature	Guideline	Baseline design	Improved design
Part orientation	Overall build	"... self-supporting angle between 45°/50°"	Excessive slope angles	Positioning angle has been improved
	Downfacing surfaces	"... avoid big overhang sections or large downfacing sections"	Excessive unsupported overhang	Overhang has been reduced by part orientation
	Shrink lines		Does not take into account the support structures	Considering the support structures provides a better orientation
Support generation	Thermal stresses reduction	"Support structure is needed to avoid warping and to keep part in position."	Inadequate thermal dissipation leading to increased stresses	Support structure with adequate sections help dissipating heat
Holes	Straight		Inadequate hole support	Block Support structures have been increases to prevent dross formation
Surfaces	Influence on quality	"... avoid big overhang sections or large downfacing sections"	See downfacing surfaces	See downfacing surfaces
TOTAL BEST PRACTICES APPLIED				6

5 Discussion

5.1 KPIs of the Quality-Oriented Decision Support Framework

The following KPIs (Table 3) have been monitored and evaluated. KPI1 is about DMQ and has been measured during the printing phase via metallographic analysis through a smart monitoring system integrating also a predictive MBD approach to monitor the production and to identify potential issues in-process. KPI2 measures the shape variation improvement by MBD by comparing the variation respect to the nominal shape of a geometry produced using a job not evaluated with MDB and the same geometry produced using an optimized process by application of MBD. KPI3 (Lead time reduction) wants to reduce the number of the quality control (mostly, volumetric) on parts once printed. KPI4 (Reduce the complexity of design by abstraction) tracks the number of design iterations (virtual and/or actual), that can be reduced thanks to the application of the features catalogue throughout the MBD.

Table 3. KPIs: baseline, target, and results achieved

KPI	Baseline	Target	Results achieved
KPI – Quality improvement 1 (DMQ)	DMQ $\approx$ 99.4%	DMQ $\geq$ 99.5%	DMQ $\geq$ 99.8%
KPI2 – Quality impr. 2 (shape variation through MBD)	Baseline model shape variation	Model shape variation improved (>20%)	200% improvement observed in simulation. Estimates have been confirmed by inspection
KPI3 – Lead time reduction	100% control on all critical printed components (32 h in 4d)	80% control on all critical printed components	80% control feasible with in-process smart monitoring. 15% processing time reduction without DMQ degradation
KPI4 – Reduce the complexity of design by abstraction	No design process checklist	$\geq$5 guideline steps applied	6 guideline steps applied to improve the baseline design

5.2 Value Proposition of the Quality-Oriented Decision Support Framework

Results end users are identified as AM process engineers and quality managers. Figure 4 presents the Value Proposition Canvas for such roles. The framework value proposition consists of an innovative metal AM production workflow, which enables the quality optimization of the project design. The proposed approach facilitates the reduction of defects occurring during the production process, such as the distortions of the component during their realization, leading to shapes that may not comply with the original design and may eventually hit the dust spreading blade, also blocking the printing job. While foreseeing these defects, the framework enables to reduce the number of attempts needed to obtain the desired component so to reach the result in less time and cost.

The framework provides AM SMEs with an innovative solution to address production defects in metal AM. Defects at the macro scale can be identified by the process

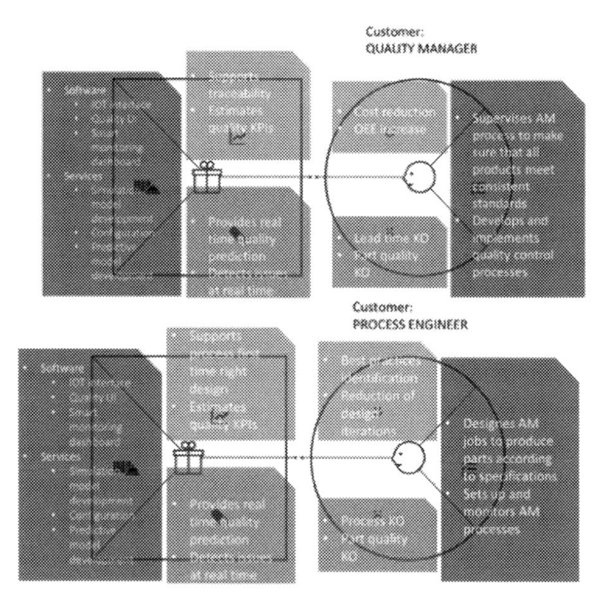

Fig. 4. Value Proposition for Quality Manager (up), and AM Process Engineer (down) user

simulation, while defect at the meso and micro scale can be analyzed, modelled, and predicted by applying advanced quality-oriented modelling to real time process data. The proposed framework provides a significant contribution for improving the metal AM production at various stages. During the MBD stage, through process simulation and a feature catalogue enriched with experimental and virtual information, the framework provides quality-oriented support to product/process design decision leading to improved product and processes, better asset utilization, improved predictability, and lower scrap rates. During the Quality Assurance stage, applying a CPS capable of detecting quality issues at real time, improved process monitoring enabling process quality conformity declaration. Finally, in the Quality Control stage, by application of validated predictive models, a smarter process control (e.g., from wide inspection areas and range of defects to targeted areas and specific defect types) enables the implementation of different policies and cost reduction (e.g., from 100% inspection to sampling).

6 Conclusions

This research demonstrated that manufacturers can improve their metal AM processes by systematically applying an innovative quality-oriented decision support framework. Such framework assists the component design providing a checklist of best practices and qualitatively links them to DfAM MBD-tools features (e.g., through ANSYS additive suite). It also provides valuable insight regarding the manufacturing process and the relationship between process parameters combinations and quality results. Once integrated in the design process, they enable the confident selection of parameters sets different from the nominal ones. Finally, it integrates a CPS by augmenting an AM asset with a real-time smart monitoring software application. This allows to guarantee and improve a set of quality (i.e., component density) and economic KPIs (e.g., lead time). The validated quality-oriented predictive models can be further enriched by expanding the DOEs and considering additional data and parameters.

Acknowledgements. This study was carried out within the MICS (Made in Italy – Circular and Sustainable) Extended Partnership and received funding from the European Union Next-Generation EU (PIANO NAZIONALE DI RIPRESA E RESILIENZA (PNRR) – MISSIONE 4 COMPONENTE 2, INVESTIMENTO 1.3 – D.D. 1551.11-10-2022, PE00000004).

References

Aboulkhair, N.T., Simonelli, M., Parry, L., Ashcroft, I., Tuck, C., Hague, R.: 3D printing of Aluminium alloys: Additive Manufacturing of Aluminium alloys using selective laser melting. Prog. Mater. Sci. **106** (2019). https://doi.org/10.1016/J.PMATSCI.2019.100578

Baijens, J., Helms, R., Bollen, L.: Data analytics and SMEs: how maturity improves performance. In: Proceedings - 2022 IEEE 24th Conference on Business Informatics, CBI 2022, vol. 1, pp. 31–39 (2022). https://doi.org/10.1109/CBI54897.2022.00011

Frazier, W.E.: Metal additive manufacturing: a review. J. Mater. Eng. Perform. **23**, 1917–1928 (2014). https://doi.org/10.1007/s11665-014-0958-z

Lamperti, S., Cavallo, A., Sassanelli, C.: Digital servitization and business model innovation in SMEs: a model to escape from market disruption. IEEE Trans. Eng. Manag., 1–15 (2023). https://doi.org/10.1109/TEM.2022.3233132

McMillan, M., Leary, M., Brandt, M.: Computationally efficient finite difference method for metal additive manufacturing: a reduced-order DFAM tool applied to SLM. Mater. Des. **132**, 226–243 (2017). https://doi.org/10.1016/J.MATDES.2017.06.058

Sames, W.J., List, F.A., Pannala, S., Dehoff, R.R., Babu, S.S.: The metallurgy and processing science of metal additive manufacturing. Int. Mater. Rev. **61**, 315–360 (2016). https://doi.org/10.1080/09506608.2015.1116649

Sassanelli, C., Terzi, S.: The D-BEST reference model: a flexible and sustainable support for the digital transformation of small and medium enterprises. Glob. J. Flex. Syst. Manag. **40171**, 1–26 (2022). https://doi.org/10.1007/s40171-022-00307-y

Slotwinski, J.A., Garboczi, E.J., Hebenstreit, K.M.: Porosity measurements and analysis for metal additive manufacturing process control. J. Res. Natl. Inst. Stand. Technol. **119**, 494–528 (2014). https://doi.org/10.6028/JRES.119.019

Song, X., et al.: Advances in additive manufacturing process simulation: residual stresses and distortion predictions in complex metallic components. Mater. Des. **193**, 1–14 (2020). https://doi.org/10.1016/J.MATDES.2020.108779

Wang, W., Ning, J., Liang, S.Y.: In-situ distortion prediction in metal additive manufacturing considering boundary conditions. Int. J. Precis. Eng. Manuf. **22**, 909–917 (2021). https://doi.org/10.1007/S12541-021-00496-Z

Examining the Influence of Business Models on Technical Implementation of Smart Services

Samuel Helbling(✉) and Felix Nyffenegger

Eastern Switzerland University of Applied Sciences, St. Gallen, Switzerland
{samuel.helbling,felix.nyffenegger}@ost.ch

Abstract. This paper explores the relationship between business models and their technical implementation of smart services in the manufacturing industry. The study employed a research methodology that involved examining real-world use cases through expert interviews with companies offering smart services. The business models were assessed by using the business model patterns proposed by Gassmann et al. while the technical implementation aspect employed generic smart service patterns based on the conceptual model outlined in ISO/IEC 30141:2018. The data analysis resulted in the identification of 11 distinct generic smart service patterns with varying properties. Furthermore, the distribution of the business model patterns among the generic smart service patterns was examined to determine potential relationships and influences on the technical implementation of the smart service. This study's findings indicate several dependencies and connections between the technical implementation of smart services and their corresponding business models. The identification of interdependencies can serve as a foundation for informed decision-making in the planning and development phases of smart service implementation for organizations.

Keywords: smart services · digitalization · servitization · business model · IoT · IT architecture

1 Introduction

The manufacturing industry is undergoing a digital transformation, with companies increasingly leveraging technology to improve their operations and customer relationships [1]. Smart services, which involve communication and cloud technologies to gather information and data from a company's installed base, are one-way manufacturers achieve this. The data obtained through smart products can be used to offer new services, such as predictive maintenance and remote monitoring, and improve customer relationships through personalized service offerings and proactive customer engagement.

However, implementing and building smart service business models is not a simple task [2]. Many manufacturing companies lack the necessary resources and expertise to do so and must acquire or purchase new know-how and skills [3, 4]. Additionally, transitioning from product-centric to service-centric business models can be challenging, as for example, the selling process of services is different from selling products

Published by Springer Nature Switzerland AG 2024
C. Danjou et al. (Eds.): PLM 2023, IFIP AICT 701, pp. 47–58, 2024.
https://doi.org/10.1007/978-3-031-62578-7_5

[5, 6]. Companies must also overcome technical and commercial challenges related to digitalization [7].

Despite the growing adoption of smart services in the manufacturing industry, there is a lack of understanding of the technical and strategic considerations involved in implementing these services. Previous studies have primarily focused on the business aspect of smart services [8–10], leaving a gap in knowledge on the technical implementation of digital services and the underlying business strategy decisions. This study addresses this gap and aims to answer the following questions: How does the business model impact the technical implementation of a smart service?

From expert interviews, specific case studies were extracted to illustrate how the technical implementation of digital services is influenced by various aspects of the organizations' business models. Based on this consideration, 11 distinct architectural patterns could be identified and set in context with their corresponding business models.

2 Literature Review

2.1 Business Models

Business models provide a structured and simplified representation of a company [11]. One of the most widely used approaches is the Business Model Canvas, developed by Osterwalder and Pigneur [12]. This framework includes nine building blocks describing a business model's key elements, customer segments, value proposition, channels, customer relationships, revenue streams, key resources, key activities, key partners, and cost structure [13].

Another approach for business models is provided by Gassmann et al. with their St. Galler Business Model Navigator. This toolkit includes a set of patterns that companies can use to create new business models. The patterns include examples of successful businesses and guide how to implement these business models. They describe business models based on four dimensions: Customer, Value Proposition, Value Chain, and Revenue Model [14].

2.2 Smart Services and IoT

Digital servitization, the amalgamation of digitalization and servitization, presents novel opportunities for value creation and capture through smart product-service systems (SPSS) [15]. Pöppelbuß and Durst define smart services as a subset of SPSS [8] and as services that leverage data from smart products to deliver enhanced value to customers [16]. Similarly, Mittag et al. posit that smart services comprise a digital service that is based on data gleaned from a physical product, which may also be augmented by an additional physical service [8]. These smart services require a connection component in addition to their physical and smart components, as emphasized by Porter and Heppelmann [17]. The Internet of Things (IoT) offers a viable solution to this requirement by connecting the physical component of smart services to the Internet [3]. Ardolino et al. assert that IoT is crucial to digitalization as it enables the collection and transmission of data, making it a vital component of any service transformation strategy implementation [18].

2.3 Reference Architecture

Effective mapping of the technical structure of smart services requires a comprehensive reference architecture, and the ISO/IEC 30141:2018 standard provides such an architecture for IoT applications. This architecture consists of four essential aspects, including (1) the characteristics of IoT systems, (2) a Conceptual Model to represent the essential concepts and relationships of the elements of an IoT system, (3) a Reference Model to describe the overall structure of the architecture, (4) and a set of relevant views to represent the architecture from different perspectives [19, 20].

However, the IoT Architectural Reference Model (IoT ARM) is another model that provides concepts and definitions for IoT architectures, developed as part of the IoT-A project up to 2013. The IoT ARM defines four sub-models, including (1) the Domain Model, (2) the Information Model, (3) the Functional Model, and (4) the Communication Model, which together provide definitions of the key functionalities and communication paradigms for connecting elements in the IoT Domain Model and provide a guide for developing IoT-A compliant functional views and building interoperable stacks [21, 22].

3 Research Methodology

The research methodology employed in this qualitative study is based on examining real-world use cases and is illustrated by Fig. 1. Given the technical complexity of smart services, a literature-based approach alone is insufficient to understand the subject matter comprehensively. Therefore, expert interviews with companies offering smart services were conducted to gain insight into these services' technical implementation and business models.

To be eligible for inclusion in the study, the use cases would have to meet the following criteria: (1) the smart service is offered to business customers, (2) the smart service consists of both a physical product and a digital service, and (3) IoT technology is used to connect the physical and digital components. A list of 32 potential use cases and corresponding contacts was compiled, and the companies were contacted. Of the 32 use cases initially identified, 9 were found not to meet the study's criteria, and no response was received from 4 companies. Expert interviews were conducted with the remaining 19 companies, focusing on a specific smart service offered by the company rather than the company's entire service portfolio.

The structured interviews aimed to gather information on the business model elements outlined by Gassmann et al., including the value proposition, value chain, revenue model, and target customers [14]. Following the interviews, the smart services were characterized by the business model (BM) patterns identified by Gassmann et al.. Additionally, the technical implementation of the smart services, discussed within the context of the value chain, was recorded using the conceptual model (CM) according to ISO/IEC 30141:2018.

The conceptual models were then standardized by uniformly naming entities, generalizing, combining elements, and limiting entities that occurred multiple times. The highest common denominator among the processed conceptual models was determined to be the base context module. Deviations from the base conceptual models were identified as generic smart service patterns or combinations of generic smart service patterns.

These generic smart service patterns were then applied to the original conceptual models and sent back to the interviewees for feedback on their applicability. The feedback received was incorporated into the conceptual models, and the IoT patterns were adjusted as necessary. The result are 11 distinct generic smart service (GSS) patterns with varying properties. Finally, the co-occurrence of business model patterns and generic smart service patterns was examined to determine potential relationships and the impact of strategic business decisions on the technical implementation of the smart service.

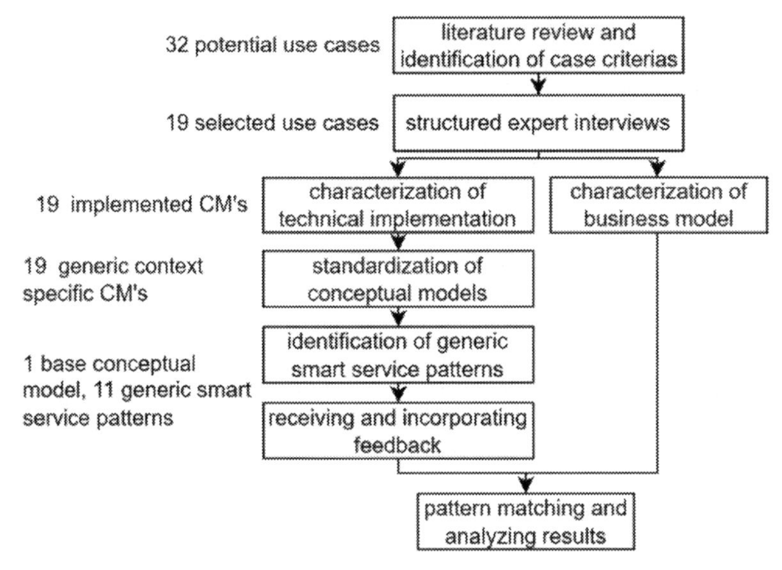

Fig. 1. Selected methodology for the qualitative study

4 Case Studies

The analysis of the relationship between the business model and the technical structure of Smart Service is based on the 19 case studies listed in Table 1. The table includes general information about the smart service provider, such as industry, the number of employees, and the age of the service. Furthermore, the GSS patterns, described in Sect. 5, and BM patterns, according to Gassmann et al., are also listed. All companies studied in this research are headquartered in a country within the DACH region, the vast majority in Switzerland.

Table 1. Case Studies

#	Sector	Employees	GSS Patterns	BM Patterns	Age Service
1	Measuring and Control Instrument Manufacturing	250+	4,8	1,27,48,57	–
2	Construction	250+	3,6,11	7,11,25,34,40,48,56,57	1 year
3	Biotechnology Research	250+	3,8,9	18,57	6 years
4	Industrial Machinery Manufacturing	250+	2,6,8,11	1,25,32,48,57,58	3 years
5	Machinery Manufacturing	250+	4,5,8,9,11	19,25,27,47,48,57	2 years
6	Construction	250+	7	11,27,48,56,57	4 years
7	Machinery Manufacturing	50–249	2	11,34,57	3 years
8	Appliances, Electrical, and Electronics Manufacturing	250+	8,9,10,11	25,48,57	9 years
9	Appliances, Electrical, and Electronics Manufacturing	250+	9,10,11	7,11,25,40,48,56,57	7 years
10	Machinery Manufacturing	–	3,4,5,6,8	23,48,57	<1 year
11	IT Services and IT Consulting	1–9	3,4,5,6	11,23,48,57	6 years
12	Industrial Machinery Manufacturing	1–9	1,5,9	11,40,48,57	6 years
13	Wholesale	250+	1,2,3,9,10,11	11,25,31,48,40,57,58	+10 years
14	Transportation Equipment Manufacturing	250+	3,8,10,11	11,25,48,57	8 years
15	Utilities	10–49	3	11,48,56,57	2 years
16	Automation Machinery Manufacturing	50–249	3,5,8,9	23,48,57	2 years
17	Information Services	1–9	2,3	11,31,48,57	1 year
18	Industrial Machinery Manufacturing	1–9	3,4,5,11	11,23,48,56,57	2 years
19	Machinery Manufacturing	250+	8,9,10,11	20,25,39,47,57	+10 years

5 Generic Smart Service Patterns

In this work, we present 11 generic smart service (GSS) patterns as technical counterparts to the business model (BM) patterns proposed by Gassmann et al. These patterns are described in accordance with ISO/IEC 30141:2018 and consist of various elements, which are detailed in Sect. 5.1. Section 5.2 examines the fundamental architecture of every Smart Service, serving as the foundation for the GSS patterns outlined in Sect. 5.3.

5.1 Entities

An entity describes an element of the CM according to the standard ISO/IEC 30141:2018. The entities of the base conceptual model (BCM) are described in more detail in Table 2. The description is based on the ISO/IEC 30141:2018 standard, the IoT-A standard, and experience from the use cases. The GSS patterns introduce additional entities. The entities can be classified into three distinct categories based on their operation and management. In Fig. 2, the entities that the smart service provider operates are shown with a hatch pattern. The dashed hatch pattern indicates that the smart service provider operates the entity in question. Finally, the crosshatch pattern indicates that a third-party organization operates the entity.

5.2 Base Conceptual Model

The base conceptual model (BCM), shown in Fig. 2, was developed by identifying the commonalities among the 19 smart services that were studied. The BCM encompasses the fundamental capabilities of a smart service, including the ability to acquire data through IoT devices, transmit data through various networks, store data in a data storage by a service, and visualize data through an application. This model aligns with the foundational requirements of smart factory use cases, as described by Budde et al., which can also be applied to smart services [23]. In addition, the Smart Service Building Blocks appearing in every case study examined by Mittag et al. are also covered by the entities of the BCM [8].

5.3 Generic Smart Service Patterns

The analysis of the normalized use cases resulted in 11 generic smart service (GSS) patterns. These can be used independently or in combination with other generic smart service patterns. Table 3 shows the GSS patterns with the corresponding definition. As an example, Fig. 2 shows the conceptual model of GSS pattern *5 Enterprise Application Integration*.

Table 2. Entities [20, 21].

Entity	Definition	Examples
Physical entity	A physical entity is a physical object, which can include living organisms and may have a hierarchical structure	–Motor –Sheep –Machine
IoT device	IoT device connects physical and digital worlds with sensors and actuators to collect data and perform actions	–Machine control –Sensor, Actuator
Local Network	A Local Network connects IoT devices to their gateway using protocols and relies on an IoT gateway for Internet access	–CAN Bus –LoRa
IoT Gateway	An IoT gateway connects local devices to the Internet and provides various functions like protocol conversion, data processing, and security	–LoRa gateway –Firewall
Internet	The Internet is a global access network that connects IoT devices and gateways with cloud applications	
Service	A service is a digital entity, typically implemented as software on a server, and provided over the Internet	–Backend –Analytics services
Data Storage	Data storage is persistently storing data in databases for various purposes, such as data analysis	–mySQL –InfluxDB
Application	An application is a digital software that executes tasks and interacts with users, hosted in the cloud and accessible through an interface	–Dashboard –App –REST API
User	A user is someone or something that interacts with a smart service, including both human individuals and non-human automation services	–Customer –3rd party service

6 Discussion

The contingency table analysis between GSS patterns and BM patterns in Fig. 3 reveals several dependencies and connections between the two. One notable example is the frequent co-occurrence of BM pattern 25, *Leverage Customer Data*, and the GSS pattern 11, *Data-Driven Insights*. The case studies demonstrate that in order for a company to use customer data for service optimization or product improvement effectively, a corresponding interface, such as dashboards or an M2M interface, must be made available to employees by the smart service, and employees must be authorized and trained to read and use the customer data. This disclosure, possibly also to third parties, must be contractually defined. These patterns occur in just under half of all the cases examined, suggesting that leveraging customer data is a key opportunity for companies to differentiate themselves from competitors. Similarly, Kowalkowski and Ulaga highlight the importance of companies collecting data from their installed bases and using it strategically [24].

Another trend observed is the increased co-occurrence of BM Pattern 11, *Digitalization*, and GSS Pattern 8, *IoT Data Platform*. This may be due to companies digitalising

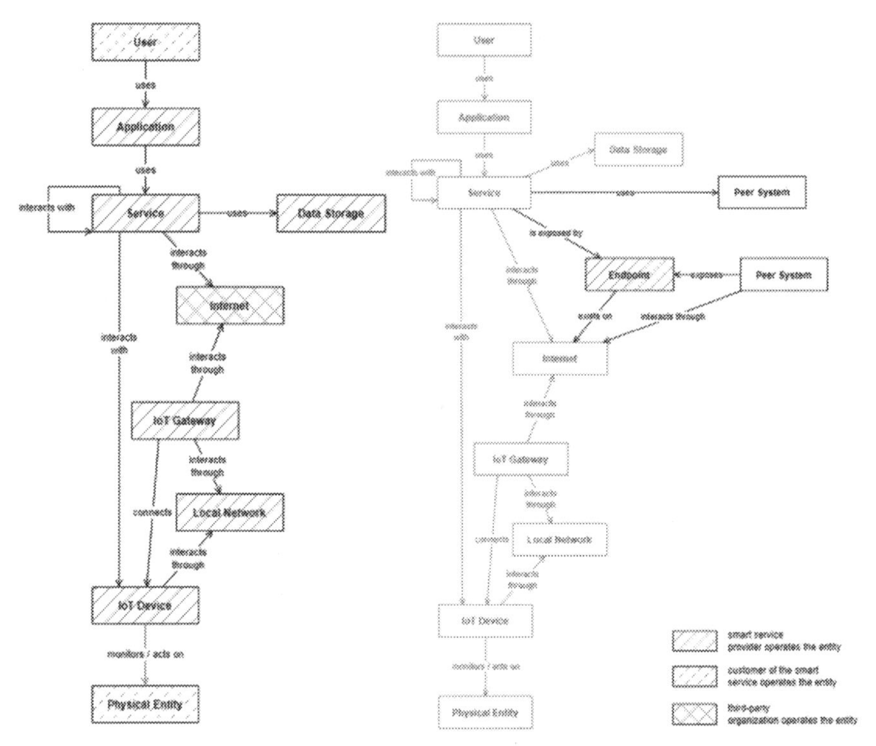

Fig. 2. Left: Base conceptual model, right: Conceptual model of pattern 5 *Enterprise Application Integration*, modelling language according to ISO/IEC 30141:2018

objects for their customers via sensors and making this data available to the customers via an interface such as REST API. Additionally, the value proposition of smart services often includes simplifying the customer's life through the virtualization of a physical process, such as reading a sensor or ordering spare parts. This is reflected in the high frequency of BM Pattern 57, *Virtualization*, in every single use case. BM Pattern 48, *Subscription*, also occurs frequently as companies with smart services aim to generate recurring revenue and therefore prefer to offer the services in a subscription model, as also pointed out by Bonnemeier et al. [9], Rabe et al. [8], and Wortmann et al. [25].

Table 3. Generic smart service patterns

#	GSS Pattern	Definition
1	Physical Object is a Human	A human user is treated as a physical object, detecting input through sensors and providing feedback through actuators
2	Partner for Connectivity	Outsourcing connectivity to partners for cost savings and scalability
3	IoT Data Platform	The smart service offers a digital interface, typically an M2M interface, for querying data to integrate into existing 3rd parties systems
4	Hybrid Data Storage	The telemetry data is stored in a special database, in addition to a database for the master data
5	Enterprise Application Integration	The service queries and uses data from internal and external corporate services
6	All-in-one	The IoT device also acts as an IoT gateway and typically communicates via the mobile network
7	Smartphone as IoT Gateway	A customer's smartphone is used as an interface between an IoT device and the cloud
8	Edge Gateway	Edge gateway is an IoT device that manages and gathers data from multiple IoT devices through a local network, optimizing communication with the cloud
9	Customer Intranet	The IoT device is connected to the firewall through the customer's intranet, which serves as a gateway to the Internet
10	User-Object Service	The smart service provider offers remote or physical services to a physical entity
11	Data-Driven Insights	The smart service provider uses data from IoT devices and other sources to optimise products, services and internal processes

It is also noteworthy that every smart service, except for Case Study 14, incorporates either BM Pattern 11, *Digitalization*, or GSS Pattern 8, *Edge Gateway*. The implementation of BM Pattern 11 implies using an IoT device to measure influencing variables such as current or temperature. On the other hand, GSS Pattern 8, *Edge Gateway*, is employed in situations where a digitalised system, such as a machine control system, already exists. Both options require an IoT device to digitalise the physical world, yet in the case of an edge gateway, the IoT device is not included in the functional scope, resulting in the exclusion of BM Pattern 11, *Digitalization*, from the smart service.

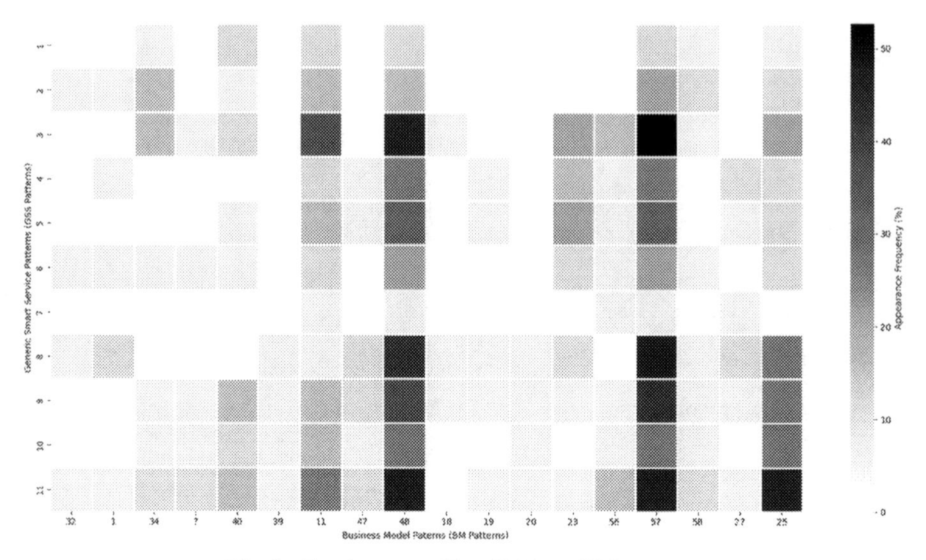

Fig. 3. Contingency table of BM vs. GSS patterns

7 Conclusion and Outlook

This study aimed to investigate the relationship between the business model of smart services and their technical structure. By analyzing 19 case studies, we determined that strategic business decisions influence the IT architecture of smart services and to what extent. The results of this study provide a basis for further research in this area.

Additionally, the generic smart service patterns identified can be utilized by companies in the development and enhancement of new and existing smart services. The connections identified between BM patterns and GSS patterns enable the creation of a well-grounded blueprint of the IT architecture of smart services at an early stage, which can assist in making fundamental decisions. This can also aid in identifying and addressing challenges and risks at an early stage. For companies that already have established smart services, this research offers the opportunity to explore alternative business models that are employed with similar technical infrastructures.

Applying the GSS patterns and BM patterns to additional case studies and projects can enhance the patterns and uncover more GSS patterns. The current study serves as a foundation for further research on the influence of the business model on technical implementation. However, the study did not examine the influence of the individual elements of a business model, such as the value proposition or the revenue model. These dependencies could be further investigated. Additionally, it would be valuable to examine the influence of smart service use cases as proposed by Budde et al. [23], and an in-depth analysis of the structure of the cloud and its relationship to business model patterns also presents potential for future research.

References

1. Herterich, M.M., Uebernickel, F., Brenner, W.: The impact of cyber-physical systems on industrial services in manufacturing. Procedia CIRP **30**, 323–328 (2015). https://doi.org/10.1016/j.procir.2015.02.110
2. Alghisi, A., Saccani, N.: Internal and external alignment in the servitization journey – overcoming the challenges. Prod. Plann. Control **26**(14–15), 1219–1232 (2015). https://doi.org/10.1080/09537287.2015.1033496
3. Borgmeier, A., Grohmann, A., Gross, S.F. (eds.): Smart Services und Internet der Dinge: Geschäftsmodelle, Umsetzung und Best Practices: Industrie 4.0, Internet of things (IoT), Machine-to-Machine, Big Data, Augmented Reality Technologie. Hanser, München (2017)
4. Baines, T., Lightfoot, H. (eds.): Made to Serve: How Manufacturers Can Compete Through Servitization and Product-Service Systems. Wiley, Chichester (2013)
5. Chalal, M., Boucher, X., Marques, G.: Decision support system for servitization of industrial SMEs: a modelling and simulation approach. J. Decis. Syst. **24**(4), 355–382 (2015). https://doi.org/10.1080/12460125.2015.1074836
6. Valtakoski, A.: Explaining servitization failure and deservitization: a knowledge-based perspective. Ind. Mark. Manage. **60**, 138–150 (2017). https://doi.org/10.1016/j.indmarman.2016.04.009
7. Gebauer, H., Fleisch, E., Friedli, T.: Overcoming the service paradox in manufacturing companies. Eur. Manag. J. **23**(1), 14–26 (2005). https://doi.org/10.1016/j.emj.2004.12.006
8. Mittag, T., Rabe, M., Gradert, T., Kühn, A., Dumitrescu, R.: Building blocks for planning and implementation of smart services based on existing products. Procedia CIRP **73**, 102–107 (2018). https://doi.org/10.1016/j.procir.2018.04.010
9. Bonnemeier, S., Burianek, F., Reichwald, R.: Revenue models for integrated customer solutions: concept and organizational implementation (in En;en). J. Revenue Pricing Manag. **9**(3), 228–238 (2010). https://doi.org/10.1057/rpm.2010.7
10. Fleisch, E., Weinberger, M., Wortmann, F.: Geschäftsmodelle im Internet der Dinge. HMD **51**(6), 812–826 (2014). https://doi.org/10.1365/s40702-014-0083-3
11. Jodlbauer, H.: Digitale Transformation der Wertschöpfung. Verlag W. Kohlhammer, Stuttgart (2018)
12. Bieger, T., zu Knyphausen-Aufseß, D., Krys, C. (eds.): Innovative Geschäftsmodelle. Springer, Heidelberg (2011). https://doi.org/10.1007/978-3-642-18068-2
13. Osterwalder, A., Pigneur, Y.: Business Model Generation: Ein Handbuch für Visionäre, Spielveränderer und Herausforderer. Campus Verlag, Frankfurt/New York (2011)
14. Gassmann, O., Frankenberger, K., Choudury, M.: Geschäftsmodelle entwickeln: 55+ innovative Konzepte mit dem St. Galler Business Model Navigator, 3rd edn. Hanser, München (2021)
15. Kohtamäki, M., Parida, V., Oghazi, P., Gebauer, H., Baines, T.: Digital servitization business models in ecosystems: a theory of the firm. J. Bus. Res. **104**, 380–392 (2019). https://doi.org/10.1016/j.jbusres.2019.06.027
16. Pöppelbuß, J., Durst, C.: Smart Service Canvas – Ein Werkzeug zur strukturierten Beschreibung und Entwicklung von Smart-Service-Geschäftsmodellen. In: Bruhn, M., Hadwich, K. (eds.) Dienstleistungen 4.0, pp. 91–110. Springer Fachmedien Wiesbaden, Wiesbaden (2017). https://doi.org/10.1007/978-3-658-17552-8_4
17. Porter, M.E., Heppelmann, J.E.: How smart, connected products are transforming competition. Harv. Bus. Rev. **92**(10), 64–88 (2014)
18. Ardolino, M., Rapaccini, M., Saccani, N., Gaiardelli, P., Crespi, G., Ruggeri, C.: The role of digital technologies for the service transformation of industrial companies. Int. J. Prod. Res. **56**(6), 2116–2132 (2018). https://doi.org/10.1080/00207543.2017.1324224

19. Holtschulte, A.: Praxisleitfaden IoT und Industrie 4.0: Methoden, Tools und Use Cases für Logistik und Produktion. Hanser, München (2021)
20. Internet of Things (IoT) – Reference architecture, ISO/IEC 30141. International Organization for Standardization (2018)
21. Walewski, J.W.: Internet-of-Things Architecture IoT-A: Project Deliverable D1.2 – Initial Architectural Reference Model for IoT (2011). Accessed 17 Jan 2023
22. Bassi, A., et al. (eds.): Enabling Things to Talk: Designing IoT Solutions with the IoT Architectural Reference Model, 1st edn. Springer, Berlin (2016). https://doi.org/10.1007/978-3-642-40403-0
23. Budde, L., Hänggi, R., Friedli, T., Rüedy, A.: Smart Factory Navigator: Identifying and Implementing the Most Beneficial Use Cases for Your Company–44 Use Cases That Will Drive Your Operational Performance and Digital Service Business. Springer, Cham (2023). https://doi.org/10.1007/978-3-031-17254-0
24. Kowalkowski, C., Ulaga, W.: Service Strategy in Action: A Practical Guide for Growing your B2B Service and Solution Business. Service Strategy Press, Scottsdale (2017)
25. Wortmann, F., Bilgeri, D., Weinberger, M., Fleisch, E.: Ertragsmodelle im Internet der Dinge. In: Seiter, M., Grünert, L., Berlin, S. (eds.) Betriebswirtschaftliche Aspekte von Industrie 4.0, pp. 1–28. Springer Fachmedien Wiesbaden, Wiesbaden (2017). https://doi.org/10.1007/978-3-658-18488-9_1

Navigating the Digital Twin: 3D Exploration for Asset Administration Shell Content

Mario Wolf(✉), Oliver Vogt, and Detlef Gerhard

Digital Engineering Chair, Ruhr University Bochum, 44801 Bochum, Germany
mario.wolf@rub.de

Abstract. The digital twin (DT) as the core for digitalization activities becomes ever more prevalent in both research and industry. One emerging interpretation of the DT concept is the Asset Administration Shell (AAS), which describes – on a semantic and machine-readable level – each asset with a structured data set for its characteristics, lifecycle status, capabilities, etc.

In complex products, such as production systems that consist of many subsystems, there are consequently also many AAS, each with individual submodel structures. Because the AAS hierarchy does not necessarily align with the physical structure of a system, assistance in accessing information and services from the digital twin is required for different processes, e.g., maintenance and service tasks.

In this contribution, the concept for a web-based visualization application is presented, which uses the association of 3D geometry and AAS to create an interactive 3D view. This allows access to AAS hierarchies, submodels, and associated information on different levels and granularity. The goal is to utilize standardized submodels to enable easy navigation through the many-faceted AAS and fast data retrieval. In this way, the AAS and respectively the DT can intuitively be queried, and its contents can be displayed to the user without prior knowledge of the structure. The first working prototype is implemented for a production system, where the focus is on accessing information about its control cabinet with certain terminals.

Keywords: Digital Twin (DT) · Asset Administration Shell (AAS) · Visualization

1 Introduction

The digital twin (DT) is a core concept for representing complex technical systems across the whole lifecycle, particularly capturing information from the use phase. One major challenge in the production sector is the growing complexity of the machines themselves, as well as their increased connectedness and the ability to instantly access operational data. The Asset Administration Shell (AAS) is one interpretation and formalization of the DT concept, as it describes each asset with structured data for its features, lifecycle status, capabilities, and so on. A smart manufacturing component (SMC) consists of

Published by Springer Nature Switzerland AG 2024
C. Danjou et al. (Eds.): PLM 2023, IFIP AICT 701, pp. 59–69, 2024.
https://doi.org/10.1007/978-3-031-62578-7_6

the asset and the associated AAS. This paper aims at enabling shop floor, operating, or maintenance personnel to intuitively access data and services generally available to an SMC through a web-based 3D browsing and visualization tool.

In the following section, the current state of the art regarding the AAS concept as well as digital twins implemented using asset administration shells will be explained, which leads into the stakeholder analysis for the proposed application. Thereafter, the concept of the Visual AAS Browser is laid out in detail to lead to the implementation, which is the first working prototype based on a production system, where the focus is on accessing information about its control cabinet with its internal terminals.

2 Related Work

2.1 Digital Twin

Through advanced communication and interaction technologies, real and virtual space are linked to form a Digital Twin (DT), enabling cyber-physical production systems (CPPS) [1], in which intelligent products are interacting and communicating with machines and within complex processes. The DT concept was coined by Grieves [2] back in 2002 and has been adopted by many different industries since then, resulting in the fact that no uniform DT definition exists [3, 4]. In a fully digitized world, every entity can collect or generate data throughout its operation. The data collection creates a digital footprint of a process or product and is an enabler for real-time data evaluation [5]. This digital footprint of operational data can be linked to CAD or simulation models, created during product development in order to create a lifecycle spanning DT of a system. The DT can be used to support various functions, such as visualization, interactions, or simulations. It can also have different hierarchies, i.e., can describe a single component, a product, or entire systems. Depending on the use case and the entity to be described, this results in different levels of complexity [6]. Overall, the common understanding is that the physical product is connected to its virtual representation and a bidirectional link between those two components exists.

2.2 Asset Administration Shell (AAS)

In order to address the different interpretations and implementations of the DT and to form a unified data schema with a focus on production systems, the concept of the asset administration shell (AAS) as a metamodel was introduced by the German Platform Industry 4.0 [7]. In this concept, an SMC consists of the physical object, the asset, and the virtual representation of the asset, the AAS [8]. The AAS is divided into a header and a body. The header contains information for identifying every AAS and the corresponding asset through a unique identifier. The body contains the actual information about the asset. The information is structured in individual submodels which in turn contain an undefined number of properties describing individual data. Each submodel refers to a certain characteristic or function of the asset [9]. Particular submodels for the AAS are already standardized and published and more standardized submodels are in development [10]. The Platform Industry 4.0 presented a first prototype of an implementation for a pick and place station, in which the AAS consists of sub-AAS and therefore contains a bill of material submodel (BOM) [11]. Three different types of AAS can be distinguished. The first type is the passive AAS where the content is represented in a file for data exchange between partners in the value chain. Re-active AAS is the second type and specifies server-based access to AAS information via an API for applications or services. The third type is the proactive AAS. A proactive AAS can interact with other AAS and exchange information using the I4.0 language [12]. The interface for exchanging or accessing information via API is also specified by Platform Industry 4.0. The interface is specified technology agnostic but also gives implementation guidelines for given technologies like REST, OPC UA, or MQTT [13]. For creating a new AAS, an open-source tool named AASX-Package Explorer was developed and provided [14]. To host the AAS on a server and access it via an API, several solutions already exist [15, 16]. The AASX Server also provides a graphical user interface for exploring all AAS hosted on the server [15].

2.3 Implementations Based on AAS

Platenius-Mohr et al. illustrated how they achieved file- and API-based information exchange for interoperable digital twins. In the paper, they proposed a mapping system to transform existing digital twins to the AAS format [17]. Schelter presents two examples to show what a prototypical AAS for industrial plants might look like and how data from these machines could be updated in the AAS. For example, AAS for electron beam machines and vacuum pumps are developed and it is shown that it is extremely difficult to link data from actual machines or plants to an AAS in a meaningful way [18]. How submodels for pumps can be constructed and converted into XML schemas for usage in an OPC-UA mapping is demonstrated by Müller and Both [19]. Cavaleieri and Salafia presented an approach of representing an IEC 61131-3 program and its relationships by using the AAS metamodel. With this approach engineers and technicians from different domains can understand the relations between the plant and its control programs better [20].

3 Concept

CPPS consist of many subsystems with varying levels of connectivity, access to operational data, or documentation. In this paper's example, a smart manufacturing system is composed of certain individual components, which are manufactured by different companies. These different components were integrated and supplied by one general contractor. When viewing the smart manufacturing system as a singular asset with its accompanying DT, which holds all relevant (meta) data about the asset as well as references to its components, the operator needs a way to understand and use this particular hierarchy. This is getting more complex if the components of themselves are considered assets that provide individual DT. It is assumed that each of these (partial) assets has an AAS as the representation of the DT, in which information about the asset and its functionalities is stored. The operator now needs to be enabled to retrieve the contents of a selected AAS as the follow-up step of understanding the asset hierarchy. Therefore, the following concept aims to develop a concept for human interaction with asset administration shells focused on exploration and depending on the user's access rights. The authors propose a Visual AAS Browser that allows browsing through an asset's AAS, as well as their accessible contents, by using an interactive version of the 3D geometry of the primary smart manufacturing system as the user interface. The concept starts with a stakeholder and use case analysis, before detailing the IT implementation concept of the Visual AAS Browser.

3.1 Basic Scenario and Stakeholder Analysis

In the given scenario, a system supplier acquires the components required for the smart manufacturing system from component suppliers. In addition to the physical components, the system supplier also receives 3D representations of the components from the component suppliers as well as the AAS belonging to the components in the AASX exchange format.

After receiving the purchased components, the system supplier can build the smart manufacturing system and simultaneously create the DT in the form of a structured AAS hierarchy. Access to the final system's AAS is to be provided via standardized (web)services, which also includes access to all other AAS belonging to the system in the lower tiers of the hierarchy. In the assumed scenario, the server is hosted by the system supplier, who is the provider of the Visual AAS Browser as a service to the customer and takes care of access rights, data, and system availability. However, different stakeholders can be considered as consumers of the service. The first stakeholder is the system supplier. The Visual AAS Browser can support the production of the smart manufacturing system's components as well as the integration of the third-party components and quality control, as all of those components offer AAS, which in the assumed process, have already been integrated and are available before physical assembly. During operation, the operator of the smart manufacturing system becomes the consumer of the Visual AAS Browser services. After handover to the customer, only the customer shall have access to the current usage and operating data of the product, while the static descriptive data of the AAS remains unchanged from the as-delivered state until changes are made to the physical system.

When configuring the Visual AAS Browser service, the provider (system supplier in this case) specifies the customer-individual server on which the AAS is provided, which is then published from the internal representations. The functionality of browsing the AAS should be as simple as exploring the 3D representation of the smart manufacturing system in a general 3D viewer.

When selecting a component, the user selects not individual physical components that would be represented by a typical BOM, but the actual assets as represented by the accompanying AAS. The Visual AAS Browser retrieves the data through the standardized web service interface from the supplier's AAS-server instance that is provided specifically for the customer. Figure 1 summarizes the basic scenario.

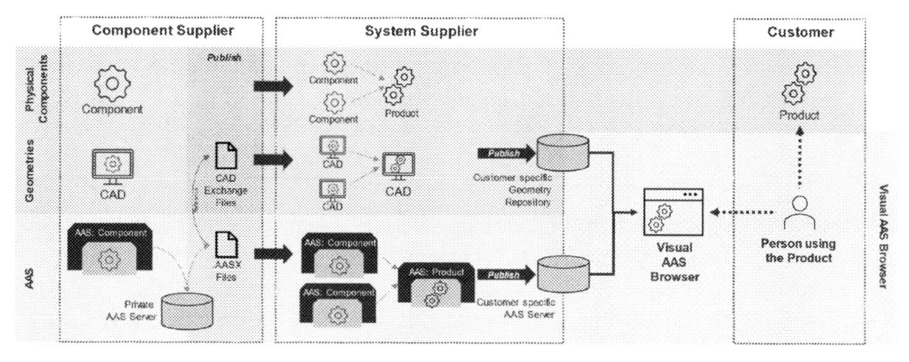

Fig. 1. Basic Scenario

3.2 AAS Creation

To deliver a SMC, one needs to provide the physical product alongside the corresponding AAS, which also includes the associated CAD files. Since a component can consist of several subsystems, the component AAS itself can be composed of several subjacent AAS. The references are stored in a specific submodel of the corresponding AAS.

In the context of the presented approach, each AAS contains (at least) the submodels *Nameplate* and *TechnicalData*. Further submodels are added for additional use cases, such as for quality control. If the component AAS refers to further AASs, a *BOM* submodel referencing these subjacent AASs must also be added to the component AAS by the component supplier. The component supplier sends the AAS to the system supplier as a file in the AASX format.

3.3 Preparation of a CAD Model for the Visual AAS Browser

To enable the correct interaction between the Visual AAS Browser and the AAS belonging to the selected element, an association must be accessible between the selectable geometry and the associated SMC, to then retrieve the correct AAS in the hierarchy of the system. For this purpose, a unique ID is required that realizes the reference to the associated AAS. In this concept, the ID of the AAS is stored in the 3D model of the component in the form of a user-defined property named *AAS_ID*. It should be noted that multiple instances of the same type of component can be installed in a system. Accordingly, the ID of the associated AAS instance must be stored in the user-defined property in the CAD model that is instance-specific.

3.4 Integration Process by the System Supplier

The system supplier's role is to create the SMC, which in the assumed use case includes the components from other suppliers. The integration happens on both the virtual layer, with the CAD and BOM information, and the physical layer, with the assembly of the real production system. Additionally, the BOM is added to the smart manufacturing system's AAS as reference. The system supplier can also add the AAS received from the component suppliers to the hierarchy of the smart manufacturing system's SMCs.

The system supplier now associates the CAD models of the BOM components with the SMCs. The AAS and the referenced 3D models are then put in repositories accessible to the Visual AAS Browser with the customer's credentials.

3.5 Visual AAS Browser

The Visual AAS Browser is the main component of the concept, which provides a user interface between the smart manufacturing system and the operator. The main functions of the viewer are displaying the 3D model and the SMC hierarchy of the complete system, enabling the selection of SMC in the 3D model and hierarchy, and retrieving and displaying the AAS data of the selected component.

For displaying the complete production system with its components, the Visual AAS Browser accesses the AAS and the associated 3D models provided by the system supplier's repositories, authenticated with the customer's credentials. After selecting the currently available systems to browse, which in this case is the smart manufacturing system only, the global 3D view is loaded and ready for the user to interact with.

Figure 2 shows a rough sketch of the planned front end of the Visual AAS Browser. The object of study in this sketch is the smart manufacturing system consisting of various machines and parts provided by the component supplier, as well as a control cabinet. The user of the Visual AAS Browser has the option to select an asset in either the 3D model, in the BOM structure, or the SMC structure. Both latter two are displayed in tree views. Either way, the associated components are highlighted in all three representations, and the contents of the associated AAS are displayed in a separate window next to the 3D model, which can display either tabular data or specific other formats, if supported.

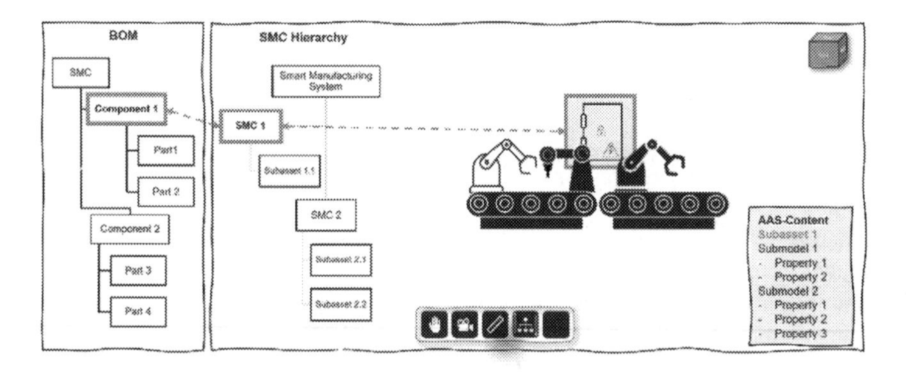

Fig. 2. Visual AAS Browser concept user interface

4 Prototypical Implementation

The aforementioned object of study is a smart production system with various components, as depicted in Fig. 3 (top). For the implementation, the drill-down using the Visual AAS Browser is shown by first selecting the control cabinet as a sub-component to the production system (Fig. 3, top) and then a profile rail with terminals (Fig. 3, bottom) as part of the control cabinet. In this implementation example, the CAD model and AAS repositories are set up in a way that the system supplier would provide to the customer as per the proposed concept. The Visual AAS Browser loads the selected global 3D geometry from the online repository with the accompanying AAS and displays it to the user, to start the browsing process. When an SMC is selected by the user, the data is retrieved from the associated AAS and visualized accordingly.

Fig. 3. Smart Production System with selected control cabinet (top) and control cabinet with selected terminal rail component (bottom)

4.1 Implementation of the AAS and the AAS-Server

The AAS required for the system and its SMC were created manually with the open-source tool AASX Package Explorer in preparation for the implementation. A new aasx-file is created for each AAS, containing the submodels *Nameplate* and *TechnicalData,* which are both templated and published by the Industrial Digital Twin Association

(IDTA). The AAS for the system itself also contains a BOM submodel. These files are published on the open-source AASX Server and accessible to the Visual AAS Browser through the REST web service Interface.

4.2 Adaptations to the CAD Model

Only one extra step is required for the preparation of the Visual AAS Viewer usage when creating the CAD model in the engineering process, as described in the concept section. The ID of the corresponding AAS must be added to the part that has its own AAS. For this implementation, Autodesk Inventor is used as the CAD software. In Autodesk Inventor, custom properties can be added to parts or assemblies via the so-called iProperties by defining the name, type, and value of the property. In this case, the name is "*AAS_ID*", the type is "*string*" and the value is the ID of the corresponding AAS created in the step before.

4.3 Implementation of the Visual AAS Browser

The first two parts of the implementation can be considered preparation for the implementation of the Visual AAS Browser. The authors chose Autodesk Platform Services as the visualization framework for the implementation [21]. The prepared CAD model is made available on Autodesk Fusion Team to be accessible online for the Visual AAS Browser. It is already possible to display 3D models and their basic component structure with the original Autodesk viewer. However, since the main intention of the Visual AAS Browser is to display and browse the contents of the AAS of SMC selected in the viewer, a JavaScript-based extension is created to enhance the base functionality and to enable the concept's functionalities.

When clicking on a component, the ID stored in the CAD model of the selected element is evaluated by the extension. The viewer retrieves the AAS from the AAS server based on the mapping of the element in the associated AAS. The response from the server returns a list of all submodels' *shortIDs* contained in the AAS. Consequently, a new request is sent to the server for each submodel *shortID*, once the user wants to access a particular submodel. The server's answer contains the content of the requested submodel.

Figure 4 shows a screenshot of the Visual AAS Browser. In this case, the part "2-Leiter-Schutzklemme" (2-wire protective terminal) inside the control cabinet is selected as a part in the lowest tier of the SMC hierarchy. The Visual AAS Browser is configured to display the data of all submodels in a tabular view, which the user can collapse based on the structure of the individual submodels. The data in this table can also be sorted by attribute name or value by the consumer.

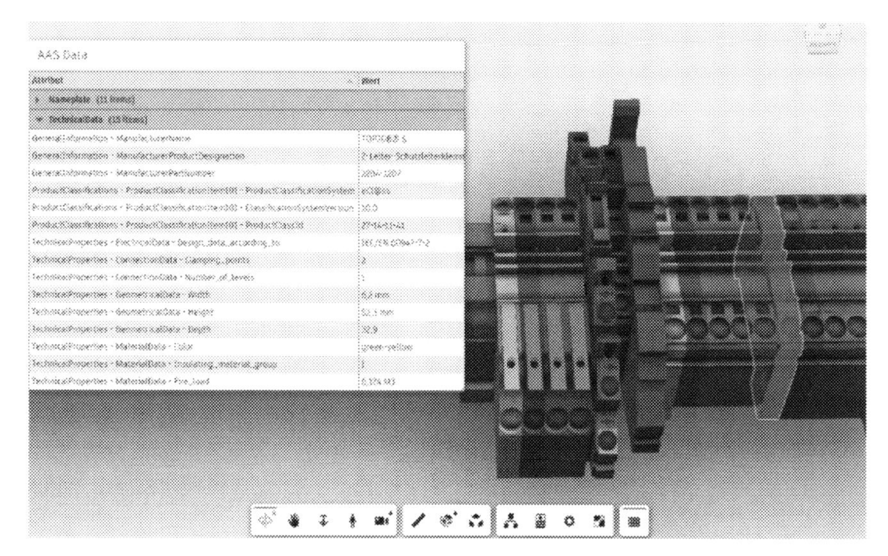

Fig. 4. Focus on the lowest tier (part level) of an SMC in the Visual AAS Browser

5 Conclusion and Outlook

The presented approach of visualizing and browsing through a complex system's AAS was implemented based on established industrial software products, with most of the planned functions of the Visual AAS Browser's concept. Firstly, it is possible to load the corresponding AAS and display its contents when selecting an SMC. This function was prototypically implemented and is functioning as intended. Contents of any AAS can be queried dynamically, independently of how many properties are stored in a submodel or how many submodels an AAS contains. One possible improvement of the developed viewer extension is the optimized display of the contents of the AAS submodel, especially for standardized submodels, while falling back to the tabular view for any non-standardized submodels with textual content and the option for both image and PDF viewing capabilities.

Future work will encompass the implementation of a second extension for the Autodesk platform services for showing the SMC hierarchy and enabling the selection of an asset from the structured hierarchy while highlighting all linked components in the 3D model and part and SMC hierarchy, as laid out in the concept.

Once the full concept is implemented, the scope of the use cases will be broadened to enable writing access to the AAS. This allows scenarios like quality inspection processes to be performed based on submodel availability and the therein-contained data fields and instructions.

References

1. Biffl, S., Lüder, A., Gerhard, D. (eds.): Multi-disciplinary engineering for cyber-physical production systems. Springer, Cham (2017). https://doi.org/10.1007/978-3-319-56345-9

2. Grieves, M.: Origins of the Digital Twin Concept (2016)
3. Kritzinger, W., Karner, M., Traar, G., Henjes, J., Sihn, W.: Digital Twin in manufacturing: a categorical literature review and classification. IFAC-PapersOnLine **51**(11), 1016–1022 (2018). https://doi.org/10.1016/j.ifacol.2018.08.474
4. Jones, D., Snider, C., Nassehi, A., Yon, J., Hicks, B.: Characterising the digital twin: a systematic literature review. CIRP J. Manuf. Sci. Technol. **29**, 36–52 (2020). https://doi.org/10.1016/j.cirpj.2020.02.002
5. Schuh, G., Walendzik, P., Luckert, M., Birkmeier, M., Weber, A., Blum, M.: Keine Industrie 4.0 ohne den Digitalen Schatten. Zeitschrift für wirtschaftlichen Fabrikbetrieb **111**(11), 745–748 (2016). https://doi.org/10.3139/104.111613
6. Stark, R., Damerau, T.: Digital twin. In: Chatti, S., Tolio, T. (eds.) CIRP Encyclopedia of Production Engineering, pp. 1–8. Springer Berlin Heidelberg, Berlin, Heidelberg (2019). https://doi.org/10.1007/978-3-642-35950-7_16870-1
7. Federal Ministry for Economic Affairs and Climate Actions (BMWK). Details of the Administration Shell - Part 1: The exchange of information between partners in the value chain of Industrie 4.0 value chain of Industrie 4.0 (Version 3.0RC02) (2022). https://www.plattform-i40.de/IP/Redaktion/DE/Downloads/Publikation/Details_of_the_Asset_Administration_Shell_Part1_V3.html
8. Hoffmeister, M.: Die Industrie 4.0-Komponente. ZVEI - Industrie 4.0 (2015). https://www.plattform-i40.de/IP/Redaktion/DE/Downloads/Publikation/zvei-faktenblatt-i40-kompon ente.html
9. Federal Ministry for Economic Affairs and Energy (BMWi). Structure of the Administration Shell: Continuation of the Development of the Reference Model for the Industrie 4.0 Component (2016). https://www.plattform-i40.de/IP/Redaktion/EN/Downloads/Publikation/structure-of-the-administration-shell.html
10. IDTA English. IDTA - AAS Submodel Templates. https://industrialdigitaltwin.org/en/con tent-hub/submodels. Accessed 24 Jan 2023
11. Plattform Industrie 4.0. AAS Reference Modelling: Exemplary modelling of a manufacturing plant with AASX Package Explorer based on the AAS metamodel (2021). https://www.pla ttform-i40.de/IP/Redaktion/EN/Downloads/Publikation/AAS_Reference_Modelling.html
12. Federal Ministry for Economic Affairs, "Verwaltungsschale in der Praxis: Wie definiere ich Teilmodelle, beispielhafte Teilmodelle und Interaktion zwischen Verwaltungsschalen? (Version 1.0) (2020). https://www.plattform-i40.de/IP/Redaktion/DE/Downloads/Publikation/2020-verwaltungsschale-in-der-praxis.html
13. Federal Ministry for Economic Affairs. Details of the Asset Administration Shell - Part 2: Interoperability at Runtime – Exchanging Information via Application Programming Interfaces (Version 1.0RC02) (2021). https://www.plattform-i40.de/IP/Redaktion/DE/Downloads/Publikation/Details_of_the_Asset_Administration_Shell_Part_2_V1.html
14. GitHub. admin-shell-io/aasx-package-explorer: C# based viewer/editor for the Asset Administration Shell. https://github.com/admin-shell-io/aasx-package-explorer. Accessed 24 Jan 2023
15. GitHub. admin-shell-io/aasx-server: C# based server for AASX packages. https://github.com/admin-shell-io/aasx-server. Accessed 24 Jan 24 2023
16. Eclipse BaSyx. About BaSyx. https://www.eclipse.org/basyx/about/. Accessed 28 Jan 2023
17. Platenius-Mohr, M., Malakuti, S., Grüner, S., Schmitt, J., Goldschmidt, T.: File- and API-based interoperability of digital twins by model transformation: an IIoT case study using asset administration shell. Futur. Gener. Comput. Syst. **113**, 94–105 (2020). https://doi.org/10.1016/j.future.2020.07.004
18. Schelter, C.: Überführung der Industrie 4.0 Verwaltungsschale in die Praxis anhand zweier industrieller Anwendungen. In: Automation 2018, pp. 225–234. VDI Verlag (2018)

19. Both, M., Müller, J.: Entwicklung einer Industrie 4.0 Verwaltungsschale auf Basis des allgemeinen Geräteprofils für Pumpen. In: Automation 2018, pp. 705–716. VDI Verlag (2018)
20. Cavalieri, S., Salafia, M.G.: Asset administration shell for PLC representation based on IEC 61131-3. IEEE Access **8**, 142606–142621 (2020). https://doi.org/10.1109/ACCESS.2020.3013890
21. Autodesk Platform Services. Autodesk Platform Services. https://aps.autodesk.com/. Accessed 24 Jan 2023

Industrial Layout Mapping by Human-Centered Approach and Computer Vision

Osmar Moreira da Silva Neto and Marcelo Rudek(✉)

Pontifical Catholic University of Paraná (PUCPR), Curitiba Paraná, Brazil
{marcelo.rudek,marcelo.rudek}@pucpr.br

Abstract. Digital twins and immersive mixed reality have been improving the human-centered digital transformation within the automotive industry context. In computer vision, photogrammetry-based techniques for reconstructing 3D environments make it possible to create virtual models of real scenery based solely on a set of digital images. This paper addresses the problem of projecting and updating shopfloor layouts through 3D mapping of the real environment, corresponding to its digital twin. Within industrial plants, we encounter objects of various sizes, shapes, and layouts, which would typically require a significant amount of time to create a digital twin model using manual measurement devices such as tape measures and clipboards to annotate length, distances, and angles. Through photogrammetry using images, we can recreate these layouts and objects with accuracy, speed, and rich details, eliminating the need for hand-drawing them in CAD software. Therefore, we propose an automated method for digitizing shop floor layouts and creating digital twins where humans can interact and make changes to the modeled equipment structure to simulate different arrangements before implementing actual changes on the shop floor in an industrial setting. This text presents the innovations to digital environments technologies, discusses the feasibility, and evaluates the expectation of the proposed method.

Keywords: Digital Twin · Computer Vision · Industrial Layout

1 Introduction

A recurrent industry problem is to configurate the shopfloor layout due constant changes to new products and their respective new production methods, devices and technologies. In the engineering sector this problem is a big challenge because nowadays this work continues to be done as handmade drawing sketch as an initial step to create a CAD model. Thus, this work delays several hours due to its entirely manual nature. Aiming to minimize the time required to execute this activity, we propose utilizing some new technological artifacts with greater efficiency, such as computer vision techniques and photogrammetry, together with CAD software, in order to create a corresponding Digital Twin (DT). Nowadays, in the Industry 4.0 scenario, as numerous industries undergo to digital transformation, the digital twin emerges as a support element. It is widely

Published by Springer Nature Switzerland AG 2024
C. Danjou et al. (Eds.): PLM 2023, IFIP AICT 701, pp. 70–79, 2024.
https://doi.org/10.1007/978-3-031-62578-7_7

recognized as an indispensable asset for competitiveness and achieving significant economic advantages over competitors. By ushering in a revolutionary wave, digital twins are reshaping industries [1]. In this context, virtual simulations allow us to evaluate of different scenarios, becoming shorten the design and analysis steps, and making the process of re-designing easier and faster. Thus, after the DT implemented, it can be applied in different parts of the design of a new product design, from the conceptual the idea of the product until testing for evaluation [2, 3]. According [4], the DT modeling follows the rule of these components: real space, virtual space, and linking mechanism to get the data/information interchange between them. From normalization ISO DIS 23247 to DT environments presented in [5], we found the four mains parts of a DT standardization, as: (i) implementation study; (ii) modelling and synchronizing, (iii) analytics, and (iv) user interaction and safety. After understanding the problem's need from the initial study, the following question to be solved is about the mapping to the virtual ambient from real process. Thus, we need to build the integrated way that replicates the real scenery. And, after that to stablish the protocols of communication and synchronization rate and to define which data will be collected. In the analytical phase, the data collected are stored, analyzed and used as simulation information. The result of simulated can suit with signal of control for a plant and/or information for user. Lastly, the user and safety phase, is the definition of a way an user communicates with the digital twin, for example, between an IHM (Men-Machine Interface) or Web application [5]. Since our ongoing research is now in the phase of modelling, we present a method to time reduction in creating a Digital Twin shopfloor layout by computer vision approach instead the traditional handmade drawing layout.

2 Background

2.1 The Problem Addressed

In this research we are addressing a recurrent industry problem, that is the alteration of the shopfloor layout. The constant changes are needed due new products and its respective new production devices and technologies. In the engineering sector this problem is a big challenge because nowadays this work was doing with paper, pencil, measure tape and clipboard to handmade drawing a sketch as an initial step to create a CAD model. The flow of manual operations is presented in Fig. 1.

As presented in Fig. 1, we are representing a real process whose operation involves the technical drawing which initially is used to represent the objects and the machines clearly and precisely identified from shopfloor [6]. This process takes several days because an employed should measure all layout area with the measure-tape and drawing it. After, the handmade sketch should be sent to the specific engineering department to be draw in a CAD system also in a handmade way. The remain of process is planning the layout update by accessing a specific equipment library and then proponing a new shopfloor layout based on the free or occupied area initially measured. This manual process is not a difficult task by itself, just a clipboard for hand drawing the layout in a sketch and after redrawing it to a CAD 2D system. However, the time consumption added to perform the drawing twice and the respective final time accumulated is a bottleneck in the process of layout reprograming. In this paper will be shown how computer vision

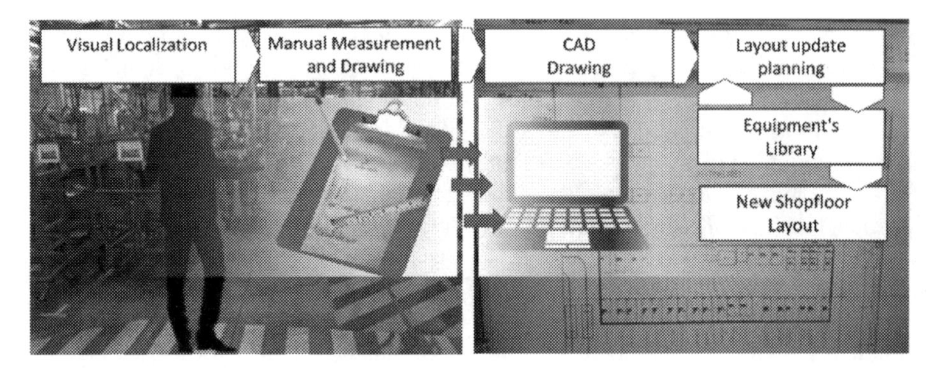

Fig. 1. The manual process steps from visual inspection to layout creation.

and 3D reconstruction become feasible a digital twin to transform a real layout in your digital representation by an automatic way from video sequence images.

2.2 The Computer Vision and the 3D Reconstruction

Computer vision is the process that intends to represent a real scenery to corresponding 2D or 3D digital formatted image trough sensors, as RGB cameras, ToF cameras, Lasers and Lidars, and so one. In order to mapping the environments, both the coordinates and distance from each image pixel to the sensor in the tridimensional space must be computed faster and accurate [7, 8]. The data captured by range sensors generates a cloud of points that represents an object or a complex scenery, and the identification of all elements compounding the scene can be computed by deep learning approaches. In order to overcome the high costs of the laser-based range sensors (LIDAR, ToF) the cloud of points from 3D space can be recovered by photogrammetry based on multiple views [9] from a single RGB camera, as in Fig. 2.

With the development of VR (virtual reality) technology and the continuous emergence of new devices, three-dimensional reconstruction has become an usual research topic in the field of computer vision [8, 10]. The main task is to build a three-dimensional model [9] of the real physical world based on the data collected by various sensors, using mathematical tools such as multiview geometry, probability statistics, and optimization theory, and building a bridge between the real world and the virtual world [11, 12]. These theories are encapsuled in the reconstruction systems, as the example in Fig. 2, the equivalent 3D model from reconstructed images sequence in the software Micro Station [13].

In order to change the manual drawing layout is needed integrate these concepts with available technologies tools, as software language like for instance the Python or C++ and it is OpenCV library to handling images and format them to be used as input to the simulation environments. Also, to build the virtual environment we addressed to apply Unity, Blender, etc. engines to recreate the similar scenery.

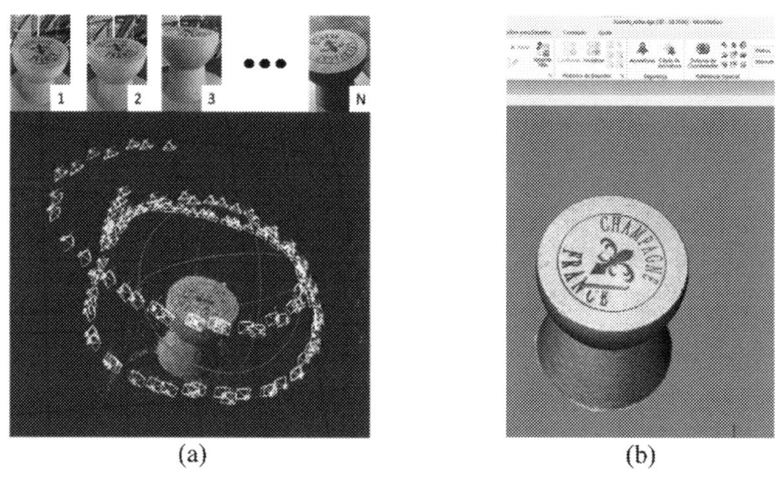

Fig. 2. (a) Camera's positions to *1,2,3, ..., N* images from the object snapshots. (b) The equivalent 3D model from reconstructed images sequence.

3 The Proposed Method

In order to overtake the handmade process as mentioned previously in Fig. 1, we proposed the automated step inside the Digital Twin that; analytical phase, the data collected are stored, analyzed and with based on desired be simulated. The result of simulated can suit with signal of control for a plant and/or information for user [5]. Described in the steps of Fig. 3 as the strategy to reduce the time and the to bring a real corresponding layout for engineering analyzes and for this analyzes is necessary verify the size of object and capture the images or create a video when the ambient has not movement people or machines because this confounding the software in the process of create a Digital Twin.

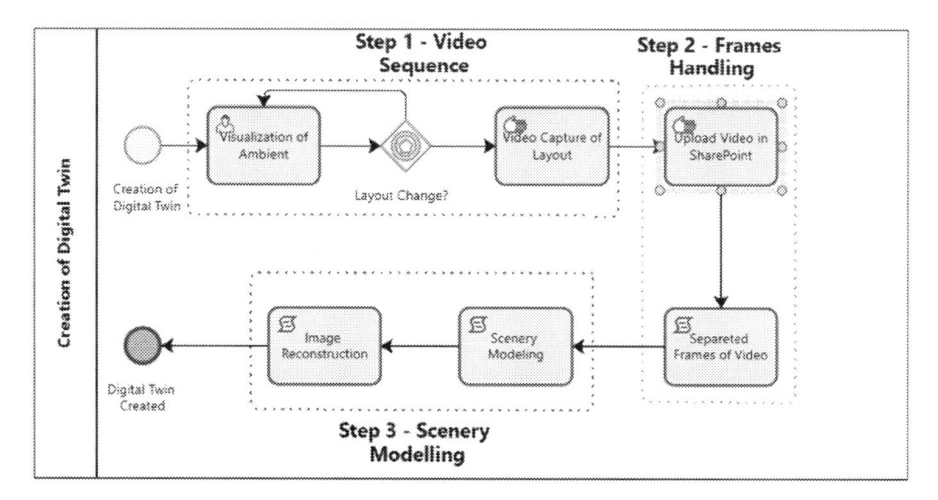

Fig. 3. The proposed method with respective steps to automated layout generation.

The proposed method is human centered because we have interaction between the system and the human operator as supervisor in all steps in order to decision maker about the best strategy to produce the video sequence. We replace the sketch drawing by a video sequence captured from a single RGB camera, like embedded in a smartphone.

3.1 Step 1 – the Video Sequence

The first step is to generate a video sequence of the factory area. The video needs to have minimum tree minutes in order to obtain enough images to data generation, and a maximum of seven minutes due the size of the number of pictures to process. Also, it is ideal that this video be doing like a "caracole" sequence around the area/objects capturing all the layout up, middle down and a little of floor for reference (see Fig. 2). For more reliability, it was experienced that on the frames sequence should appear the one meter of floor, one meter of the extreme side left, one meter of extreme right side and one meter of up the captured area. The video sequence, as presented in Fig. 4, can be taken by a human operating a smartphone or piloting a drone around the objects area. By this way we can prepare the multiple views from the object/scenery in the following step.

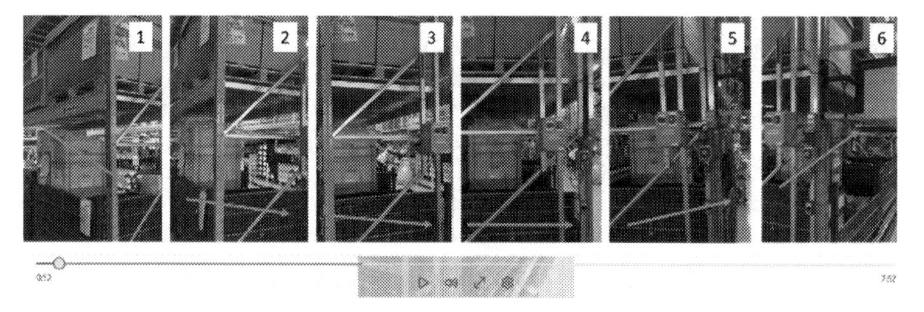

Fig. 4. Multiple views taken by a RGB camera. The arrows represent the observer viewpoint.

3.2 Step 2 - Separation of Frames

With the video created, it is necessary to load it to a computer with Python language [14–16] and OpenCV library. Here we developed the code for separation of the video frames setting the FPS accordingly, as for instance 4 fps or 30 fps, looking for the influence in quantity of frames to improve the image processing. Once performed this step, we have as resulting a set of image files to be used as input to the next step.

3.3 Step 3 - Scenery Modelling

After the image uploaded in computer it is needs execute the sequential operations steps, as presented in Fig. 5 and described as following:

1. Images Upload – It is the stage to upload the sequenced images files to be utilized in the reconstruction process.
2. Feature Extraction – After upload the images it is necessary find and extract the pixels that remains equals between the images irrespective of rotation, translation, and scale. [18].
3. Image Matching – This section is for to find the distance of pixels equals inside the images that utilized the same scene [19].
4. Feature Matching – After find the pixel in common and the distance between pixel in images its possible separated the bad images for the good images utilizing the discretion that find in Feature Extraction and Image Matching [18].
5. Structure From Motion - The objective of this step is starting the construction the 3D object utilizing the geometric relationship utilizing the internal calibration of all the images capture and created a model if position and orientation with point clouds [20].
6. Prepare Dense Scene – In this part the software created a folder in PC where images will begin the Texturizing in scene that is transformed in another set of images [21].
7. Depth Map and Depth Map Filter – In this point each image is considered how camera as in Fig. 2. Item A and its selection the best cameras will create a volume W, H, Z with many candidates that share pixels. Considering that distance between cameras is not uniform so the software utilized a filter for help in creation of volume of object [22].
8. Meshing and Mesh Filter - The objective of this step is to create a representation of scene utilizing mesh triangular that show how this object is, but in this step the object does not color yet just your format with triangles [23].
9. Texturing – In this section is painting of mesh triangular utilizing colors of images for this is applicate a filter UV in mesh [24]. In each mesh is applicate a color of accord your pixel [25], for this process be more fast its applicate a filter of low frequency and this approach is in the concept than [26] and [27].

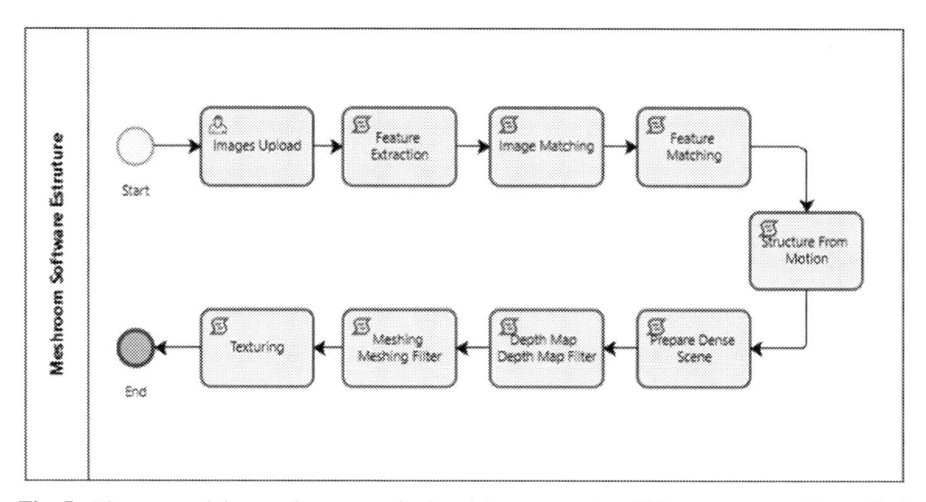

Fig. 5. The sequential operations steps to virtual image creation. This sequence can be applied, for instance, by Meshroom software structure [28].

The software Context Capture can be utilized for produce 3D models utilizing videos or photos by using these step sequences [17].

3.4 Image Reconstruction Processing

Here we need an environment to process the 3D reconstruction of scenery, for instance, using the software Context Capture. The steps performed here are the generation of the clouds of points, triangulation and reconstruction. The time of this process depends on the quantity of image frames. Recent approaches to 3D object detection exploit different data sources, and camera based approaches utilize either monocular [29] or stereo images [30]. With the introduction of digital photogrammetric techniques, certain errors associated with the subjectivity of human operators during triangulation may be avoided [31].

The main steps are:

1. Frames/Point Clouds – In this step insert the images taken with the same camera configuration (image dimension, focal length and sensor size). [32].
2. Camera Properties – Keep the knowledge of the configuration of camera that was utilized in project.
3. Aero triangulation – In this step it is the selection of all images with points in common between your characteristic is views from above. [33, 34].
4. Reconstruction Settings – Configure the parameters about reconstruction of 3D model (depends on which software is used).
5. Production – In this step is generated a model, with error feedback, progress monitoring and a possible option for retouching. In this point its possible export the file for different formats how: 3D meshes, point clouds, orthophotos/DSM [32].

After all operations we can visualize the resulting virtual image corresponding with real environment, as presented in Fig. 6.

Fig. 6. Resulting 3D layout image after reconstruction.

4 Results

The process of creation for one industrial layout how showed in example of Fig. 6 if to utilize the today methods that be manuals methods wait around the tree business days of work utilizing the tools: paper, pencil, measure tape and clipboard. The propose of solution in this paper it is reduce the time for create the layout utilizing a cell phone it will do a movie of layout about the 7 min after in the computer wait more 1 min for code separated frames in the movie and with this photography get in the software context capture that late 35 min for processing these images and generated the Digital Twin with a 3D model of layout. The time difference between these techniques was the 16h:20m – 0h:43min = 15h:37m. Thus, the proposed method saves time to around 2 full workdays as presented in the Table 1.

Table 1. Comparing of Manual Process X Automated Process for Creation of One Digital Twin

Task/Activities for One Layout	Manual Process for One Activity (current)	Automated Process for One Activity (Proposed method)
Separated Tools	20 min	1 min
Drawing Layout in Clipboard	8 h	0 min
Capture Layout	No applicable	<=7 min
Drawing Layout in Software CAD 2D	8 h	0 min
Drawing Layout in Software CAD 3D	No applicable	35 min
Total time consumption	16:20 h	43 min

The contribution of this research is that the process is centralized of human in the loop concept just of in this process anything object can captured and transformed in a Digital Twin all depends on the time of video.

5 Conclusion

In this paper we investigate some digitalization techniques can be applied to virtual modelling of an environment. This environment can be mapped for a 3D digital scenery representation creating a simulation scenario, from that we can identify which equipment are present in the image. Once equipment is identified, a human operation can reallocate it in a virtual sketch to evaluate the processes, (for example, like some sequential operation is working or to define a new layout configuration to improve a factory process). Once analyzed it in virtual space, the real space can be rearranged with a new devices' positioning. This paper shows the digitalization phase and the interaction between real and virtual is the ongoing research. For now, we can evaluate the process to build the virtual model by analyzing the tools as the meshroom and context capture software and evaluate the gain of time in the process. The next part of the research is to finish the second step of a DT creation, by including the synchronization activity between real and virtual worlds. For the future we will test how to utilize virtual reality for walk inside the layout utilizing an oculus VR. Also, we can embed Intelligence Artificial routines to define better layout configurations.

References

1. Tao, F.: Applications of digital twin. In: Digital twin driven smart manufacturing, pp. 29–62. Academic Press, Cambridge, MA, USA (2019)
2. Tao, F., Cheng, J., Qi, Q., Zhang, M., Zhang, H., Sui, F.: Digital twin-driven product design, manufacturing and service with big data. Int. J. Adv. Manuf. Technol. **94**, 3563–3576 (2018)
3. Grieves, M.: Origins of the Digital Twin Concept. https://www.researchgate.net/publication/307509727_Origins_of_the_Digital_Twin_Concept. Accessed 9 Jan 2022
4. Grieves, M.W.F.: Product lifecycle management: the new paradigm for enterprises. Int. J. Prod. Dev. **2**, 71–84 (2005)
5. Bittencourt, V., Carvalho, L., de Barros, R.C.: Evolução e estudo da aplicação do digital Twin no setor industrial. (The evolution and study of the application of the digital Twin in the industrial sector) (in Portuguese) (2020)
6. Dosovitskiy, A., Beyer L., Kolesnikov A., Weissenborn D., et al.: An image is worth 16 × 16 words: transformers for image recognition at scale. In: International conference on learning representations ICLR 2021, pp. 1–21,(2021)
7. Ardeshiri, T., Larsson, F., Gustafsson, F., Schön, T.B., Felsberg, M.: Bicycle tracking using ellipse extraction. In: 14th international conference on information fusion, pp. 1–8 (2011)
8. Silva, R.L., Canciglieri Junior, O., Rudek, M.: A road map for planning-deploying machine vision artifacts in the context of industry 4.0. J. Indust. Product. Eng. **39**(3), 167–180 (2022)
9. Kurka, P.R.G., Rudek, M.: Three-dimensional volume and position recovering using a virtual reference box. IEEE Trans. Image Process. **16**(2), 573–576 (2007)
10. Ding, Y., Xiaofeng, Z., Zhili, Z., Wei, C., Nengjun, Y., Ying, Z.: Semi-supervised locality preserving dense graph neural network with ARMA filters and context-aware learning for hyperspectral image classification. IEEE Trans. Geosci. Remote Sens. **60**, 1–12 (2022)
11. Xin, S., Adam, M., Bahram, J.: Three-dimensional profilometric reconstruction using flexible sensing integral imaging and occlusion removal. Appl. Opt. **56**(9), D151 (2017)
12. Lei, J., Liu, Q.: Three-dimensional temperature distribution reconstruction using the extreme learning machine. IET Signal Proc. **11**(4), 406–414 (2017)
13. MicroStation CONNECT Edition® Design, Model, and Manage Infrastructure Homepage. https://www.bentley.com/wp-content/uploads/PDS-MicroStation-LTR-EN-LR.pdf. Accessed 22 Jan 2023
14. Reffold, C.N.: Teaching and learning computer aided engineering drawing. Int. J. Eng. Ed. **14**, 276–81 (1998)
15. Hai-bo, W.: Computer Aided Industrial Design Journal of Anhui University of Technology, pp. 23–26 (2005)
16. TIOBE Software Index F.: TIOBE Programming Community Index Python. pg1 (2011)
17. Bentley Systems, Homepage. https://www.bentley.com/wp-content/uploads/PDS-ContextCapture-LTR-EN-LR.pdf. Accessed 15 Feb 2023
18. David, G., Lowe, F.: Distinctive image features from scale-invariant keypoints. Int. J. Comput. Vision **60**, 91–110 (2004)

19. Nister, D., Stewenius, H.: Scalable recognition with a vocabulary tree. In: IEEE Computer Society Conference on Computer Vision and Pattern Recognition (CVPR'06), pp. 2161–2168 (2006)
20. Cheng, J., Leng, C., Wu, J., Cui, H., Lu, H.: Fast and accurate image matching with cascade hashing for 3D reconstruction. In: 2014 IEEE conference on computer vision and pattern recognition, CVPR (2014), pp. 1–8 (2014)
21. Meshroom Homepage. https://meshroom-manual.readthedocs.io/en/latest/feature-documentation/nodes/PrepareDenseScene.html. Accessed 15 Feb 2023
22. Hirschmüller, H.: Accurate and efficient stereo processing by semi-global matching and mutual information. In: IEEE conference on computer vision and pattern recognition, CVPR (2005)
23. Jancosek, M., Pajdla, T.: Exploiting visibility information in surface reconstruction to preserve weakly supported surfaces. Hindawi Publish. Corpor. Int. Scholar. Res. Notices **2014**, 1–20 (2014)
24. Lévy, B., Petitjean, S., Ray, N., Maillot, J.: Least Squares Conformal Maps for Automatic Texture Atlas Generation, pp. 1–10. ISA (Inria Lorraine and CNRS), France (2002)
25. Burt, P.J., Adelson, E.H.: A multiresolution spline with application to image mosaics. ACM Trans. Graph. **2**(4), 217–236 (1983)
26. Baumberg, A.: Blending images for texturing 3D models, pp. 404–413. BMVC (2002)
27. Allene, C., Pons, J., Keriven, R.: Seamless image-based texture atlases using multi-band blending. In: 2008 19th international conference on pattern recognition, pp. 1–4. Tampa, FL, USA (2008)
28. Chen, X., Kundu, K., Zhang, Z., Ma, H., Fidler, S., Urtasun, R.: Monocular 3d object detection for autonomous driving. In: Proceedings of the IEEE conference on computer vision and pattern recognition, pp. 2147–2156 (2016)
29. Chen, X., Kundu, K., Zhang, Z., Ma, H., Fidler, S., Urtasun, R.: 3d object proposals using stereo imagery for accurate object class detection. IEEE Trans. Pattern Anal. Mach. Intell. (2017)
30. Helava, U.V.: Digital comparator correlator system (DCCS). ISPRS J. Photogram. B Remote Sens. **44**(1), 37–47 (1989)
31. Bentley Homepage.: Context Capture User Guide Homepage. https://docs.bentley.com/LiveContent/web/ContextCapture%20Help-v17/en/GUID-B9804464-3D42-420F-ACFF-7C61C1C028BD.html. Accessed 22 Jan 2023
32. Jones, A. D.: American Society of Photogrammetry (ASP), Manual of Photogrammetry, Fourth Edition, Church Falls, VA, (1980)
33. Moffitt, F.H., Mikhail, E.M.: Photogrammetry. Harper & Row, New York (1980)
34. Sallam, K.F., Abdulhadi, A., Hawas. K.: 2023 Bentley Systems, Incorporated. Homepage. https://www.bentley.com/software/contextcapture. Accessed 20 May 2023

A Data Management Approach for Modular Industrial Augmented Reality Applications

Jan Luca Siewert(✉), Matthias Neges, and Detlef Gerhard

Ruhr-University Bochum, Bochum, Germany
{Jan.Siewert,Matthias.Neges,Detlef.Gerhard}@ruhr-uni-bochum.de

Abstract. Modular architectures for Industrial Augmented Reality (IAR) applications allow more flexibility and can lower the barrier of entry, compared to custom-built applications dominant today. This contribution presents a data management approach for such architectures. Work plans are converted into a presentation-independent format that can be consumed by the system. Based on available modules and capabilities, related data, like CAD models, images, or descriptions, are automatically converted into a format suitable for the modules responsible for presenting the content. This can include converting a CAD module into a tessellated format, but also generating a rendering in cases where 3D models cannot be display, e.g., in a projection-based AR experience.

In IAR applications, the position of the displayed content in the users surrounding is an important aspect. For industrial contexts, this mostly relates to presenting content at specific parts of larger assemblies. This contribution shows how different combination of existing AR tracking technologies, in addition to CAD data of such assemblies, can be used to setup flexible IAR systems.

Keywords: Industrial Augmented Reality · Data Management · Product Lifecycle Management

1 Introduction

With growing maturity of available Augmented Reality devices, data management approaches become increasingly important. Industrial Augmented Reality (IAR) has various usecases across the whole product lifecycle [1]. Research has shown advantages of using AR: during service tasks workers are reliable guided through unknown procedures [2]; precise spatial information can be transmitted for remote maintenance on complex equipment [3]; process reliability is increased in quality control applications. Nonetheless, IAR applications have not found widespread adoption in the industry, yet [4]. Today's IAR applications mostly fall in one of two main categories: they are custom-built using a Software Development Kit or are provided as-a-service for a specific application. While the first approach results in highly customized systems tailored to the specific need for the specific use-cases, both their initial creation and later customization requires highly trained professionals, making this approach only suitable for

Published by Springer Nature Switzerland AG 2024
C. Danjou et al. (Eds.): PLM 2023, IFIP AICT 701, pp. 80–90, 2024.
https://doi.org/10.1007/978-3-031-62578-7_8

larger companies. As-a-service products mostly cover single applications, e.g., maintenance or remote support IAR systems that are flexible enough that they can be adopted and expanded upon by smaller companies as well need to be composed of modular, reusable, and expandable parts.

This contribution presents a novel data management approach for such a modular IAR architecture. Instead of creating a monolithic IAR application from scratch or trying to adapt an as-a-service product, this architecture proposes the use of stand-alone modules, that can be composed, adapted, and reconfigured to fit a specific use-case. After the architecture is introduced, it is shown hot it can adapt to a specific assembly use-case.

2 Prior Works

While various studies reported on various advantages of IAR systems, there remain obstacles before a widespread use can be achieved. IAR systems should be integrated into existing data management infrastructure, like PLM and IoT systems. To achieve that, these systems need to communicate using standardized interfaces and data formats [5]. Because there is no "one-size-fits-all" solutions to support all tasks, both the content and the application logic needs to be adapted to fit a given use-case. Today, this often requires expert knowledge, making the process time-consuming and expansive [6]. To implement IAR systems cost-effective and with minimal set-up time, an expandable and adaptable approach needs to be taken [7, 8]. To achieve expandable and flexible IAR systems, various modular IAR architectures have been proposed.

MacWilliams et al. identified a set of common components in AR applications. These include an application subsystem, containing the actual, use-case dependent application logic. The interface to the user consists of interaction subsystems on the one hand and presentation subsystems on the other hand. Tracking subsystems are responsible to display the content at a fixed position in space. All data is part of a world subsystems, while user specific content is part of a context subsystem [9].

Kuster et al. propose a service-based architecture for IAR applications. Services register their capabilities at a central service registry. Exemplary services are extracting payloads from QR codes or asking for input from a user. These services are then composed to business process using the BPMN graphical modelling language. The actual display and interaction concepts used by the various AR devices used, however, is not specified [10].

To support assembly tasks with IAR applications, numerous approaches for an (semi-) automatic generation of work instructions based on CAD assembly documents exist. Neb et al. extract high-level "assembly features" from form features using a macro in a CAD system. These are then the basis for an automatic generation of an assembly sequence as well as additional information, like assembly distance and a suitable animation to use in a visualization. The resulting data is exported in a structured format together with a tessellated geometry. Combining these, an AR Head-Mounted-Display (HMD) is then used to visualize the assembly instruction [11].

Gors et al. show that a fully automatic content generation approach is not suitable for real-world CAD models, because elastic parts, like springs or cables, are only added as static entities in the assembly model. Here, candidates for the next assembly step are found by identifying parts that are unobscured by the already assembled parts of the structure. However, manual intervention is necessary by an operator when the algorithms cannot find a next part because of the stated limitations. After an assembly sequence is identified, instruction content is created. This includes text describing the part and the required assembly direction, image sequences, 3D models of the various intermediate assembly steps, and animations [12].

3 Concept

The proposed modular architecture is roughly based upon a classic 3-layer architecture, while incorporating the most relevant components of AR systems as discussed by MacWilliams et al. [9]. The view layer can be divided into multiple interaction and presentation systems. These can be developed independently from any special use-case, allowing them to be reused in different settings. An application logic module reacts to events from the interaction systems and forwards content to suitable presentation systems.

3.1 System Architecture

The overall system architecture (Fig. 1) is roughly based on a classic 3-layer architecture. The view layer consists of the afore mentioned presentation and interaction systems. They communicate with a runtime through a message broker. The runtime is responsible for executing the actual application logic. It receives events from the interaction systems and reacts by sending data to be presented by the presentation systems. This is based upon the data prepared by the planning module. This module takes the source data from the data layer and brings it in a format that is understood by the runtime. Thereby it considers which interaction and presentation systems are available, as well as the requirements of the use-case. This process is presented in more detail in the next section.

On the data layer, three modules for data preparation are responsible for creating the content based on input data, convert the data to suitable formats, and act as a data proxy to simplify access. Data source can include different systems, like PLM for CAD data and assembly information, ERP systems for work plans, or IoT data for telemetry. The necessary adapters are implemented as small services. All available services are registered at a central service registry.

Content creation can happen manually by an operator, for example in a stand-alone web-application or as part of a broader ERP system. It relates the input data, like CAD models or assemblies, to the specific work plan that will be executed by the worker.

Data conversion includes converting the same data type into different format, e.g., different formats for images. It can also mean transforming the data in a new format. For example, a 3D model can be transformed into an image by creating a thumbnail.

Sometimes, a direct connection between an AR device and other software system might be undesirable. For example, a device in a production shopfloor should not have direct access to the supporting PLM system. In such a case, the data can be transferred through a data proxy, which also supports caching data and forwarding new data back to the source systems.

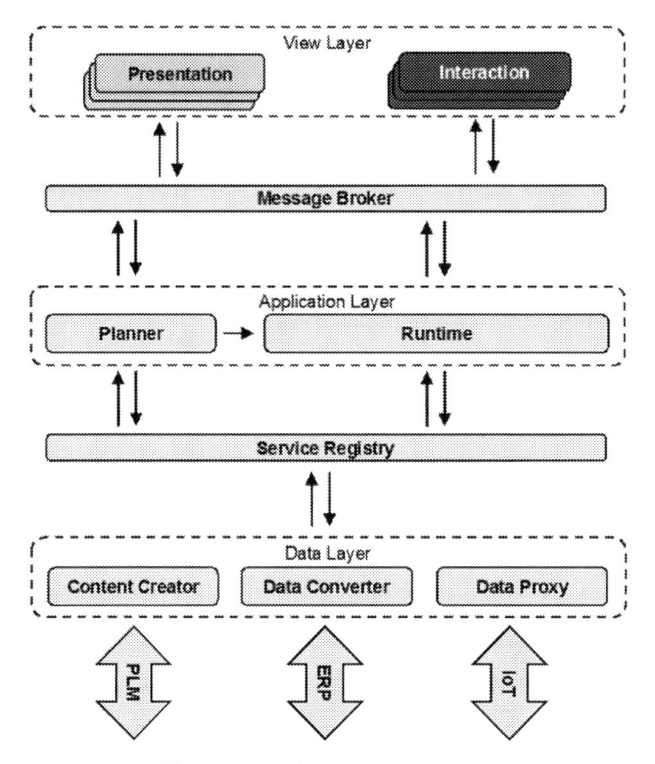

Fig. 1. Overall System Architecture

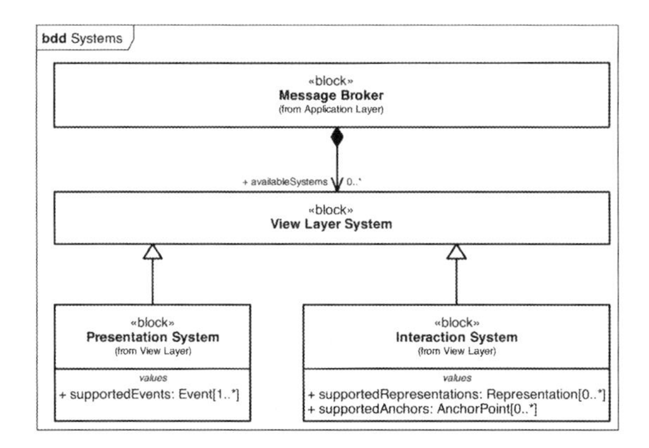

Fig. 2. Block Diagram with the requirements of the Presentation and Interaction Systems.

The details of the presentation and interaction systems are shown in the Block diagram in Fig. 2. An interaction system is described by the different event types it supports. A presentation system is described by both the data it can represent, like images, or 3D models, and the anchor points it can attach data to.

3.2 Implementing a Use-Case

To meet changing requirements from different users or different use-cases, the assistant system must be composed from the different available subsystems and their capabilities. Most of the required steps can run automatically. Some, however, may require input from an operator who is responsible for maintaining the system. These are shown as manual actions in the BPMN diagram in Fig. 3.

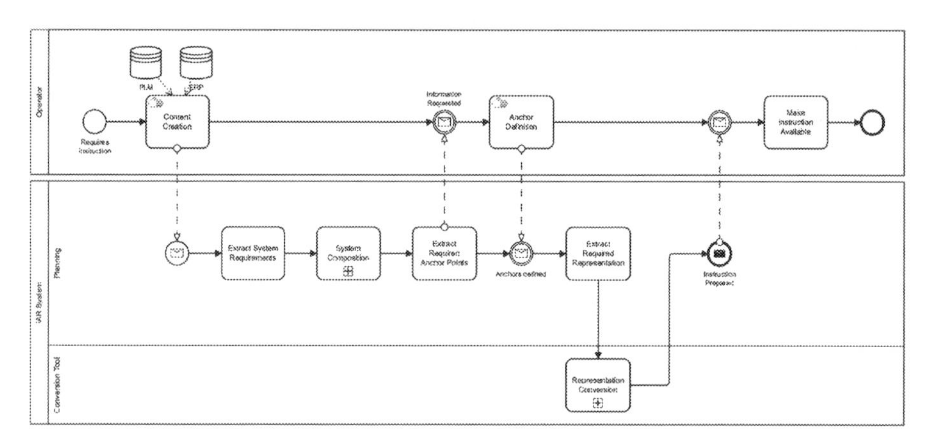

Fig. 3. Process for creating content for an IAR system in the proposed architecture.

Work Plan Creation. In a first step, the work instruction is created. Depending on the source data, this step can be performed automatically, the operator can create them manually, or he is assisted in the creation. After this step, a standardized description of the work to be performed is created. The work is divided into individual steps that are connected through transitions. Each transition is tied to a specific event. For example, one event might always start the next step in a list of steps. For every step, one or more pieces of information are displayed at a given anchor. The same information can be conveyed to the worker in different ways, like with an animation, a static 3D model, or a combination of a description and a rendering. A set of minimal system requirements in the form of required events, anchors and minimum adequate representations is derived from the work plan.

Not every IAR application needs a concrete, step-by-step work plan. However, the architecture requires a standardized definition of the displayed data to derivate the required anchor points and content representations. For those use-cases, the "work plan" associates available data-points with their real-world location and implies arbitrary transitions between each entry. This enables a worker to freely select displayed content.

System Composition. In the next step, subsystems are selected to match these requirements. Because all subsystems register themselves with their capabilities at a central registry. From there, the optimal set of subsystems is selected to support the use case.

Anchor Definition. More so than the display of three-dimensional content, an AR application is characterized by being able of presenting content at a specific point in space. This point remains static even when the user moves and is typically referred to as an anchor. While content might be shown at an arbitrary position in space, IAR applications mostly present information tied to a specific position at a machine or an individual part in it. A presentation system might support one or more of anchor types, like model targets or image markers. Some might be capable of supporting any anchor of a given type, e.g., any QR code independently of its actual payload. Mostly, supporting a specific anchor requires additional work that must be performed automatically or manually by the operator. For model-based tracking algorithms, some preprocessing is necessary based on either available CAD models, or by scanning a physical part before the worker can start his work. In some cases, the operator might assign static position to a given anchor. An example might be an assembly station where work pieces are always stored at the same location. Here, no special anchors are needed to be capable of highlighting those.

Representation Conversion. A system usually has to support required anchors to convey the necessary information to the worker. A system, however, is more flexible with the specific representations it uses therein. When the required anchors are provided by the overall system, different data representations are automatically created from the source data. The presentation subsystems describe their capabilities. This can include the type of representation, like 3D models, images, or texts, as well as supported file formats. The resulting representations are permanently saved so that there are available later on.

Representations can be created by converting different file formats of the same type, e.g., creating a tessellated OBJ file from a step file, as well as by transforming then, e.g., derivate a 2D rendering from a CAD file. These conversion and transformations are done by stand-alone services that register their capabilities at the registry. They are then tasked on demand to create the representations to best fit the requirements and content of the use-case.

System Expansion. Creating an IAR application for a given use-case with the proposed architecture does encourage modularity and reusability. Individual subsystems, like converters, but also interaction and representation systems, are implemented only once. When their functionality is required for another use-case, they work out of the box. To support other use-cases with different requirements, the system needs to be adapted.

The work plan conversion is strongly tied to both the specific use-case as well as the source data. For different use-cases, this module is therefore different. It is, however, indifferent to the actual output to the user. The type of output devices, like a projection-based AR setup or AR glasses, the work plan creation process stays the same. The work plan creation has to transform the source data into discrete states with representations attached to each and transitions between them.

Representation and interaction systems are independent from the actual use-case. When a use-case requires special interactions, like sending a measured value in a quality assurance process, a new event type has to be introduced. Then, an existing system gets expanded with the functionality to emit these events. These events should be defined as general as possible to be able to reuse them in other contexts as well.

When the requirements of a use-case cannot be fulfilled even after attempting to convert representations, additional systems need be created and registered. This can be new representation systems that support new anchors or representation types, or additional converters or transformers that support different output types. In this case, the operator gets a list of unfulfillable requirements as well as suggestions on how to solve them.

4 Implementation

To demonstrate the system, several prototypes have been implemented and tested. To show how different presentation systems can convey the same information, an assembly use-case is assisted by an in-situ projection system on the one hand. On the other hand, a worker has access to the same information in a HoloLens-based HMD system. Because IAR systems based on the proposed architecture are easily adaptable, the HMD system is later expanded to support a simple Quality Control (QC) use case as well. The implemented modules and the data flow between them are shown in Fig. 4.

Both systems are controlled through a runtime module written on-top of Node-JS, while Eclipse Mosquitto and the MQTT protocol were chosen for the Message Broker functionality. A Fusion Data Adapter module written in Python connects to Autodesk Fusin 360 Manage as a PLM system and Autodesk Platform Services to perform data conversions. It makes the methods available as a REST-API.

4.1 Supporting an Assembly Use-Case

Work Plan Creation. The input data is already in a structured format. The individual assembly steps are converted to states. Each state has the textual description associated to it. Where available, each step also has a 2D rendering of the current assembly state. Between steps, transitions are introduced corresponding to the events "next" and "previous".

System Composition. Each subsystem describes its capabilities using a defined JSON structure. For interaction systems, these include the types and names of the events it supports. For presentation systems, the capability includes the type and file formats of supported representations. Furthermore, they supply a list of anchors that are supported without manual work by the operator. Based on this information, the subsystems suitable for assisting the specific applications are selected.

For the projection system, presentation modules to display text and images have been implemented, respectively. They are only able to present information without any anchors. For interaction, the setup can project buttons on the workbench. Through a depth sensor mounted next to the projector, the users can interact with these by hovering their hand above them. Similarly, the system can highlight storage containers attached to the workbench and recognize when the user takes a part out of these.

The HoloLens has similar functionalities. Instead of images, it can render complete and interactive 3D models. The simple button functionalities are supported through a more sophisticated menu that is attached to the user's hand.

Anchor Definition. Each specific anchor point has a type and a unique identifier. For anchor points describing parts, this should be same identifier as in the Autodesk Fusion 360 Manage PLM system. Additionally, a system can declare a list of anchor types it

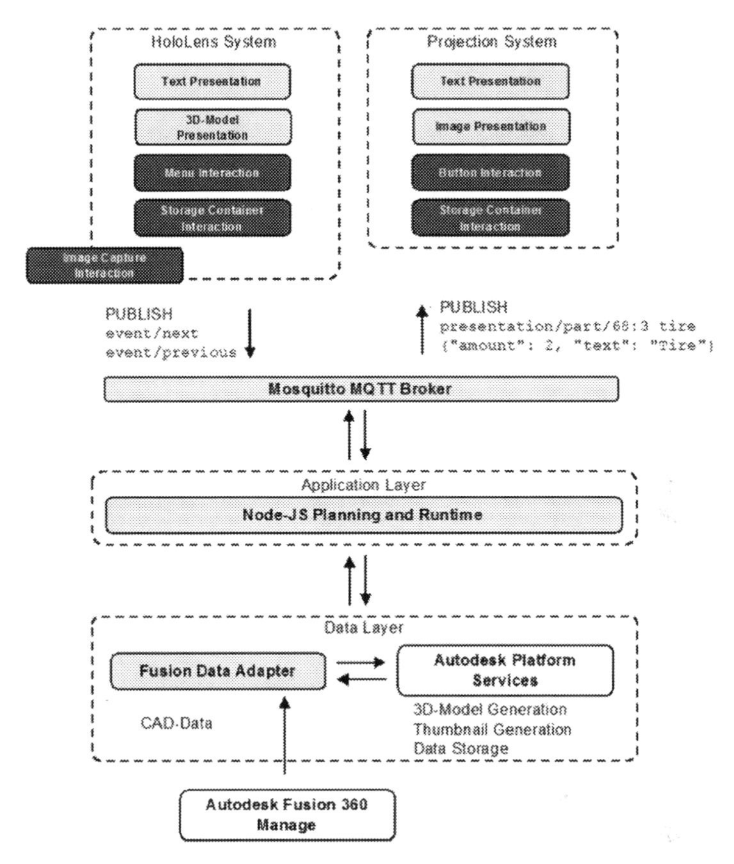

Fig. 4. Implemented modules and data flow between them.

might support. When a use-case requires an anchor that is not already supported but with a type that might be, the system gets request indicating the specific anchor to support.

How the subsystem handles such a request depends on its implementation. In this implementation, the operator setting up the system must create all anchors manually by defining the storage containers. Both the projection setup and the HMD make suggestions for potential storage containers based on their depth map and environment sensing capabilities, respectively. Then, the operator associates the container area with the specific part by selecting it from a dropdown menu with the previously undefined parts.

When a new anchor is supported by a subsystem, it updates its registered capabilities by updating the retained message at the MQTT broker. This restarts the planning process.

Representation Conversion. After all anchor requirements are fulfilled, the runtime converts data into representations supported by the presentation systems. Through the Autodesk Platform Services Platform, various types of 3D-CAD data formats can be converted into tessellated descriptions of the geometry based on the OBJ file format. A Fusion Data Adapter module uses this API to convert the CAD data and to store the results for subsequent executions. Additionally, the CAD data can be described by creating thumbnail images for the various models.

4.2 Supporting a Quality Control Use-Case

To showcase the **System Expansion** step of the proposes architecture, the HoloLens application is expanded with an additional interaction system to ask the user for numerical inputs. Furthermore, the user can capture images and annotate them to document issues. These support an additional event that sends the data gathered from the user to the runtime. Source data for this Quality Control use-case is again described as a state machine with new transitions where user input is required. Because additional modules can subscribe to these events on the MQTT broker, a new reporting module stores inputted data outside of the architecture and writes it into a quality control report.

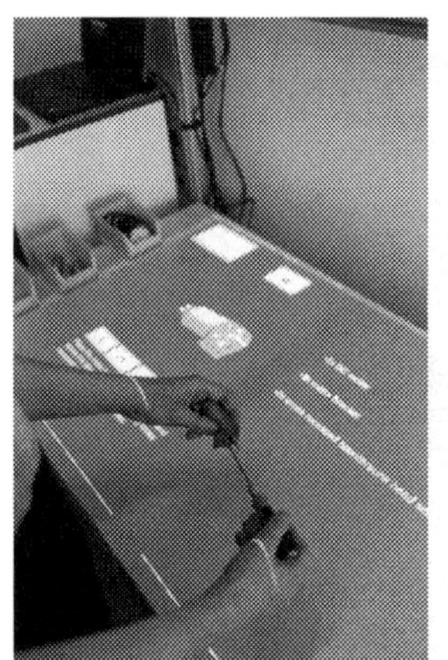
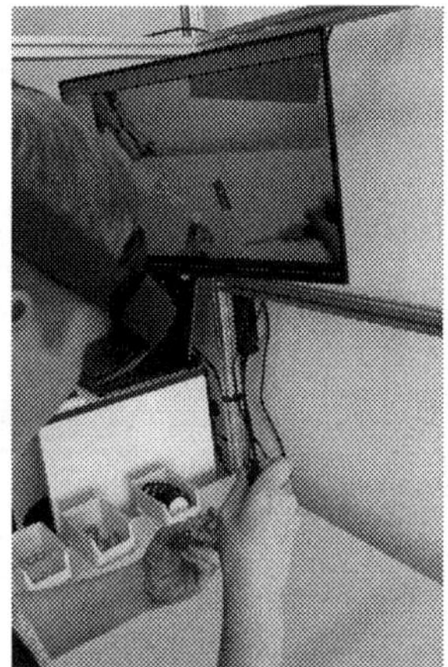
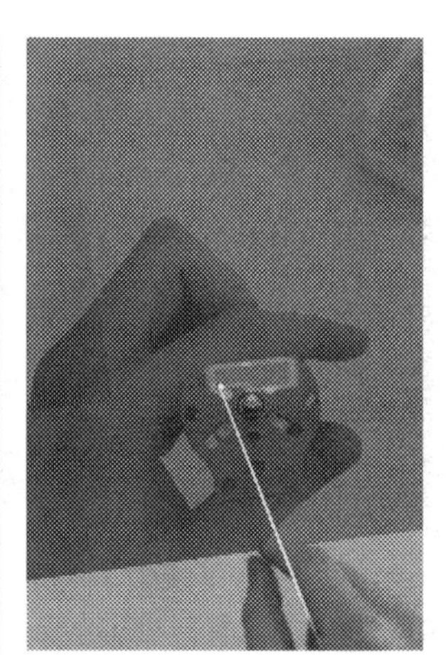

Fig. 5. The projection (left) and HMD system for assembly (center). The worker can take and annotate images for quality control in the QC application. in use.

Figure 5 shows the implemented systems in use. First the projection setup for assembly of robot models, then the HoloLens application supporting the same use-case, and finally the additional quality control use-case on the HoloLens.

5 Conclusion

The proposed modular IAR architecture enables a more flexible approach to create IAR applications. By using encapsulated modules, the system is easily expandable to more use cases. This contribution presents an approach for data management and preparation that takes into account the specific requirements for a given use-case as well as the capabilities of the different modules that are available in a given setting. By enabling operators to easily configure their systems for new use-cases, adaption can be quicker and more cost-effective. Through automatic conversion of different source data, the systems easily integrate into existing processes and data management software.

The architecture imposes a certain overhead, both during setup and during execution. Therefore, it might not be suitable for bigger companies that want to create highly specialized assistant systems to a lot of workers. At the same time, the architecture must be further tested to validate the possible repurposing and adaption of an existing system to a completely different use case.

In future works, the proposed service-based architecture could be the foundation for new business models for AR modules. Because the individual modules are standalone and reusable for a right range of use-cases, software vendors might offer specific subsystems, e.g., for model tracking, that can then be added to the specific IAR system on demand. One example could be the use of Artificial Intelligence to automatically detect errors during an assembly task. This could be added to the system as a new type of interaction subsystem that emits specific events, like going to the next step because the assembly is correct or showing an error message.

References

1. Fite-Georgel, P.: Is there a reality in industrial augmented reality? International symposium on mixed and augmented reality, science and technology proceedings. IEEE (2011)
2. Lorenz, M., Knopp, S., Klimant, P.: Industrial Augmented Reality: Requirements for an augmented reality maintenance worker support system. In: Adjunct proceedings of the 2018 IEEE international symposium on mixed and augmented reality (Ismar), pp. 151–153 (2018)
3. Masoni, R., et al.: Supporting remote maintenance in industry 4.0 through augmented reality. Procedia Manufac. **11**, 1296–1302 (2017)
4. Martinetti, A., Marques, H.C., Singh, S., van Dongen, L.: Reflections on the limited pervasiveness of augmented reality in industrial sectors. Appl. Sci. **9**(16), 3382 (2019). https://doi.org/10.3390/app9163382
5. Egger, J., Masood, T.: Augmented reality in support of intelligent manufacturing – a systematic literature review. Comput. Ind. Eng. **140**, 106195 (2020)
6. Masood, T., Egger, J.: Augmented reality in support of Industry 4.0—implementation challenges and success factors. Robot. Comput.-Integrat. Manufact. **58**, 181–195 (2019). https://doi.org/10.1016/j.rcim.2019.02.003
7. Quandt, M., Knoke, B., Gorldt, C., Freitag, M., Thoben, K.-D.: General requirements for industrial augmented reality applications. Procedia CIRP **72**, 1130–1135 (2018)
8. de Souza Cardoso, L.F., Mariano, F.C.M.Q., Zorzal, E.R.: A survey of industrial augmented reality. Comput. Indust. Eng. **139**, 106159 (2019)
9. MacWilliams, A., Reicher, T., Klinker, G., Bruegge, B.: Design patterns for augmented reality systems. In: Proceedings of the international workshop exploring the design and engineering of mixed reality systems (MIXER), Funchal, Madeira, CEUR Workshop Proceedings, Stycze (2004)
10. Kuster, T., et al.: A Distributed architecture for modular and dynamic augmented reality processes. In: 2019 IEEE 17th international conference on industrial informatics (INDIN) (2019)
11. Neb, A., Brandt, D., Rauhöft, G., Awad, R., Scholz, J., Bauernhansl, T.: A novel approach to generate augmented reality assembly assistance automatically from cad models. Procedia CIRP **104**, 68–73 (2021) (54th CIRP CMS 2021 - Towards Digitalized Manufacturing 4.0.Johannes Egger, Tariq Masood. Augmented reality in support of intelligent manufacturing – a systematic literature review. Comput. Indust. Eng. **140**, 106195 (2020))

12. Gors, D., Put, J., Vanherle, B., Witters, M., Luyten, K.: Semi-automatic extraction of digital work instructions from cad models. Procedia CIRP **97**, 39–44 (2021) (8th CIRP Conference of Assembly Technology and Systems)

A Method to Interactive Simulations of Industrial Environments Based on Immersive Technologies

Richard Valandro[1,3], João Cláudio Nogueira[2], and Marcelo Rudek[3](✉)

[1] Serviço Nacional de Aprendizagem Industrial - SENAI, Curitiba, Brazil
[2] Facilities Management, Robert Bosch LTDA, Curitiba, Brazil
[3] Pontifical Catholic University of Paraná PUCPR, Curitiba, Brazil
marcelo.rudek@pucpr.br

Abstract. Mixed reality technologies and digital twins have been widely applied when the objective is to create immersive virtual environments, especially when it comes to the industrial context. Industries are increasingly looking for ways to update the new 4.0 reality and, with the use of scenario's virtualization technologies, it is possible, for example, to train employees or people responsible for the development and creation of new solutions to digitalization in production systems for stakeholders. In this context, the article presents a project for the application of mixed reality and digital twin to create an immersive virtual environment for training employees at a company focused on the electromechanical area, more specifically on the maintenance of machines and equipment for the electrical substations at the headquarters located in the city of Curitiba - Brazil. The application's environment, in the diary activities, cannot be accessed with a certain frequency due to risks and the impossibility of stopping the manufacturing process. Thus, the work proposed a method to create immersive training sceneries based on Microsoft Hololens technologies in order to prepare the employees to be able to act on maintenance after virtual training.

Keywords: Digital Twin · Mixed Reality · Industry 4.0

1 Introduction

The constant technological advances allow us to look for adaptative tools faces the need to find the successive solutions and improvements to avoid the recurrent problems in the industry and services. The competitiveness has always been an engine for the companies to adopt resources that allow them to achieve a greater advantage over their main competitors. There is an innovative human-centered evolution by digital transforming and the virtual environments play the role. The main idea of implementing immersive technologies in the industry is to help employees to comply with their work processes carried out on the shopfloor, as well as ensuring their safety. This implementation has been happening by Virtual Reality (VR), Augmented Reality (AR), Mixed Reality (MR) and

Published by Springer Nature Switzerland AG 2024
C. Danjou et al. (Eds.): PLM 2023, IFIP AICT 701, pp. 91–101, 2024.
https://doi.org/10.1007/978-3-031-62578-7_9

Extended Reality (XR) technologies [1]. They are increasingly accessible due several aspects, such as: cost reduction, ergonomics, usability, less technical barriers to implementation, and maintenance facilities. Through AR and VR, employees can simulate its diary situations, as well as predict occurrences (accidents, equipment breakdowns, etc.). In addition, the "digital twins" are another technological area that have enhanced the experience of both collaborating students and the company, since it is possible to virtually simulate the behavior of equipment in real time through the operation carried out. As presented in Fig. 1 by the NTL Institute for Applied Behavioral Science [2], the learning pyramid points out that the practice reaches the level of 75% of content fixation, while the audiovisual reaches 20%; and reading and lectures reach around 10%.

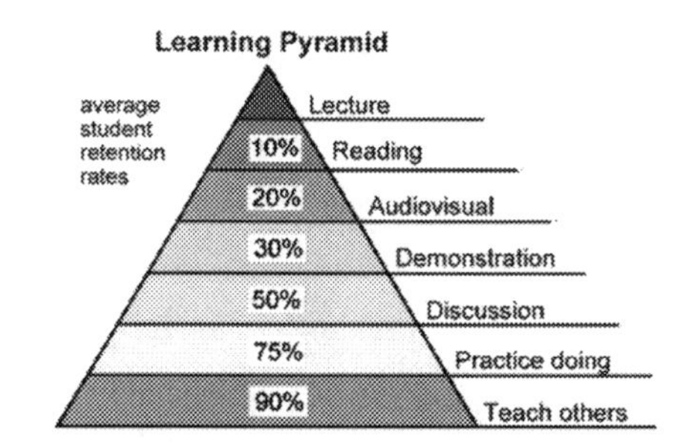

Fig. 1. Adapted of The NTL Learning Pyramid [2, 19].

We expect that a realistic simulation environment achieves the similar rate as the real practice. A simulation training offers a controlled virtual scenario for the professional to practice activities with risks involved, encouraging him to make an assertive decision. The main objective of simulation is precisely to offer interactive and intelligent systems, based on visual graphic tools, for the user to immerse and explore reality under stressful conditions. Simulation based learning breaks the learning paradigm by allowing the student to have a more active participation than traditional theory. Simulation learning makes use of software, VR and AR devices (3D glasses, swivel chair), graphic (games), sound (and even olfactory, depending on the training) resources. Our objective with this research project, is to create a method to virtual training focused on maintenance. We explore an example application to be developed in a German multinational company, in its Brazilian branch. Among the problems listed by the company's managers, the most critical one involved the plant's main substation: the "SD-01 69kV Substation" (Fig. 2). The substation is responsible for distributing electricity to 14 other existing substations, and based on this care, it's clear that it is a critical point for the plant's operations. Every year there is a scheduled shutdown (lasting 3 days) for maintenance of the substation and verification of components. However, unscheduled downtime for maintenance interrupts the operation of machines and equipment responsible for the factory's production processes, resulting in a very high financial loss estimated at a few million dollars per hour/stop. The causes of unscheduled stoppages are diverse: pieces of objects/animals that fall into the environment causing a short circuit, heavy rain in a short period of time, burning of an electrical component, overload in the system, among others. Due to the risks involved in carrying out face-to-face training (in the environment), employees end up not being trained, resulting in a longer period of unscheduled downtime (when it occurs). Another point is the time to detect the root cause. As much as the maintenance technicians know the basic operation of the substation, regardless of the cause of the shutdown, for safety reasons the entire area is de-energized for inspection. Another critical factor is the absence of a formalized history of unscheduled shutdowns and their respective causes, with no "lessons learned": making it difficult to predict failures.

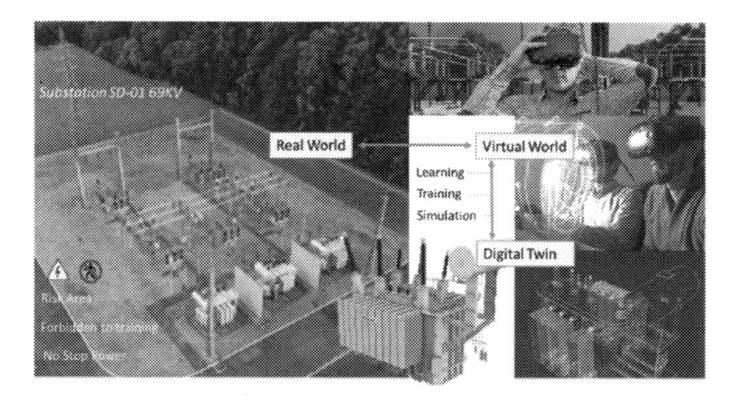

Fig. 2. Conceptual features to digital twin.

2 Background

For some learning situations, especially when they involve skills development, the practical contact with equipment, environments and phenomena are the best way to learn. Using real environments is not always feasible or safe. Virtual reality (VR) and augmented reality (AR) are excellent alternatives in these cases. Furthermore, the natural evolution of computational interfaces should transform environments, currently graphic and two-dimensional, into immersive and three-dimensional ones, increasing the demand on the part of industries for interfaces of this type. Regarding the literature review, having already defined the research theme, we identified the main keywords such as: *interactive simulation, virtual reality, augmented reality, mixed reality, and digital twin.* We addressed the SCOPUS platform to search for relevant articles, reviewed by pairs and from the last five years. We also defined/applied, for the selection of articles, some inclusion/exclusion criteria such as: publication phase, type of source, thematic areas, keywords and classification of articles according to the Scimago Journal & Country Rank (h index). In addition, the articles were analyzed considering the conversion of titles and abstracts in order to select articles that would be read in full and used as a basis for the research. According to literature review, we can apply immersive technologies because the following characteristics:

The convergence between the physical and virtual worlds. The [3] addresses the impossibility of visual inspection and the complexity of combined loads make it difficult to quantitatively assess (difficulty of accessing a certain environment) and the [4] presents the challenges of high-quality scanning as it requires a comprehensive description and real-time rendering of the system. Also, model-driven digital twin system as presented by [4], is based on the descriptions of geometric, physical and sequential rules to produce an accurate and real-time simulation, as well as, according [5] a catalog of components and intelligent data models as a solution for the development of any digital twin. This disruptive convergence tends to augment operations and services traditionally restricted to physical spaces with new virtual-based capabilities.

- The issue of human motor skills. Once the traditional hands-on training is resource and labor intensive, and the virtual training based on video demonstrations have the lack of interactivity, the [6] addresses the importance of developing remote training methods providing effective sensory feedback. For Buchmann [7], interaction consists of the user's ability to act in the environment in order to promote changes and reactions to their actions, which can bring performance gains as the user interacts more intuitively with the system.
- Also, the [8] propose the use of a virtual reality digital twin of a physical layout as a mechanism to understand human reactions. Finally, in the [9], simulation is approached as a technological tool for the development of exploratory and planning models to optimize decision-making.

2.1 The Problem

The training currently carried out at the study site does not use any immersive technology. The improvement of employees in electromechanical maintenance is carried out by instructors/technicians who have been working in the area for a longer time, as in Fig. 3.

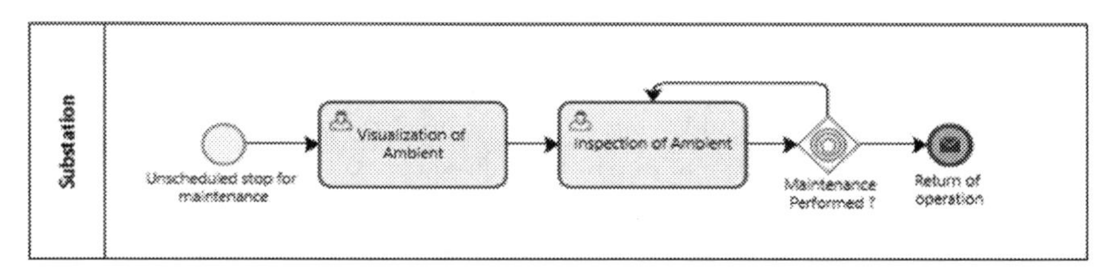

Fig. 3. Workflow (no immersion)

An example of training that we can approach is for detecting electromechanical failures in circuit breakers. In 2020 the company reported that, after completing a scheduled maintenance shutdown, a circuit breaker was not resetting. However, it took the technicians approximately two hours to discover that the problem was caused by this circuit breaker and that it was due to a mechanical failure in its mechanism. Furthermore, after performing a fault analysis, another probable cause for the tripping of the circuit breaker could be the existence of a current value above that specified in the device. With that in mind, we thought about the possibility of carrying out training aimed at measuring current in the substation electrical network using ammeters or multimeters (for example). Positive points in carrying out training using a virtual environment:

- Work at height simulation
- Simulation of electrical measurement with the network powered/energized.
- Technical analysis of devices belonging to the environment.
- More time for employee training.

2.2 Immersive Technologies

According to [10], the immersive technologies refer to technology that attempts to emulate a physical world through a digital or simulated world, creating a surrounding sensory feeling, thus creating a feeling of immersion. The system uses VR to simulate virtual operation scenarios and interactions to provide an immersive operation experience. In this same context, the mixed reality connects and combines the physical and digital worlds in virtual and augmented reality applications. In visual terms, mixed reality can be defined as a creative space that exists between the extremes of the physical and digital world. Experiences range from superimposing virtual content on objects in the physical world, as in augmented reality apps, to a fully immersive experience where the user has no real-world participation, as in virtual reality.

To increase the sensation of immersion, mixed reality systems use dedicated hardware that is more powerful than traditional cell phones and tablets, in addition to having sensors capable of rapidly scanning the environment around the user in 3D and making virtual objects interact with the identified geometry of the site. The most complete systems use special glasses, with semi-transparent lenses, where virtual objects are projected in perfect alignment with the environment, which causes a very realistic illusion of the presence of these objects (Fig. 4).

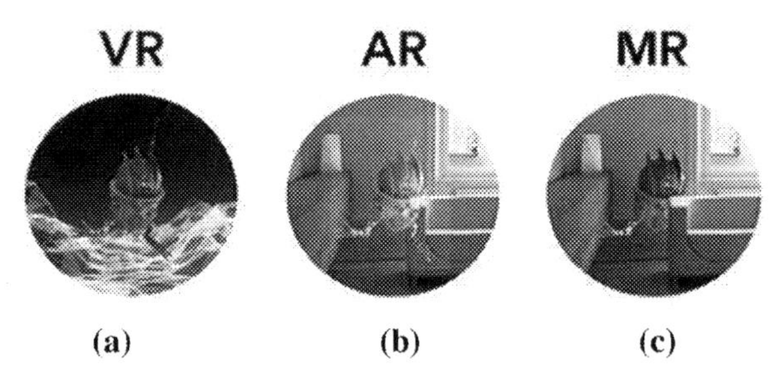

Fig. 4. (a) Digital environments that shut out the real world. (b) Digital content on top of your real world. (c) Digital content interacts with your real world [10].

Mixed reality environments are created by combining objects into a single environment. Real objects can be captured by various means, such as: 3D scanning or photogrammetry. Meanwhile, virtual objects can be created using computer-aided design (CAD) software. Photogrammetry is a technique that allows the study and definition of shapes, dimensions and positions of objects in space, using measurements obtained from photographs or digital images. According to Kempter [11], 3D laser scanning and photogrammetry technologies are the digitization technologies most used for surveys of buildings and, despite differences in cost of equipment and processes of detection, are three-dimensional automated and wireless acquisition systems, contact with the analyzed object, which use sensors based on waves of light for the measurement, direct or indirect, of the object.

These processes generate files that can be migrated and manipulated using augmented reality glasses such as, for example, Microsoft's Hololens 2. Using these glasses, the user can interact with virtual objects (holograms) and place them on something physical, like a real table, to visualize parts, for example. A point to consider regarding Hololens 2 is the possibility of connecting to Microsoft Teams, in which the person can use the glasses to talk to a professional who is in another distant environment, and who will give instructions for maintenance on a machine, also through Teams. This device creates [12] high definition virtual environments, allowing interactions between augmented reality and the user.

Directing to our problem, we can use these glasses for the interaction between the students (collaborators) and the virtual environment in which it will be possible, for example: the measurement of electrical variables, exchange of mechanical components, continuity tests, simulation of failure situations electromechanical and etcrealistic illusion of the presence (Fig. 5).

Fig. 5. User view through Hololens 2, adapted from [13].

However, before importing the files (generated through photogrammetry or 3D scanning), it is necessary to use some software that prepares the 3D models (modeling, rendering, animation, creation, and visualization of interactive 3D content and etc.) for use on the glasses. Among them, we researched: the Blender and Unity.

With Blender it is possible to crop and configure the captures and smooth textures and edges to obtain the best model. From Unity, we can turn 3D scans into animations or create scenarios. In addition, we can edit and combine them however we want. Both software communicates making it possible to export files between one platform and another.

2.3 Digital Twin

We can define digital twins as virtual models of a product or asset connected to the physical prototype or instance via the Internet of Things (IoT). Digital Twin is a relatively new trend which comprises building a simulator of a product, processor or machine, being powered by real sensors and (in real time) the existing device. The advantage of this concept is not only to monitor its operation, but to simulate different usage situations, improvements, failures, etc. Thus, when necessary, intervention is much more effective [14, 15].

Essentially, digital twin is a hybrid approach, built into four levels [9]: geometry, physics, behavior, and rule. The first two levels involve mainly kinematics and geometric simulation, also referred to as continuous simulation.

Using Hololens 2 [16], mentioned in item 2.2, we will apply the concept of digital twim, considered an efficient intelligence solution for digitization and automation that uses [3, 21] digital models to monitor, describe, simulate, predict and control physical counterparties in a timely manner.

3 Proposed Method

The research project proposes the development of a digital twin for training employees who are responsible for the maintenance of the substation in question. The development of maintenance skills involves several high-risk activities such as: work at height and high voltage. Activities that end up not being realistically trained on a daily basis due to the existing risk. Through a digital twin, it is possible to simulate/virtualize the scenarios and, logically, the possibility of carrying out employee training without offering risks to the integrity of people. Along with the digital twin, augmented reality and virtual reality features will be used. In Fig. 6, we briefly describe the steps of the proposed method.

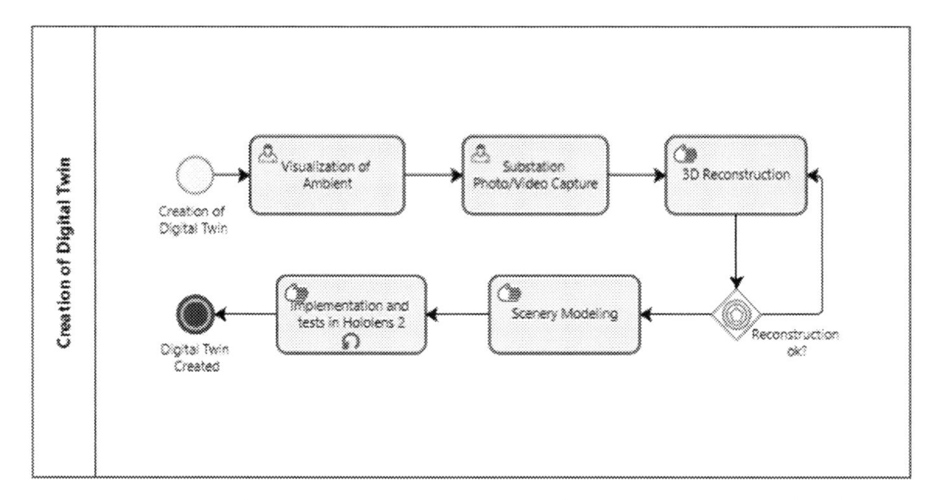

Fig. 6. The proposed method with the respective steps for training through the digital twin.

3.1 Visualization of the Ambient

As for the implementation of the proposed method, initially a face-to-face visit will be made to the study environment (substation) for space analysis and planning for the next step: substation photo/video capture according image processing requirements [17, 18]. A next step is the analysis of physical variables present in the real environment with the aim of implementing them in the virtual world.

3.2 Substation Photo/Video Capture and 3D Reconstruction

In the stage of capturing photos and videos of the substation, we will use an Iphone model 13 pro through the Polycam application. We have already carried out some preliminary tests and obtained satisfactory results. To create the virtual environment (mixed reality), it is necessary to reconstruct the physical environment in 3D, based on the captured images. A solution for 3D reconstruction is the use of some software or application that performs the process and provides us with the 3D environment for the next steps. We checked and chose two softwares for a more detailed analysis in practice: Polycam and Meshroom.

Meshroom is capable of producing sharp and accurate models, but requires a lot of processing power and storage, generating heavier files. While Polycam creates high quality 3D models from photos with any mobile device. It is possible to view the 3D captures directly on the device, and it is also possible to export them in various file formats. A point of attention is that Meshroom is only available for Windows and Linux, while Polycam only supports mobile devices (IOS/Android).

3.3 Scenery Modeling

After the 3D reconstruction, we will use the Unity software, which is a tool that allows you to create video games for different platforms (PC, consoles, mobile, VR and AR) using a visual editor and programming through scripting, offering users professional tools, capable of fulfilling the requirements of any game. Some factors related to the real environment are important with regard to virtual modeling: risk sequence, kinematics, physical simulation (if possible) and, finally, special effects to show the consequences of the electromagnetic fields existing in the environment in question (substation).

3.4 Implementation and Tests in Hololens 2

With the scenario modeled, we will carry out the implementation and tests using Hololens 2 [14, 20]. We will have the support of the company's own employees for this step. From this, we will have feedback regarding usability, reliability, practicality, etc.

4 Results

From proposed method we simulated in laboratory the 3D reconstruction process, from acquisition of the image of an electrical transformer. Despite the scale difference, we intended here to validate the 3D image acquisition. As shown in the image of the Fig. 7a, we extract the measures by manual way as reference to validate the object size.

In Fig. 7b is presented the respective reconstructed image after all steps executed of proposed method. It is observed distortions on extreme points of the objects. We observed that we have the same pattern in sizing, however, the distorted parts of object changes with rotations in the virtual environment as presented in Fig. 8.

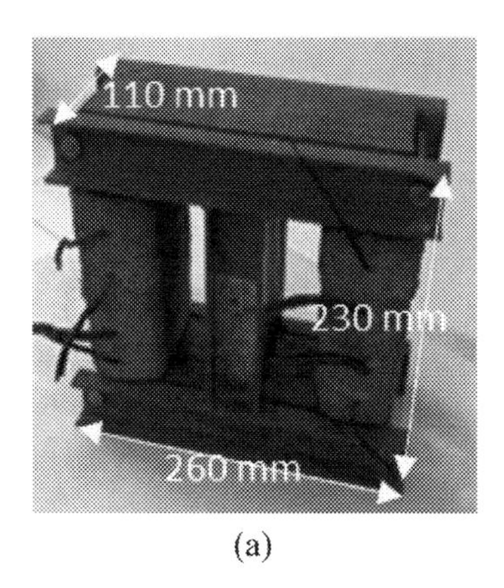

(a)

(b)

Fig. 7. (a) A 2D image of a real electrical transformer (260 mm × 230 mm × 110 mm). (b) The respective 3D reconstructed image.

(a)

(b)

(c)

Fig. 8. The three views of the reconstructed image. (a) top view. (b) front view. (c) side view.

As this article deals with a proof of concept, in which we evaluate the technologies that can be applied in the creation of the environment, we do not have quantitative technology evaluation data. However, qualitatively we can say that the use of the Polycam application was satisfactory because it gave us expressive results with regard to image quality and fidelity with the real object.

5 Conclusion

As presented in this article, we select eight main articles that used some of virtual technologies for study and/or practice. The examples found deals about the impossibility of visually inspecting underground objects by applying augmented reality to create a corresponding digital twin. This kind of problems has in common with our research, the impossibility of accessing the environment due to external risk factors.

Also, we are concerned to prevent the humans' actions in sharing a common workplace due the impossibility of access by employees to the environment while the equipment is in full operation; The creation of a digital twin connecting to a 3D model arose too because [4] from the 3D graph generated by the 3D model it is possible to extract the various information relevant to the physical twin, such as: shape, size and positions of the objects present. In addition, it is possible to simulate physical properties describing the state and behavior of entities.The solution for us will in future use the mixed reality and a digital twin concept to train the maintenance teams. Here in this paper, we shown the initial step of image acquirement and respective 3D reconstruction by a simulated operation. The next step is the implementation (through Blender, Polycam and Unity) using an AR/VR headset: initially Hololens 2 (Microsoft). The 3D reconstruction test we did aimed to understand the digitization process to insert it in a virtual environment (Blender, Unity, etc…) and to understand what functionalities Hololens 2 (for example) can offer. These next step will be focused on the visual part and on the virtual graphical environment, with the creation and implementation of the scenarios that will be used for training the company's employees. In parallel, we will carry out the analysis and studies to obtain the kinematics and risks of the real environment. We will also verify the possibility of including, in the virtual environment, the simulation of electromagnetic effects existing in the real environment.

References

1. Meta Homepage. https://discoverycommerce.valor.globo.com/novas-tecnologias-imersivas-transformam-o-presente-e-o-futuro-das-vendas/. Accessed 23 Feb 2023
2. ResearchGate Homepage. https://www.researchgate.net/figure/The-learning-pyramid-from-NTL-Institute-for-AppliedBehavioral-Science_fig4_221801860. Accessed 14 Feb 2023
3. Li, M., Feng, X., Han, Y.: Brillouin fiber optic sensors and mobile augmented reality-based digital twins for quantitative safety assessment of underground pipelines. Autom. Construct. **144** (2022)
4. Li, X., He, B., Zhou, Y., Li, G.: Multisource model-driven digital twin system of robotic assembly. IEEE Syst. J. **15**(1), 114–123 (2021)
5. Rivera, L.F., Jimenez, M., Villegas, N.M., Tamura, G., Muller, H.A.: The forging of autonomic and cooperating digital twins. IEEE Internet Comput. **26**(5), 41–49 (2022)
6. Mokhtari, F., Imanpour, A.: A digital twin-based framework for multi-element seismic hybrid simulation of structures. Mech. Syst. Sign. Process. **186** (2023)
7. Buchmann, V., et al.: FingARtips: gesture based direct manipulation in augmented reality. In: Graphite ´04: Proceedings of the 2nd international conference on computer graphics and interactive techniques in Australasia and South East Asia, pp. 212–221. ACM, New York (2004)

8. Oyekan, J.O., et al.: The effectiveness of virtual environments in developing collaborative strategies between industrial robots and humans. Robot. Comput.-Integrat. Manufac. **55**(Part A) (2019)
9. Paula Ferreira, W., Armellini, F., De Santa-Eulalia, L.A.: Simulation in industry 4.0: a state-of-the-art review. Comput. Indust. Eng. 1–17 (2020)
10. Lam, T.: TEDxQueensU Homepage. How immersive technologies (AR/VR) will reform the human experience, https://www.youtube.com/watch?v=Fi97-DAcGMk. Accessed 14 Feb 2023
11. Medium Homepage. https://medium.com/tend%C3%AAncias-digitais/realidade-mista-mixed-reality-d406ed0bf923. Accessed 14 Feb 2023
12. Dezen-Kempter, E., Soibelman, L., Chen, M., Muller, A.V.: Escaneamento 3D a laser, fotogrametria e modelagem da informação da construção para gestão e operação de edificações históricas. Gestão e Tecnologia de Projetos, São Paulo **10**(2), 113–124 (2015)
13. Globo Homepage, https://g1.globo.com/tecnologia/noticia/2023/03/22/hololens-2-microsoft-lanca-no-brasil-oculos-de-realidademista.ghtml?utm_source=meio&utm_medium=email. Accessed 29 May 2023
14. Experimentable Digital Twins—Streamlining Simulation-Based Systems Engineering for Industry 4.0, http://www.reconcell.eu/files/publications/ii_2018.pdf. Accessed 8 Feb 2023
15. Conde, J., Munoz-Arcentales, A., Alonso, A., López-Pernas, S., Salvachua, J.: Modeling digital twin data and architecture: a building guide with FIWARE as enabling technology . IEEE Internet Comput. **26**(3), 7–14 (2022)
16. Microsoft Hololens 2. https://www.microsoft.com/pt-br/hololens/hardwareAccessed 22 Feb 2023
17. Silva, R.L., Canciglieri Junior, O., Rudek, M.: A road map for planning-deploying machine vision artifacts in the context of industry 4.0. J. Indust. Product. Eng. **39**(3), 167–180 (2022)
18. Kurka, P.R.G., Rudek, M.: Three-dimensional volume and position recovering using a virtual reference box. IEEE Trans. Image Process. **16**(2), 573–576 (2007)
19. Lalley, J.P., Miller, R.H.: The learning pyramid: does it point teachers in the right direction? Education **2007**(128), 64–79 (2007)
20. Microsoft Homepage. https://learn.microsoft.com/pt-br/training/modules/intro-to-mixed-reality/2-what. Accessed 31 Jan 3023
21. Forbes Homepage. https://forbes.com.br/forbes-tech/2022/09/digital-twin-como-funciona-a-tecnologia-que-espelha-o-mundo-real/. Accessed 28 Jan 2023

Organisation: Knowledge Management, Change Management, Frameworks for Project and Service Development

How to Support Knowledge Exchange in a Multi-division Manufacturing Firm Using a Prototype Platform?

Mélick Proulx[1(✉)] and Mickaël Gardoni[1,2]

[1] École de Technologie Supérieure, Montréal, Québec H3C 1K3, Canada
melick.proulx.1@ens.etsmtl.ca, mickael.gardoni@etsmtl.ca
[2] INSA Strasbourg, 670000 Strasbourg, Grand Est, France

Abstract. Knowledge management and intellectual capital are essential factors in a company's success. An intrinsic element of knowledge management is knowledge exchange. Obtaining knowledge outside of a firm's boundaries, also known as open innovation, is necessary. Although these topics are well described in the literature, they seem less developed in a manufacturing context. The current economic context is favourable to the firm's merger and acquisition. This paper aims to understand how knowledge exchange can be supported in a multi-division manufacturing SME. During three months, thanks to a prototype platform, part of the knowledge exchanges and collaboration were tracked and evaluated in a Quebec aerospace multi-division manufacturing firm. Through this period, knowledge exchanges' quantity, success and collaboration symmetry were monitored. Semi-structured interviews were conducted at the end of the experimentation with the members to gather the foundations and limitations of the prototype platform. This one proposes a supportive organizational structure and incentives which enhance knowledge exchanges among involved SMEs. This prototype platform approach should provide the foundations of more structured knowledge exchanges.

Keywords: Knowledge exchange · Open Innovation · Collaboration · SME

1 Introduction

Knowledge is one of a firm's most important intangible assets, which resides in its employees [11, 32]. The company's internal knowledge is a limited resource and obtaining knowledge from outside is necessary to renew the knowledge pool [9]. Innovation cycles are getting shorter and shorter, and managing this knowledge is essential if firms want to remain competitive [27]. Many companies will face a knowledge management problem across potentially new subsidiaries in a merger and acquisition trend.

Knowledge management in subsidiaries belonging to the multinational firm is well documented [26], and there is also literature available in knowledge sharing in SMEs [5, 8, 13, 28]. However, when SMEs are divided into several units, managing the knowledge held by the units might be challenging. Therefore, the question that needs to be

Published by Springer Nature Switzerland AG 2024
C. Danjou et al. (Eds.): PLM 2023, IFIP AICT 701, pp. 105–115, 2024.
https://doi.org/10.1007/978-3-031-62578-7_10

asked is: How to support knowledge exchanges in a multi-division manufacturing firm? This paper aims to present opportunities and barriers to knowledge exchange among a group of SMEs by using a prototype platform.

The paper is structured as follows: Sect. 2 will present the state of the art on the knowledge exchange inside a multi-division firm and the open innovation in an intra-firm context by using a prototype platform. Next, Sect. 3 will present the context in which this study has been conducted. The methodology will be presented in Sect. 4. The opportunities and the barriers will be covered in the next two sections. The last section will be the conclusion.

2 State of the Art

2.1 Knowledge Exchange Inside a Multi-division Firm

Literature shows that knowledge exchange can be supported by using a prototype platform in a multinational group having subsidiaries. However, there are fewer examples in the literature of how knowledge exchange can be supported in a group of SMEs.

Key benefits have been observed by multinational firms which perform knowledge management and knowledge exchanges across their subsidiaries. The subsidiaries' knowledge management capabilities, such as favouring knowledge exchange across the group, showed a higher innovative performance [12]. This can be explained by selecting and leveraging the new knowledge more effectively across the firm [12]. The subsidiaries' knowledge absorption and diffusion can also positively affect the employees and internal processes [12].

To support knowledge exchange across the subsidiaries, several methods are used by larger firms. Moving employees from the headquarter to the subsidiaries, and vice-versa is one of the most used methods [10]. Multinational firm managers', which allocate resources to subsidiaries, can localize specific knowledge to better exchange knowledge across subsidiaries [22]. Moving resources between subsidiaries have proven to reduce the costs of knowledge exchanges by enhancing the building of networks and promoting knowledge exchanges across subsidiaries [22]. Larger firms can employ around more than 10 workers to ensure knowledge exchange across all subsidiaries [10]. Their knowledge exchange team oversees the dissemination of the best processes and auditing the actual processes to ensure the firm reaches operational excellence [10]. This team is also in charge of employee training and coordinating the annual event to foster knowledge exchanges [10]. Technology and IT tools, such as a prototype platform, also seem to be used by multinational firms to coordinate their operations with their subsidiaries, to describe firm projects or suggest the best practices [10, 16].

Knowledge exchange between subsidiaries at the multinational level can be fraught with barriers. The motivation of the workers performing knowledge exchange and their absorptive capacity can affect the effectiveness of the knowledge exchange [12, 10].

However, studies on knowledge exchange in SMEs suggest that it is more than just a scaled-down model of the literature available on multinational firms [5]. Few kinds of literature are available about knowledge exchange in SMEs, which its subsidiaries also consist of SMEs. On the other hand, literature is available on favouring knowledge exchanges inside a single SME, which can be a starting point.

Knowledge exchange was highlighted to be very important and beneficial to SMEs [8]. Study shows that knowledge management, thus knowledge exchange, is directly linked to SMEs manufacturing performance [28]. SMEs employ fewer people than large firms, which can favour knowledge exchange by making them work closely together [13]. Therefore, knowledge in a manufacturing sector regrouping SMEs represents a significant asset and provides a competitive advantage [24].

A study on Malaysian SMEs showed that knowledge exchange could be largely influenced by technology, motivation, and a reward system [8]. The employee's motivation to perform knowledge exchange will directly impact the result [8]. These items need to be stable and in place to foster knowledge exchange across SMEs [8]. Up-to-date technology also plays a vital role in supporting knowledge exchange in SMEs [8]. Utilizing an intranet or networking application was considered beneficial [8]. Another study on SMEs in Italy showed that SMEs prefer using e-mail, videoconference, or an ERP to share knowledge rather than data mining or social media [5]. In Iceland, a study on two SMEs from the financial and food sector showed that SMEs could perform knowledge exchange. Using e-mail, intranet and social media can diffuse new knowledge and possible solutions to encountered problems [13]. The studied SME also planned a twice-a-week meeting to communicate actual issues and find answers [13]. The second firm studied also used social media to share knowledge and planned competitions to develop new products, which is on a voluntary basis. These competitions are linked to a reward program which can be translated into a salary increase [13].

Knowledge exchange in SMEs can seem costly. Therefore, it would be imperative for SMEs to maximize the potential outcomes of knowledge exchanges [8]. However, SMEs often lack financial and human resources, limiting knowledge exchange [5, 8, 13, 29]. Another element that can limit knowledge exchanges is the need for more training on the subject in SMEs [13]. A survey on Indian SMEs revealed several barriers to knowledge exchange: lack of high management commitment, misunderstanding of the knowledge exchange concept, lack of time to share and lack of motivation [1]. A study on Albanian SMEs identified employees' motivation, lack of top management support and recognition as knowledge exchange barriers [29]. A precise knowledge management strategy is essential to standardize the collection, dissemination and use of knowledge and to reduce the risk of losing it [13].

The literature presented above shows that achieving knowledge exchange within an SME is possible. However, it will be interesting to observe whether it is similar when several SMEs perform knowledge exchange together. Knowledge exchange has been shown to be an essential aspect of successful collaborations in SMEs [7].

2.2 Open Innovation in the Intra-Firm Context

A parallel can be drawn between knowledge exchange and open innovation. As a definition, open innovation is characterized by two types of knowledge exchanges which are outside-in and inside-out. Opening innovation processes to the external environment inputs defines outside-in flow [6]. Diffusing the knowledge developed internally, which is little or not used, and making it accessible outside the company represents the inside-out flow [6]. From its beginning, the focus of open innovation has been on large companies. Intra-firm knowledge has been proven to provide a firm with an acceptable quantity of new knowledge [9]. Literature suggests that larger firms using open innovation with their subsidiaries remark that knowledge inflow benefits the subsidiaries [22]. In comparison, knowledge outflow helps the other subsidiaries of the group [22]. Multinational subsidiaries showed a high involvement in the different knowledge flows, from a high outflow and inflow profile to a low outflow and inflow profile [16].

SMEs have proven to be supporting more prominent firms providing products or services [7]. Recent years have shown an increased interest in using open innovation in SMEs. Obtaining knowledge from outside the company is a trend that has grown in popularity recently [9]. The frame on which SMEs are based makes them more open to external sources of knowledge [19]. Literature has shown that SMEs can benefit from open innovation, such as access to new knowledge or minimizing their new product time to market [19]. Study shows that in contrast to larger firms, SMEs practice open innovation more on the commercialization of new product [30]. A study realized on Italian SMEs shows that technology, globalization, and organizational culture are among the characteristics that favour open innovation in smaller firms [30]. On the other hand, SMEs can encounter some challenges using open innovation. Lack of financial resources, internal resistance and human resources are among them [7, 30]. A study on Belgian firms reveals that technology-intensive subsidiaries are less open to external knowledge [9]. To be effective, knowledge flows need to be supported by personal interactions [22]. Again, the literature shows that it is possible for a single SME to use open innovation. It would be interesting to observe the utilization of open innovation in a group of SMEs.

3 Context

3.1 Knowledge Capitalization Problematic

Few manufacturing companies perform knowledge management instinctively. This does not mean that there is no knowledge within the firm. Knowledge examples in manufacturing firms are multiple, manufacturing processes, and daily production notes are among them [31]. It will not be easy to diffuse knowledge if it is not integrated into the firm's processes [31]. Knowledge can refer to as intellectual capital for firms, representing an essential aspect of a firm's creativity and innovation and directly impacts its success [1]. Knowledge management also positively impacts the firm performance [1]. The knowledge of a firm depends on the people inside it [8]. Nowadays, human resources are highly volatile, which can complicate intellectual capitalization by the firm. The knowledge capitalization problem remains as long as the company does not control the knowledge held by its employees [22].

3.2 Studied Group

The studied group is a Tier 2 firm from the Quebec aerospace cluster. The group has five divisions, all located in the Quebec region. The divisions are mainly Tier 3 firms, with between 50 and 300 employees. In addition to the head office, which provides common services such as finance and human resources, each division has its own manufacturing expertise. In the current context, the studied group has an active merger and acquisition strategy. This strategy will highlight the need for more knowledge exchanges, open innovation and collaboration between divisions. Applying these concepts would continue to add value to current and future acquisitions.

4 Methodology

Most studies have used surveys of targeted companies to collect data [2, 8, 29]. The Design Research methodology was used in this study. Acquiring knowledge through the act of experimentation is the aim of this methodology [18]. Its application consists of five distinct steps [4, 18], which are described below.

The first step is awareness of the problem, which is described as choosing a problem to solve. As shown above, knowledge capitalization is a problem in aerospace SMEs. Supporting knowledge exchange is essential to help SMEs capitalize on their knowledge. Therefore, the chosen problem is how to support knowledge exchange in a multi-division SME.

Suggesting the key elements needed to solve the problem is the second step. It was necessary to choose a technological tool to evaluate the exchange of knowledge between the group's SMEs. The use of quantitative, qualitative data and regular follow-up also seems important in drawing conclusions.

The third step, based on the available knowledge and the specific element identified in the suggestion step, is to develop a potential solution. The prototype platform was using some functionalities of Microsoft Teams. The planner feature allows the use of tiles which are called cards. These cards are used to keep track of the different knowledge exchange topics. Two criteria - role and knowledge - were used to select the participating members of each division. A meeting was held every two weeks to monitor the knowledge exchange. The agenda included the update of the collaboration criteria, the review of the cards and the blocking point to knowledge exchange.

Evaluation of the developed solution according to certain criteria is the fourth step. This step may involve iteration between the development and evaluation steps to converge on a solution. Only one kind of prototype platform was tested for this study. Data was collected over a period of three months. Quantitative data such as the number of cards, the rate of knowledge exchange and the symmetry of the collaboration were collected to evaluate the knowledge exchanges [25]. It was then collected through Microsoft Excel and analyzed with graphs and descriptive statistics. A total of 19 collaboration topics were placed on the platform by the different divisions. Of these, approximately 52% had knowledge exchange. Opportunities and barriers were the focus of the qualitative data collection. Meeting notes were gathered in Microsoft OneNote. A semi-structured interview was conducted at the end of the experiment to gather the opportunities, barriers, and managerial implications. Content analysis was used to understand data from the meeting's notes and the semi-structured interview. The interview consisted of 13 questions. The questions were grouped by topic: three questions on barriers to knowledge exchange, three questions on the impact of culture, six questions on the impact of management and one question on the impact of technology. This study covered the knowledge exchange of a manufacturing firm and its divisions. The sample used for this study is a non-probability sample, which means that it is not representative of all populations of manufacturing multi-division firms.

The last of the five steps is the conclusion which is selecting the solution to be implemented following the development and evaluation phase. Although the chosen solution is the use of a prototype platform to promote knowledge sharing within the company, the opportunities and barriers associated with the experiment are presented below.

5 Opportunities

According to Razmerita and Nielsen [26], there are three categories to present barriers, which are individual, organizational, and technological. These categories can also be used to classify the opportunities. These opportunities are the basis for knowledge exchanges between SMEs in the same group.

In terms of individual opportunities, the introduction of a reward system seems to be a good opportunity, according to some studies [8, 14, 15, 21]. Although there was no reward system in this experiment, this addition could motivate people to share knowledge. This system could work by accumulating points or a salary increase for a certain number of knowledge exchanges. However, it is important to note that not every employee can be motivated by a reward system; in this experiment, most of the people involved simply enjoyed helping others.

As for the organizational aspect, several opportunities have emerged as a result of the experiment. Including knowledge exchange in the group processes, such as project management or inter-division project postmortem, represents some opportunities [28]. Building on the studies carried out by Demeter and Losonci and Meyer, Li and Schotter [10, 22], exchanging resources between divisions is a good opportunity. Instead, unlike multinationals, where the exchange period is mostly in years [10], the period should be adapted to SMEs and perhaps reduced to a few months. The introduction of a knowledge exchange program is another opportunity. This program could include periodic meetings between the division to exchange knowledge and resource to spread best practices among the group [10]. This experiment demonstrated the value of appointing one person to foster knowledge exchanges across the divisions [21]. This leader will need specific soft and hard skills, such as being a good communicator and having a good ability to synthesize. Finally, this person will need to act as a bridge between the divisions to favour knowledge exchanges and open innovation across the group. Having an adequate corporate structure also seems to be an opportunity. This corporate structure would favour knowledge exchanges among divisions, as this task could also be included in the job descriptions of certain employees. These opportunities could improve knowledge exchanges, communication between divisions and innovation.

Creating communities seems to be a starting point for technological opportunities. Setting up a community around a specific topic can favour knowledge exchanges between the divisions [23]. Once again, appointing a community leader seems unavoidable. Several studies suggest implementing adequate IT tools to favour knowledge exchange [3, 10, 13, 31]. An IT tool such as a prototype platform seems appropriate to foster knowledge exchange across the divisions. A prototype platform where ideas and topics are documented and then easily accessible would benefit SMEs in exchanging knowledge.

The number of collaboration topics placed on the prototype platform per week during the experiment is shown in Fig. 1. The trending line shows an increasing number of collaboration topics every two weeks, and this corresponds with the timing of the meetings. Most of the knowledge exchange and use of the prototype platform took place during these meetings. The project leader organized these meetings, highlighting the importance of appointing a knowledge management leader or community manager to facilitate more knowledge exchanges. As mentioned in Sect. 4, for the 19 collaborative topics placed on the prototype platform, the knowledge exchange rate during this experiment was 52%. While the current outcome may not seem very impressive, it's worth considering that as individuals continue to interact and meet more frequently, there will likely be a greater exchange of knowledge.

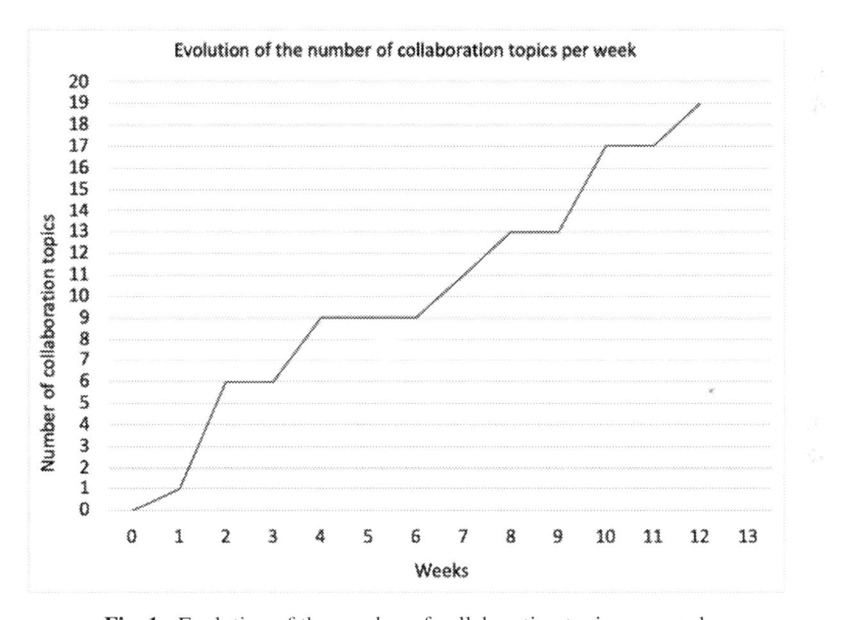

Fig. 1. Evolution of the number of collaboration topics per week

After the experiment, and based on the work of Sokoh and al, a roadmap to support knowledge exchange in a group of SMEs can be suggested [27].

1. Executive management decides to prioritize knowledge management.
2. Appoint a knowledge management leader to develop a knowledge exchange strategy, including the critical knowledge which needs to be shared.
3. Create a knowledge committee with members from each division.
4. Set up appropriate IT tools, such as an intranet and social media.
5. Create a reward and knowledge exchange program.
6. Monitor knowledge exchanges and calculate the benefits.

The knowledge exchange efficiency will depend on how well the knowledge management leader is supported by other departments, such as IT, human resource management and executive management. Management will have to allow time for their employees to perform knowledge exchange, and on their side, employees will have to be motivated and enjoy performing knowledge exchanges [15]. They also need to be trained to ensure an effective knowledge exchange.

6 Barriers

As described above, according to Razmerita and Nielsen [26], it is possible to classify barriers into three categories which are individual, organizational, and technological. These categories will be used here to develop the barriers that are encountered in the monitoring of the knowledge exchange between a group of SMEs.

Starting with the individual aspect, the first barrier to be revealed by this study was the different vocabulary used by the divisions, which was also identified by Meyer and al. [22]. Although the employees have been working for the same firm for several years and spoke the same language, this project found differences at the technical level such as manufacturing methods, specific tool names or process equipments. This project also revealed differences in administrative and technological vocabulary. In the long term, it might be good to standardize the vocabulary used to describe a concept or an object. By introducing a standardized vocabulary, communication between employees will be enhanced, and therefore knowledge exchange will be more valuable. Several studies have shown that the lack of time allocated to the resources for knowledge exchanges is a significant barrier [12, 20, 26, 27]. One of the five divisions was unable to attend most of the meetings, limiting its knowledge exchange benefits. One person from each division was appointed to the project team for this project. However, knowledge exchange and open innovation are not a single-person job. Knowledge exchange is limited by the absorptive capacity of the people involved [12, 20, 22]. In order to be efficient and add value to the group of SMEs, knowledge exchanges and open innovation have to be performed by every employee level, from the shop floor to the executive management. Sometimes, workers don't know what exactly knowledge exchange and open innovation are. On certain occasions, they do it without being aware of it. Providing training on these two concepts to every new hire employee could be an excellent solution to train employees and promote a knowledge exchange culture across the divisions. Emphasizing on employees performing knowledge exchanges to continuously create and apply new knowledge is key [11].

The corporate structure does not favour knowledge exchange and open innovation and represents an organizational barrier. This project shows that the opportunities for knowledge exchanges among the divisions are many and varied. As mentioned earlier, appointing a leader to promote and monitor knowledge exchange across the division is essential. Compared to multinational companies where a team of up to 30 employees is dedicated to knowledge management [10], this number needs to be reduced to be effective for SMEs. Selecting one leader and creating a committee of employees would appear more appropriate to begin with. The group culture needs to change to promote more knowledge exchange among people. All levels of management need to be involved in the knowledge exchange process, allocate time and resources to knowledge exchange and communicate a clear vision of knowledge management [1, 17, 21]. Holding conferences within SMEs on a specific topic, such as manufacturing methods or inspection processes, also seems to be a solution to the corporate structure [10].

In a digital era, the project team felt the technological tools available were sufficient to promote knowledge exchanges. A planner in Microsoft Teams was used for this project which seemed to be an inhibitor of knowledge exchange. There is probably a better tool to perform knowledge exchange and open innovation. However, the barrier remains in deploying the available tools to favour knowledge exchange. In addition, there is often a lack of training on the tools used, which limits the knowledge exchange across the divisions.

7 Conclusion

This paper aimed to identify opportunities and barriers to support knowledge exchange in a multi-division firm by using a prototype platform. Data was gathered from a group of five SMEs, all from the Quebec aerospace cluster. The opportunities reside in making knowledge exchange a priority for the firm and appointing a knowledge management leader. Allowing time for employees to perform knowledge exchange and using adequate technological tools are also opportunities. The main barriers are the vocabulary used by employees and the lack of training in knowledge management.

Regarding the limitations, the size of the group studied and its sector can reduce the generality of opportunities and barriers. The external environment, such as the group's region, might also limit the findings. Conducting this study in Europe or Asia, for example, might have led to possible different conclusions. The research approach used also limits the findings, using a survey to collect data on SMEs would probably have revealed different opportunities and barriers.

In terms of future perspectives, developing a model to evaluate the benefits of implementing a knowledge management system in SMEs seems appropriate. Observing how it is possible to support collaboration in another industrial sector that is clustering SMEs or using another platform to support knowledge exchange are other future perspectives.

References

1. Abualoush, S., Bataineh, K., Alrowwad, A.A.: The role of knowledge management process and intellectual capital as intermediary variables between knowledge management infrastructure and organization performance. Interdiscip. J. Inf. Knowl. Manag. **13**(2018), 279–309 (2018)
2. Anand, A., Kant, R., Singh, M.D.: Knowledge sharing in SMEs: modelling the barriers. Int. J. Manage. Enterprise Develop. **12**(4–6), 385–410 (2013)
3. Bazrkar, A.: The investigation of the role of information technology in creating and developing a sustainable competitive advantage for organizations through the implementation of knowledge management. J. Spatial Organ. Dyn. **8**(4), 287–299 (2020)
4. Carstensen, A.K., Bernhard, J.: Design science research – a powerful tool for improving methods in engineering education research. Eur. J. Eng. Educ. **1–2**, 85–102 (2019)
5. Cerchione, R., Esposito, E.: Using knowledge management systems: a taxonomy of SME strategies. Int. J. Inf. Manage. **37**(1), 1551–1562 (2017)

6. Chesbrough, H.: The future of open innovation: the future of open innovation is more extensive, more collaborative, and more engaged with a wider variety of participants. Res. Technol. Manage. **60**(1), 35–38 (2017)
7. Choudhary, A.K., Harding, J.A., Tiwari, M.K., Shankar, R.: Knowledge management based collaboration moderator services to support SMEs in virtual organizations. Product. Plan. Control **30**(10–12), 951–970 (2019)
8. Cyril Eze, U., Guan Gan Goh, G., Yih Goh, C., Ling Tan, T.: Perspectives of SMEs on knowledge sharing. VINE **43**(2), 210–236 (2013)
9. De Beule, F., Van Beveren, I.: Sources of open innovation in foreign subsidiaries: an enriched typology. Int. Bus. Rev. **28**(1), 135–147 (2019)
10. Demeter, K., Losonci, D.: Transferring lean knowledge within multinational networks. Product. Plan. Control **30**(2–3), 211–224 (2019)
11. Doan, T.N.T., Nguyen, H.H., Nguyen, T.K.V.: The importance of knowledge management in organizations: an application of SECI model and suggestions for Vietnamese enterprises. Int. J. Educ. Soc. Sci. Res. **4**(6), 312–320 (2021)
12. Ferraris, A., Santoro, G., Dezi, L.: How MNC's subsidiaries may improve their innovative performance? The role of external sources and knowledge management capabilities. J. Knowl. Manag. **21**(3), 540–552 (2017)
13. Grimsdottir, E., Edvardsson, I.R.: Knowledge management, knowledge creation, and open innovation in Icelandic SMEs. SAGE Open **8**(4), 2158244018807320 (2018)
14. Hadeeba, S.O.S.B., Yusoff, W.F.W.: Proposed framework for the usage of information technology tools to enhance knowledge management process of organizations. J. Organ. Manage. Stud. **2022**, 1–7 (2022)
15. Hau, Y.S., Kim, B., Lee, H., Kim, Y.G.: The effects of individual motivations and social capital on employees' tacit and explicit knowledge sharing intentions. Int. J. Inf. Manage. **33**(2), 356–366 (2013)
16. Jankowska, B., Di Maria, E., Cygler, J.: Do clusters matter for foreign subsidiaries in the Era of industry 4.0? The case of the aviation valley in Poland. Eur. Res. Manage. Bus. Econ. **27**(2), 100150 (2021)
17. Kahrens, M., Früauff, D.H.: Critical evaluation of Nonaka's SECI model. In: The palgrave handbook of knowledge management, pp. 53–83. Palgrave Macmillan, Cham (2018)
18. Kuechler, B., Vaishnavi, V.: On theory development in design science research: anatomy of a research project. Eur. J. Inf. Syst. **17**(5), 489–504 (2008)
19. Lee, S., Park, G., Yoon, B., Park, J.: Open innovation in SMEs—An intermediated network model. Res. Policy **39**(2), 290–300 (2010)
20. Lim, S.Y., Jarvenpaa, S.L., Lanham, H.J.: Barriers to interorganizational knowledge transfer in post-hospital care transitions: review and directions for information systems research. J. Manag. Inf. Syst. **32**(3), 48–74 (2015)
21. Margilaj, E., Bello, K.: Critical success factors of knowledge management in Albania business organizations. Eur. J. Res. Reflect. Manage. Sci. **3**(2) (2015)
22. Meyer, K.E., Li, C., Schotter, A.P.: Managing the MNE subsidiary: advancing a multi-level and dynamic research agenda. J. Int. Bus. Stud. **51**(4), 538–576 (2020)
23. Nisar, T.M., Prabhakar, G., Strakova, L.: Social media information benefits, knowledge management and smart organizations. J. Bus. Res. **94**, 264–272 (2019)
24. Obradović, T., Vlačić, B., Dabić, M.: Open innovation in the manufacturing industry: a review and research agenda. Technovation **102**, 102221 (2021)
25. Proulx, M., Gardoni, M., Farha, S.: Structuring SMEs collaborations within a cluster. In: Noël, F., Nyffenegger, F., Rivest, L., Bouras, A. (eds.) Product lifecycle management. PLM in transition times: the place of humans and transformative technologies. PLM 2022. IFIP Advances in Information and Communication Technology, vol. 667, pp. 35–44. Springer, Cham (2023). https://doi.org/10.1007/978-3-031-25182-5

26. Razmerita, L., Kirchner, K., Nielsen, P.: What factors influence knowledge sharing in organizations? A social dilemma perspective of social media communication. J. Knowl. Manag. **20**(6), 1225–1246 (2016)
27. Sokoh, G.C., Okolie, U.C.: Knowledge management and its importance in modern organization. J. Public Admin. Finan. Law **20**(2021), 283–300 (2021)
28. Tan, L.P., Wong, K.Y.: Linkage between knowledge management and manufacturing performance: a structural equation modeling approach. J. Knowl. Manag. **19**(4), 814–835 (2015)
29. Vajjhala, N.R.: Key barriers to knowledge sharing in medium-sized enterprises in transition economies. Int. J. Bus. Soc. Sci. **4**(14), 90–98 (2013)
30. Verbano, C., Crema, M., Venturini, K.: The identification and characterization of open innovation profiles in Italian small and medium-sized enterprises. J. Small Bus. Manage. **53**(4), 1052–1075 (2015)
31. Wang, P., Tian, X., Geng, J., Key, F.H.: The method of manufacturing knowledge management based on manufacturing resource capacity. In: 2010 2nd IEEE international conference on information management and engineering. pp. 457–460. IEEE (2010)
32. Zelenkov, Y.A: The effectiveness of Russian organizations: The role of knowledge management and change readiness. Российский журнал менеджмента **16**(4), 513–536 (2018)

Contextualization for Generating FAIR Data: A Dynamic Model for Documenting Research Activities

Osman Altun[1], Marc Hinterthaner[2], Khemais Barienti[1], Florian Nürnberger[1], Roland Lachmayer[1], Iryna Mozgova[3](✉), Oliver Koepler[2], and Sören Auer[1]

[1] Leibniz University Hannover, Welfengarten 1A, 30167 Hannover, Germany
[2] Leibniz Information Centre for Science and Technology (TIB), Welfengarten 1B, 30167 Hannover, Germany
[3] Paderborn University, Warburger Street 100, 33098 Paderborn, Germany
iryna.mozgova@uni-paderborn.de

Abstract. The digitization of technologies in product manufacturing results in the availability of large amounts of process and product data. To gain knowledge from this data and fully leverage its potential, its structuring and semantically annotation is essential. This allows preserving the context of data generation and makes the data machine-readable and interpretable. Contextualization is the key to generating FAIR (Findable, Accessible, Interoperable, Reusable) data. The documentation of research activities and provenance of generated data is usually achieved by protocols. However, there is often a tension between the desire to document data generation in a structured, semantically rich form and the need to design research and process parameters flexibly as experimental conditions change.

To resolve these contradictions, a dynamic model is described that allows to document research activities and implemented into a knowledge and research data management system to resolve these contradictions. The model allows a formal, semantic representation of research steps, parameters and gathered data, while also providing flexibility in the generation of protocol templates and individual experiments through the reuse of semantic building blocks. The approach is carried out within the context of a large collaborative research center, showcasing its use in managing and providing data for heterogeneous research tasks, documentation, and data types across interdisciplinary projects.

Keywords: FAIR Data · Semantic Annotation · Data Management · Knowledge Management

1 Introduction

The development of novel production technologies is often associated with complex, interdisciplinary research questions that are investigated collaboratively across several project teams and from different perspectives in order to be able to penetrate and describe

Published by Springer Nature Switzerland AG 2024
C. Danjou et al. (Eds.): PLM 2023, IFIP AICT 701, pp. 116–126, 2024.
https://doi.org/10.1007/978-3-031-62578-7_11

processes to be researched in their entirety. During the research activities, experiments, simulations or observations generate large heterogeneous data sets that increasingly merge with data from other activities into joint analysis processes. A comprehensible and reproducible documentation of the methods, materials and tools used in the execution of research activities as well as documentation of measurement results is therefore of great importance.

Documentation in the form of protocols enables contextualization of the research and the data generated, the why, how, when, where, with what and by whom. Logging is manifold and increasingly digital. Realized as locally stored text documents, Excel spreadsheets, or supplementary ReadMe files to data files, such documentation is loosely linked to the corresponding data, weakly structured and not sufficiently annotated in a machine-readable way. Software tools like digital log books, Electronic Lab Notebooks (ELN) or Knowledge Management Systems improve the structured documentation of research activities. Nevertheless, we observe few accepted standards for documenting research activities in engineering. With the increasing availability and acceptance of public data repositories, at least the standardized provision of data is becoming easier. The structured deposition and description of research activities together with generated data is yet another key to enable reproducibility and reusability of data. Looking at increasing amounts of data available, semantic annotation of data is a prerequisite for machine-readable and -interpretable data enabling machine learning and data-driven research.

However, there is a tension between making the documentation process easy to use and the need to have fixed interfaces to capture structured data. On the one hand, researchers require a user-friendly interface that enables them to quickly and easily document their constantly changing procedures, experimental conditions and parameters without disrupting their research workflow. On the other hand, a fixed interface is necessary to ensure that data is captured in a structured and consistent manner that can be easily analyzed and shared. In the following an approach is described to develop a flexible and extensible framework that allows researchers to document their procedures in a structured and semantically rich way, while also providing the necessary flexibility to accommodate changes in experimental conditions. This can be achieved by a "generic protocol structure" applying semantic building blocks that can be customized to specific experimental procedures and easily assembled to form a structured representation of experiments. The implementation is carried out within the knowledge management system of the Collaborative Research Centre (CRC) 1368 „Oxygen-free production“ [1].

2 Research Data Management in Engineering Sciences

Data is the foundation of research and science, providing the raw material from which knowledge and understanding are derived. It is used to test hypotheses, develop theories, apply scientific methods, and gain insights. The availability of research data is an important factor for the transparency and reproducibility of scientific results. The FAIR principles describe the framework of how data should be prepared and made available [2, 3]. In recent years, both the availability of data repositories and the portfolio of standards and tools for comprehensive descriptions of data generation in engineering have improved [4]. With the increasing data available the comprehensive documentation of data generation comes into focus as it provides essential context about the research methods and procedures, enabling other researchers to evaluate the validity and reliability of the data and derived results. In addition, documentation of how data was generated can be used to identify potential sources of error or bias, and to improve the design of future experiments. Therefore, the availability of data alone is not sufficient for ensuring reproducibility, as it must be accompanied by detailed documentation of how the data was generated and processed. Software tools such as ELN or knowledge management systems are also increasingly used in the engineering sciences to document the context of data generation in the form of protocols [5, 6]. E. g., the ELN Software elabFTW provides a generic documentation of research activities due to its flexible way to capture experimental procedures, process steps and parameters in less-structured text documents. Additional tools can be applied on such protocols to increase the structure and semantics of information and data [7]. When using semantic tools such as Semantic MediaWiki (SMW) [8] for the documentation of research activities, information and data are basically captured in a structured and semantically annotated form from the very beginning due to the native functionality of SMW [9–11]. Research data can be described applying the generic DataCite metadata schema [12] or using a domain-specific ontology [13]. For the description of the provenance of digital objects the PROV ontology can be used [14]. Terms from PROV have also been imported into the Metadata4Ing ontology for engineering. Metadata4Ing provides a model for documenting research activities and research data in engineering [15]. It has been developed in the context of the NFDI4Ing consortium within the National Research Data Infrastructure (NFDI) in Germany [16]. It enables researchers to document the origin and path of data created or modified during the research process. For this purpose, Metadata4Ing applies a generalized process model centered on the Class *Processing Step.*

3 Problem Analysis and Research Overview

Research problems in production technologies are often complex and require interdisciplinary collaborations among researchers from various subdisciplines within the engineering sciences. The merging and analysis of data and intermediate results from multiple research projects create the need to harmonize not only the data itself, but also the documentation of its generation and processing, in order to fully capture the context of the data. Although engineering subdisciplines have distinct standards and practices for Standard Operating Procedures (SOPs), documenting research activities can be challenging due to the constantly evolving experimental parameters and the absence of predefined standard procedures. To address this challenge, electronic lab notebooks and knowledge management systems are increasingly used to support a fully digitized documentation of experiments and data generation. These tools offer templates that can be used to structure and even semantically describe experiments and gathered data, facilitating the documentation of research activities in a harmonized and structured manner in accordance with the FAIR data principles. In the aforementioned interdisciplinary research projects, standardized templates for protocols are only partially suitable for documentation purposes. Instead, flexible documentation tools are needed that can be easily adapted and reused in parts. However, such documentation should still be structured and semantically annotated, enabling machine-readability and facilitating the combined analysis with large data collections. This allows the documentation to be effectively integrated with other data, enabling joint analysis and interpretation of interdisciplinary research questions. Adapting these ideas to fully digitized processes and linked data it is intended to create a framework of semantically annotated data and documentation of data generation. To address these competing demands, a dynamic model is being developed to describe and document research activities, data generation, and data provenance. This model enables the development and combination of semantic building blocks to create generic protocols. The implementation of this model and the generic protocols is facilitated by a knowledge management system that utilizes the SMW software.

4 Contextualization of Data Generation

4.1 Requirements

Large Collaborative Research Centres (CRCs) with multiple sub-projects are a good example of how overarching research problems are broken down into smaller research activities. For example, in the development of new or adaptation and modernization of existing production technologies, various sub-projects may examine the same prototype (referred to as "specimen" below). Different parameters (referred to as "variables" below) of both the specimen and the production process are considered using different methods. Sub-projects partially capture, measure or modify the same variables. Data and insights generated in sub-projects should be structured, networked, and made available for a comprehensive analysis.

The documentation of research activities, such as experiments, should be available across all sub-projects through digital protocols. Experiments are often repeated while varying conditions such as process variables or specimens. The basic description of the steps, specimen, and variables of an experiment should be represented by generic protocol types, which can be flexibly composed of reusable semantic building blocks for the representation of specimens and variables. Protocol types serve as an easy-to-use template for creating a protocol to document a specific experiment by capturing specific values for the variables assembled in the protocol type. Creating and adapting protocol types should be easy and intuitive so that both documentation and its structure can be quickly and independently adjusted to changes in experiment design. Semantic building blocks can be reused, so that a new declaration is not necessary when creating a new protocol type. The declaration and definition of variables is done once, separated from the protocol types, with a unique naming convention and is subsequently referenced in protocol types.

4.2 Modeling of a Generic Protocol Structure

With the aim to achieve reproducibility, reusability, and interoperability of data generation research activities are modeled reusing parts of the Metadata4Ing ontology. The Metadata4Ing ontology enables a description of data generation processes, associated artifacts, and procedures for data manipulation [15]. It implements concepts of inheritance and modularity, making it ideal for the modular approach of semantic building blocks described here. The names of entities may differ from the terms in Metadata4Ing to maintain the terms used in the joint project or to avoid reserved terms in the software.

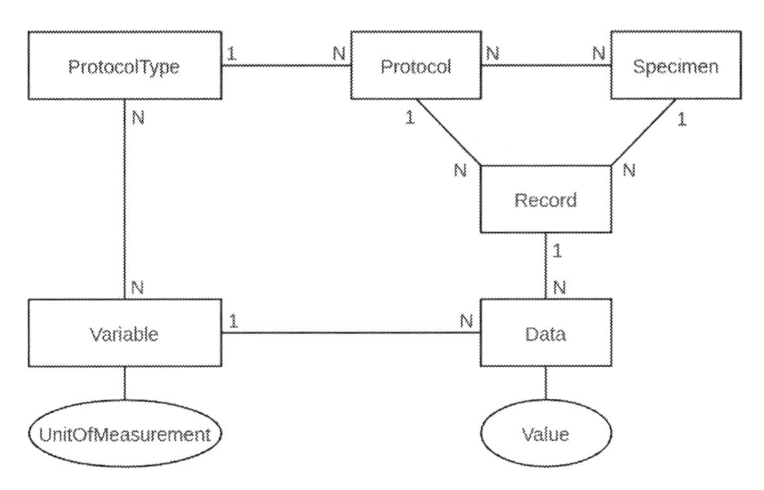

Fig. 1. Entity–relationship model of the generic protocol structure

At the center of our model (Fig. 1) is the entity *Protocol*, whose instance documents the performance of an individual experiment. A protocol type describes the basic steps and conditions. The same process variables are always used or variables are measured

Table 1. Mapping of the model to concepts in the Metadata4Ing ontology.

Entity	Category in SMW	Class Metadata4Ing	Description
Specimen	Specimen		specimen to be investigated
Variable	Variable	*PIMS-II:Variable*	variable of the specimen
Protocol	Protocol	*ProcessingStep*	documentation of an experiment
Protocol type	ProtocolType	*PROV:Activity*	type of experiment
Record (Dataset)	Record	*PIMS-II:Assignment*	container for metadata of a specimen within a protocol
Key-value-pair	Data („Value" is reserved by the system)	*PIMS-II:Value*	assignment of a value to a variable in a record
Quantity	UnitOfMeasurement ("Quantity is reserved by the system)	*EMMO:MeasurementUnit*	measured quantity including the definition of admissible units

during the execution of these steps, but with varying values and from varying specimens. In addition, multiple specimens can be considered in a protocol.

The *Variables* are separate entities to ensure consistent naming and reusability, including a unique definition with associated units. The entity *Record* serves as a container for measured values of variables in the relation between *Protocol* and *Specimen*, represented as key-value pairs. An example of a variable in the model is the oxygen partial pressure or the diameter of a specimen. During the execution of an experiment, the measured value, along with other variables and their values, is summarized and documented in a dataset within a protocol. Table 1 shows the mapping of the model to concepts in the Metadata4Ing ontology and the corresponding categories for implementation in SMW as described in the following Sect. 4.3.

4.3 Implementation into the Knowledge Management System Semantic MediaWiki

The generic protocol structure is implemented in the knowledge management system SMW [8]. SMW is a semantic extension of the software MediaWiki, known from Wikipedia. With SMW, data and information can be semantically annotated. Semantic statements about data in the form of subject, predicate, object can be modeled and represented on wiki pages using a specific syntax through categories and so-called properties.

For each entity in the model, an equivalent SMW category with required properties was created in SMW. An instance of a category and its associated data, as well as their assignment to semantic properties, are captured as key-value pairs with a form and displayed in a structured manner through templates. In general, static forms are created in SMW, so that the input form of a category always captures the same fields. Therefore, for a static protocol structure, it is necessary to define a separate form for each protocol type. In the presented approach of the generic protocol structure, the process variables and the variables to be recorded for each specimen of a protocol should be specified by the corresponding protocol type. For this purpose, instances of the variable category are created for all required variables, linked to a unit of measurement that describes the measured quantity and determines its units.

The form for the protocol type is extended with a field for a list of variables to be examined in the experiment. In order to capture exactly these variables in records of a corresponding protocol instance for each specimen, the record-form is dynamically generated in an outsourced template. This template receives the instance of the parent protocol, through which the list of variables to be examined is available through a query. Using this list, a Wiki-syntax with escape sequences is generated, which, when included in the record-form, dynamically generates the desired fields for the user. Thus, the fields in the form are dynamically generated and do not need to be implemented statically for each protocol type. The representation of the data thus captured is done similarly in the template of the record page.

5 Use Case - Knowledge Management in the Collaborative Research Centre "Oxygen-Free Production"

5.1 Status Quo of Research Activity Documentation

In practice, research involves a wide range of experiments and detailed documentation. In the context of the oxygen-free production application example, experiments as well as the outcomes are very heterogeneous. Typical experiments for material investigation like atomic emission spectroscopy, x-ray diffraction and scanning electron microscopy lead to high resolution images including the corresponding value tables. Experiments like hardness tests and tensile tests lead to large data series. Virtual experiments such as casting simulations are often also documented additionally in form of video content. Consequently, in addition to subjective or organization-internal documentation, this leads to a further lack of structured documentation that is traceable across projects. Figure 2 shows the status quo of data collection and documentation starting with the initial documentation of experiments by handwritten notes.

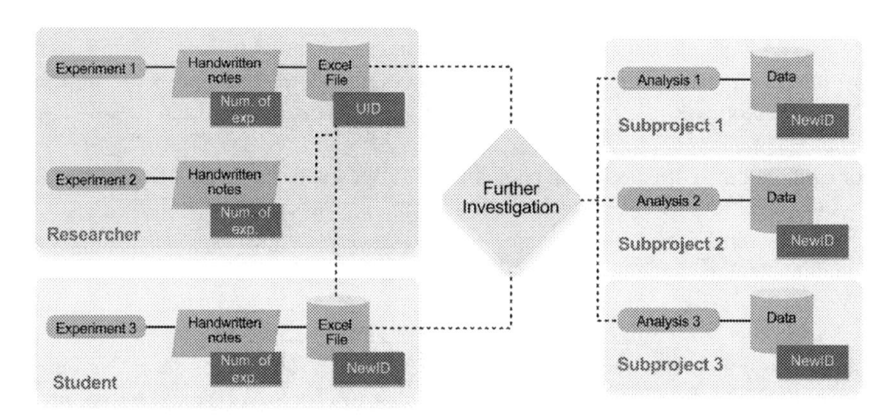

Fig. 2. Flowchart describing a common way of data documentation done by a researcher involving interaction with students and other subprojects.

For practical reasons, the specimens generated are temporarily given an ascending numerical designation, but not yet their unique identifier (UID). The handwritten protocols are transferred to a summary database, such as an Excel file. This process assigns each specimen its UID, but retains its temporary numbering from the hands-on experiment, as the specimens are often already labeled with this temporary ID. As an additional challenge, experiments may also be carried out by students using their own slightly different way of documentation, which may vary from those of the researcher. This may result in minor discrepancies in the data structure and additional efforts in merging data. In addition, specimens may be passed on to other subprojects for analysis. After successful analysis, the additional data are returned to the researcher without any information about the parameters used during the analysis, making it difficult to transfer these relevant parameters into one's own database. In the end, it is often only documented that a particular analysis was carried out.

5.2 Documentation with the Generic Protocol Model Implemented in Semantic MediaWiki

The implementation of the generic protocol structure in SMW as described in Sect. 4.3 leads to the following workflow from a researcher's point of view. For each specific experiment, a protocol type needs to be created once (Fig. 3.).

Afterwards the researcher has to choose from the list of existing variables in the system, exactly those variables for the protocol type that are used for the experiment. In case of a missing variable, the user can create and describe it beforehand, so that a unit of measurement is linked accordingly. Each time an experiment is conducted, a new protocol must be created, selecting the protocol type and the specimen to be analyzed.

The process is illustrated in Fig. 4.

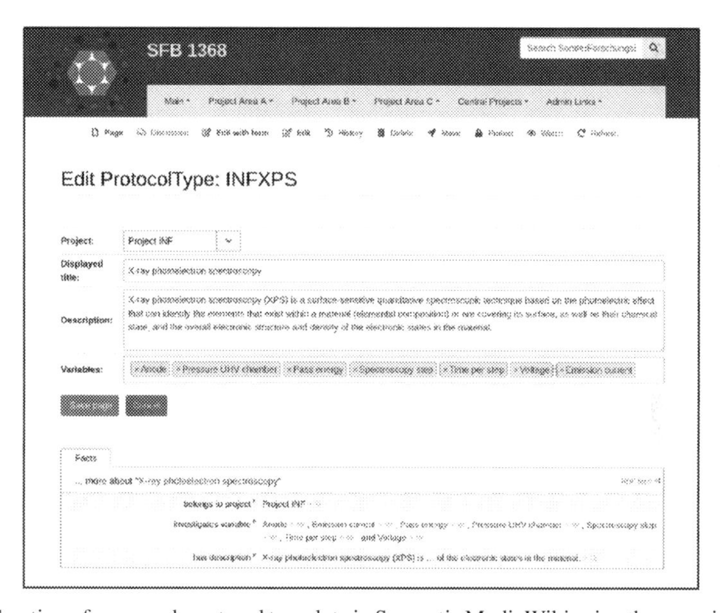

Fig. 3. Creation of a research protocol template in Semantic MediaWiki using the generic protocol structure

A link to the corresponding data record is located on the protocol's wiki page for each specimen. This link provides access to a form that can be used to enter the specific variables for the specimen, which were selected when the protocol type was created.

Thus the process of documenting research activities has not changed significantly. The user can flexibly create protocols himself and fill them out. With these generated data, the history of each specimen with the changes of variables through all subprojects can be viewed, traced and semantically queried. In addition, specimens and protocols are assigned UIDs that are generated uniformly by the system and are to be used automatically by all subprojects. This further standardization in the documentation increases the findability of the data sets and their documentation. In order to evaluate the effectiveness of the generic protocol structure and its implementation into the knowledge

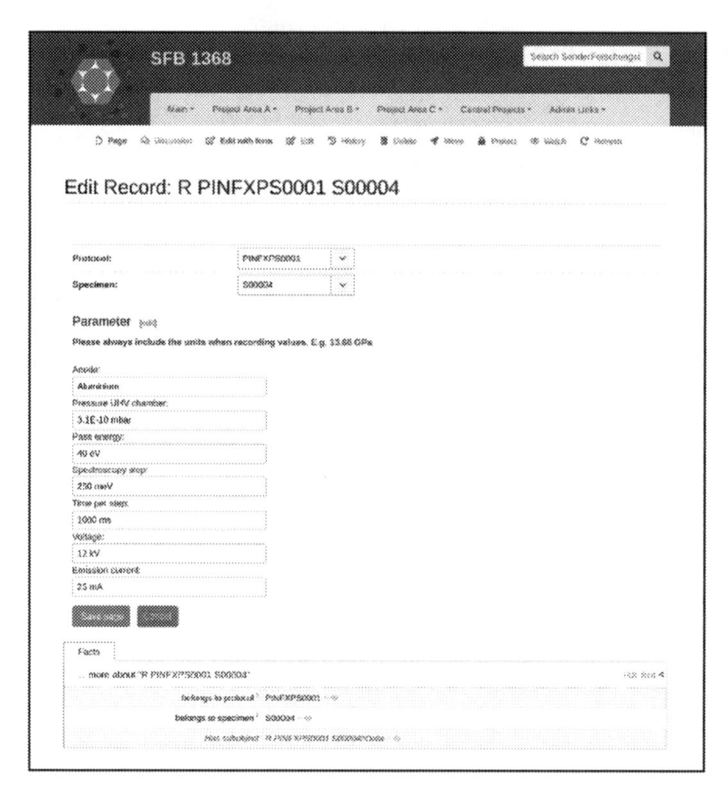

Fig. 4. Entering values of a protocol record

management system in documenting experimental procedures and parameters, an initial test with researchers from the Collaborative Research Centre was conducted. Three Participants were assigned tasks to model and document their experimental procedures, parameters, materials and instruments used with the SMW system, either by designing semantic building blocks for reuse in their protocols or by embedding already existing building blocks. The results showed that the researchers were able to independently transfer their existing documentation, which was in the form of spreadsheets, logbooks, and text documents, into the SMW system without any significant loss of information. They were also able to create their own variables and new specimens for protocols. However, the navigation between forms and templates of different categories was not clear enough, requiring too many interactions, and users indicated a desire for simpler usage, such as creating multiple specimens at once or having a better flow through multi-step forms. Additionally, users requested new features, such as personal specimen numbers and a relation of specimens to a materials database. These requirements can be implemented with reasonable effort from the current development stage and are compatible with our model and the generic protocol structure. The results highlight the importance of prioritizing user experience.

6 Conclusion and Future Work

In this paper, we introduced an approach for modeling research activities using semantic building blocks, enabling the flexible generation of structured, semantically annotated protocol templates to document research activities in a discipline-specific manner. The approach was implemented and evaluated using the SMW knowledge management system within the CRC 1368, which facilitated the documentation of experiments conducted in subprojects. Our approach shows that protocols can be created by researchers with reasonable efforts providing the necessary flexibility while generating structured documentation about data generation. In the next steps we will apply the generic protocol structure to more subprojects of the CRC creating a collection of semantically interlinked research data and data provenance. Future work will include analysis of relations between research activities, data and data provenance, enabling logical inferences from research data to support decision-making and ensuring the accessibility and interpretability of large amounts of complex structured information.

Acknowledgments. Funded by the Deutsche Forschungsgemeinschaft (DFG, German Research Foundation) – Project-ID 394563137 – SFB 1368.

References

1. Maier, H.J., et al.: Towards dry machining of titanium-based alloys: a new approach using an oxygen-free environment. Metals **10**(9), 1161 (2020). https://doi.org/10.3390/met10091161
2. Wilkinson, M.D., et al.: The FAIR Guiding Principles for scientific data management and stewardship. Sci Data. **3**, 160018 (2016). https://doi.org/10.1038/sdata.2016.18
3. Devaraju, A., et al.: From conceptualization to implementation: FAIR assessment of research data objects. Data Sci J. **20**(4), 1–14 (2021). https://doi.org/10.5334/dsj2021-004
4. Amorim, R.C., et al.: A comparison of research data management platforms: architecture, flexible metadata and interoperability. UAIS **16**, 851–862 (2017). https://doi.org/10.1007/s10209-016-0475-y
5. Grönewald, M.: Forschungsdatenmanagement mit elabFTW im SFB/TRR 270. Zenodo (2022). https://doi.org/10.5281/zenodo.6772197
6. Brandt, N., et al.: Kadi4Mat: a research data infrastructure for materials science. Data Sci J. **20** (2021). https://doi.org/10.5334/dsj-2021-008
7. Gohlke, H.: LISTER: Semi-automatic metadata extraction from annotated experiment documentation in eLabFTW. Chemotion/NFDI4Chem Stammtisch; 2022 Nov 25. https://www.nfdi4chem.de/index.php/abstract-gohlke/. Accessed 25 Feb 2023
8. Krötzsch, M., et al.: Semantic MediaWiki. In: The Semantic Web - ISWC (2006). Lecture Notes in Computer Science, vol. 4273 (2006). https://doi.org/10.1007/11926078_68
9. Mozgova, I., et al.: Research data management system for a large collaborative project. DS 101: Proceedings of NordDesign 2020, 12th - 14th August 2020, Lyngby, Denmark (2020). https://doi.org/10.35199/NORDDESIGN2020.48
10. Mozgova, I., et al.: Knowledge annotation within research data management system for oxygen-free production technologies. In: Proceedings of the design society, pp. 525–532 (2022). https://doi.org/10.1017/pds.2022.54
11. Altun, O., et al.: Integration eines digitalen Maschinenparks in ein Forschungsdatenmanagementsystem. In: Proceedings of the 32nd symposium design for X (DFX2021), pp. 1–10 (2021). https://doi.org/10.35199/dfx2021.23

12. DataCite Metadata Working Group: DataCite Metadata Schema Documentation for the Publication and Citation of Research Data and Other Research Outputs. Version 4.4. DataCite e.V. (2021). https://doi.org/10.14454/3w3z-sa82
13. Bruno, G., et al.: Efficient management of product lifecycle information through a semantic platform. Int. J. Product Lifecycle Manage. **9**(1), 45 (2016). https://doi.org/10.1504/ijplm.2016.078864
14. Moreau, L., et al.: The rationale of PROV. Web Semant.: Sci. Serv. Agents World Wide Web **35**, 235–257 (2015). https://doi.org/10.1016/j.websem.2015.04.001
15. Fuhrmans, M., Iglezakis, D.: Metadata4Ing - Ansatz zur Modellierung interoperabler Metadaten für die Ingenieurwissenschaften. Zenodo (2020). https://doi.org/10.5281/zenodo.3982367
16. Schmitt, R.H., et al.: NFDI4Ing-the national research data infrastructure for engineering sciences. Zenodo (2020). https://doi.org/10.5281/ZENODO.4015201

Gamification as a Knowledge Management Tool

Pierre Miroite(✉) and Mickaël Gardoni

École de Technologie Supérieure, 1100 Rue Notre Dame Ouest, Montréal, QC H3C 1K3, Canada
pierre.miroite.1@ens.etsmtl.ca

Abstract. Knowledge management drives innovation within a group or organization. Once implemented through codification and personalization strategies, it becomes possible to add an additional set of tools to improve it. This set comes from the principles of gamification which aim to engage the members of the organization and motivate them to participate more effectively in the culture of knowledge transmission and sharing. Several forms of gamification already exist, such as gamified Learning Management Systems, serious games or systems based on Yu-Kai Chou's Octalysis. But gamification is not sufficient on its own and requires to be based on already existing content. It is therefore to be seen as a real tool used to achieve the goals and objectives initially defined by an organization. To integrate gamification into a knowledge management process, it is proposed to follow the Design Science Research method and iteratively create artifacts. In this way, it is possible to test the effectiveness of gamification applied to knowledge management within an organization that may or may not already have an extensive knowledge base.

Keywords: Knowledge Management · Gamification · Collaboration

1 Introduction

An organization's knowledge is considered an essential resource for its survival and competitiveness. It therefore needs to be treated and managed as such in order to make the most of it, especially considering its intangible and sometimes hard-to-recover nature [1]. The difficulty of facilitating the sharing and management of knowledge within an organization is proportional to its importance, and one of the main obstacles is the willingness of employees to share their knowledge [2]. This is why gamification is seen as a solution to influence this willingness and make the management activity more enjoyable. Gamification is based on the ludic principles of games, which are now an integral part of today's society, where we enjoy the feeling of victory by earning points, receiving rewards, or the feeling of autonomy. The aim here is to propose a method for integrating gamification into a company's knowledge management process.

Published by Springer Nature Switzerland AG 2024
C. Danjou et al. (Eds.): PLM 2023, IFIP AICT 701, pp. 127–135, 2024.
https://doi.org/10.1007/978-3-031-62578-7_12

2 Knowledge Management

2.1 Fundamentals of Knowledge Management

Knowledge management is a process that allows any individual, group or organization to identify, share, use and store its knowledge and skills. However, knowledge is different from data in the sense that it is attributed to a given context, and from information since it is attributed an interpretation. But it is different from know-how, which is achieved through the recognition of one's peers in a field.

There are two types of knowledge: explicit knowledge, which could be described as theoretical or explicable as formal procedures; and tacit knowledge, which is more abstract and is born from expertise or from repeated use of knowledge already acquired. This tacit knowledge is sometimes compared to know-how since it is similar to it, but it is above all the most complex knowledge to capitalize on.

To do so, knowledge management is based on precise processes that allow an organization to develop, manage and use its knowledge: the acquisition of knowledge inside or outside an organization, the storage of this knowledge followed closely by its updating, if necessary, the sharing of knowledge within the organization, and finally its use in problem-solving or decision making.

In addition to being a driver of innovation, knowledge management ensures a response to the demands of potential customers [3] and adds value to an organization through the knowledge it possesses. This is why it is essential to implement knowledge management strategies to enable the processes presented earlier. Thus, allowing for increased efficiency and reduced duplication of effort.

2.2 Knowledge Management Strategies

There are two main strategies of knowledge management: the strategy of codification and the strategy of personalization. The strategy of codification has a focus on explicit knowledge since it is a question of working on the documentation of this knowledge [4]. It is a question here of allowing a better access as well as a better update of knowledge thanks to what we call "operators". Their role is to set up knowledge management processes. However, this strategy is not sufficient on its own because it is mainly explicit knowledge that is capitalized or transmitted, or because it is only information bases that are created. It is therefore essential to implement a personalization strategy.

The personalization strategy involves the sharing and transmission of knowledge informally between members of the organization [5]. The idea here is to encourage human exchanges and interactions to allow the transmission of tacit knowledge, which is more difficult to document. We are not trying to predict a result, but rather to engage and diversify interactions, tacit knowledge being more complex to document. One of the most common approaches to personalization is Takeuchi and Nonaka's SECI matrix, which identifies four knowledge-creation processes:

The knowledge management cycle includes Socialization, Externalization, Combination and Internalization, without necessarily starting with the same component, but always following the same direction. Initially, socialization allows certain knowledge to be transferred from one person to another, without losing its tacit form. This can be

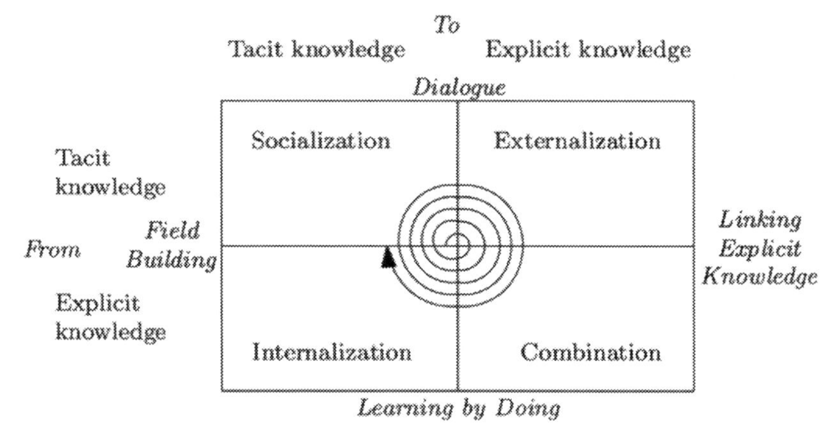

Fig. 1. SECI spiral of knowledge creation based on the Nonaka and Takeuchi model [5]

accomplished through observation, exchange or simply reproduction, without requiring the use of language. Since socialization is an essential component of this cycle and requires human interactions, it seems logical to associate this matrix with the strategy of personalization. Second, Externalization converts tacit knowledge into explicit knowledge. This represents the greatest challenge in knowledge capitalization, since it involves writing down practices or skills that are difficult to explain. Then, Combination allows the new explicit knowledge to be linked to previously acquired knowledge to reinforce learning. This creates new knowledge or improves understanding of it through previous experience. Finally, Internalization converts explicit knowledge into tacit knowledge through repetition or practice. This acquired tacit knowledge can be improved or shared through Socialization and then enter a new cycle of the spiral. The constant conversion of tacit knowledge into explicit knowledge allows the creation of new knowledge or the improvement of existing knowledge. However, this can only be accomplished if individuals are not alone in the cycle. This does not necessarily require human interaction; exchanges with other existing content can sometimes be enough.

3 Gamifying Knowledge Management

3.1 Basic Gamification Principles

Gamification is a tool that uses game mechanics and sometimes aesthetics to engage and motivate participants to achieve predefined goals [6]. However, its main goal is to make activities that are considered serious more "fun" to perform. It is to meet the demand of millennials who have grown up with games that gamification has made its entry into the workplace by being integrated into human resources processes [7] and employee training, for example. Gamification is therefore used to capture the attention of the users to whom it is proposed, and thus to create interactions with what we will call the system that serves as a medium (it can be of a random nature). But the principles of the game from which it is inspired also allow to propose direct feedbacks on actions performed, or a clear follow-up of its progress. The fact of having a clear follow-up of one's progress allows, just like direct feedback on action, to keep a dynamic of interest among users, a better retention of information, but also to allow a dynamic of performance over time, not necessarily guaranteed.

Moreover, gamification must be based on essential components. The question is no longer whether it is effective, but rather what makes it effective. Part of the answer lies in the following components: a clear objective to accomplish and a precise goal to reach; rules and a context that are clear enough to frame the gamification; a demonstration of the impact of the actions performed; rewards to gratify the users; and finally, the creation of motivation. It is also important to keep in mind that the purpose of using gamification is not to create a game as such, which is meant to entertain, but to create engagement and to incite its users to adapt a desired behavior [7]. In this case, to obtain more knowledge, or to promote the creation and transmission of knowledge. Karl Kapp [8] also specifies that the gamified system must be sufficiently qualitative, adapted to the users to whom it is proposed, but especially appropriate to the context in which it is brought. The idea is to propose relevant content, adapted to the skills of the participants in the game or to the technology used. And when we talk about the quality of the system, it is a question of making sure that it will meet the expectations that we have of it and that it will define the objectives that it allows to reach.

3.2 Gamification Alone is not Enough

For gamification to be effective, it must be understood that it is not sufficient on its own, and that it must be based on content, processes, or on an existing strategy. Just as knowledge is different from information, gamification must be contextualized. It is not the other way around. It can therefore easily be integrated into a knowledge management strategy. In the case where an organization wishes to motivate its members to implement a knowledge management culture, it becomes possible once the strategy has been chosen and implemented to think about the gamification mechanisms to be put in place. However, it is important to keep in mind that: the gamification solution is already not guaranteed, the fact that an activity is made "fun" does not guarantee the success of the gamification [8]; and secondly, the gamification mechanisms are not defined in advance. This means that as the objectives of the organizations are not necessarily the same, the chosen strategy and the processes put in place will not be the same either. Thus, the tools and techniques of gamification will not be the same either.

3.3 Several Gamification Forms

There are several forms of gamification that are similar to derivatives of it. They are all based on the same principles of the game but offer a different approach for a different purpose. One of the best known and associated with the so-called LMS (Learning Management System) platforms is what we could call the gamified platform. In the form of a desktop or smartphone application, these solutions are often very varied because they can be customized.

The Serious Game. Another form of gamification, the serious game is a particularly relevant tool in the context of a personalization strategy. It is a form situated between simulation and learning through play, since we will first play and get out of the real world, then thanks to feedback become aware of the learning received. These actions are done with full awareness in order to make the participants realize that they have mobilized real knowledge and skills that were useful during the game. The feedback stage is essential for this; it is this return to reality that closes the serious game and consolidates the knowledge acquired. We were talking about personalization strategies earlier, since the serious game lends itself perfectly to the context of group learning and makes it possible to gather different types of player profiles at the same time. Moreover, the fact that the participants are supposed to be gathered together, it becomes immediately easier to transmit knowledge in a tacit way or to associate it with the Socialization stage of Nonaka and Takeuchi's matrix (Fig. 1).

Yu-Kai Chou's Octalysis. This well-known framework is used mostly to be a source of inspiration for those who seek to adapt their gamification to an already thought situation. This model has been democratized mainly due to its modularity and the different action possibilities it allows. The model offers different mechanics distributed around eight motivational dynamics. These dynamics are based on four families of motivations that are at the heart of gamification. All the actions we undertake are linked to one or more of the dynamics presented in Fig. 2. The model assumes that each action performed is driven by a motivation considered as intrinsic (self-initiated), extrinsic (influenced by the external environment), positive or White Hat (fulfillment, personal satisfaction and control of one's actions), or negative or Black Hat (obsession, anxiety and dependency). These dynamics are therefore seen as drivers of gamification since they create engagement and motivation in users, gamification being human-centered. Each dynamic proposes a list of tools and associated techniques that allow each organization to choose the aspects to focus on according to their needs. Thanks to this tool, it is possible to be much more flexible and to adapt at all times to one's situation thanks to the numerous mechanisms that are mentioned. It is possible to change the dynamics that one chooses to propose according to the profile of a user, thus allowing a modularity even more advanced.

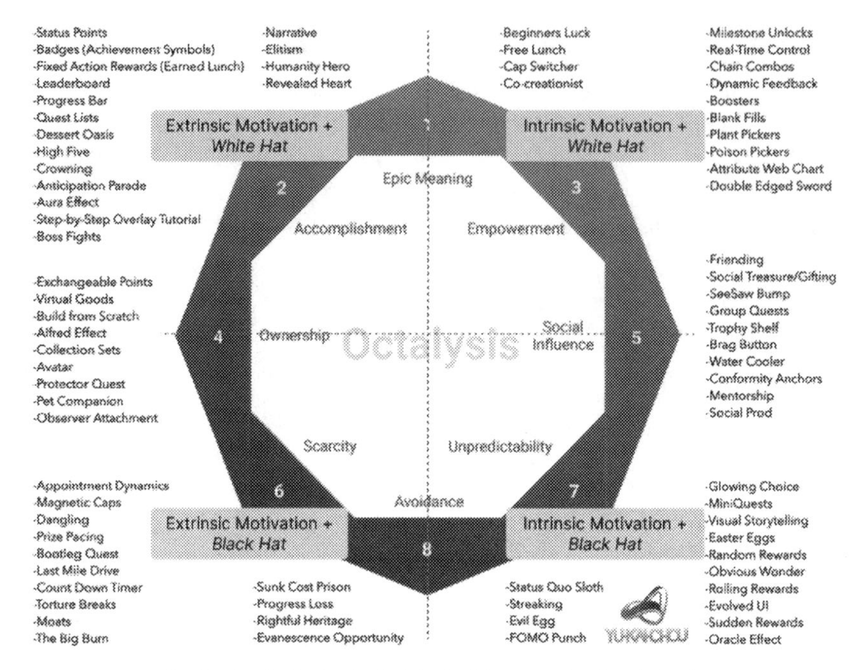

Fig. 2. Yu-Kai Chou's Octalysis [9], with additional details

4 Integrating Gamification into Knowledge Management

Again, it is necessary to remember that there is no known form of gamification that can guarantee success or increased effectiveness when added to knowledge management. It is also important to focus on the content and the context on which gamification is based, before choosing the form of gamification. Hence the importance of having a strategy in place that is adapted to the predefined objectives of the organization concerned. As a reminder, we are in a learning context, with knowledge management demonstrating an organization's ability to innovate and respond to potential needs and demands [3]. Moreover, it is not recommended to opt for linearity from a motivational point of view, since it becomes riskier to lose the interest of the users concerned. Although repetition of knowledge is important to maintain its relevance, relying solely on monotony is not the answer, especially in the context of gamification. Therefore, all potential solutions should be carefully considered and thoroughly evaluated before being dismissed. The best approach is to implement, test, and assess the impact of these solutions to determine their effectiveness.

It therefore seems appropriate to propose the Design Science Research (DSR) method, which enables us to test and gradually implement a new knowledge management system based on the mechanics of gamification. The aim of this process is to create artifacts that extend the limits of organizational capabilities. These artifacts can be of various kinds, and can therefore take the form of software, a process, a model or a method. The DSR process also aims to generate knowledge about how things can and should be built, arranged or designed, usually by humans, to achieve a set of desired goals [10].

This is known as design knowledge. A conceptual representation of the process is as follows (Fig. 3):

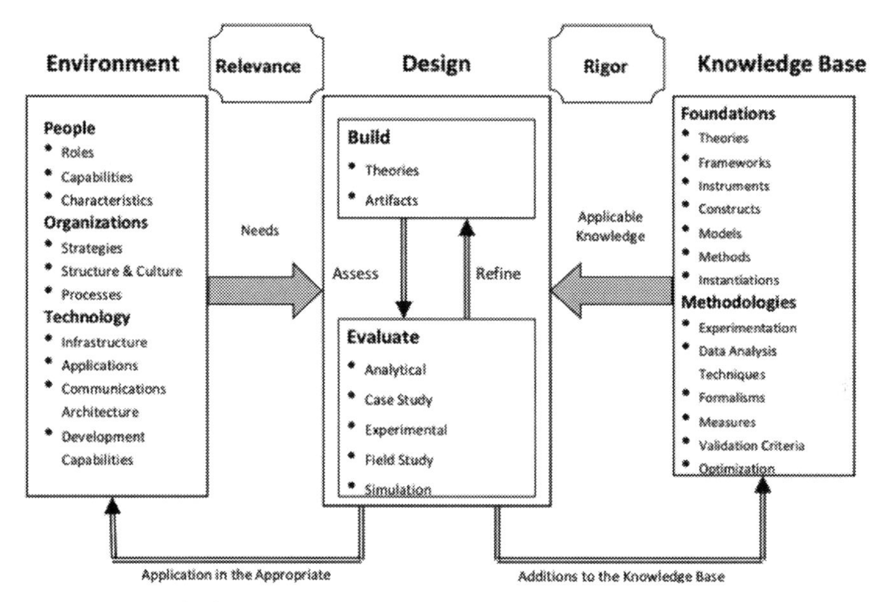

Fig. 3. DSR Model, adapted from Hevner et al. (2004) [10]

On the one hand, the environment where the problem arises, and which is made up of elements already encountered, i.e. the organization, its members and the technolo-gy in place. On the other hand, the knowledge base, which provides the material for the DSR. This is where we find the knowledge already acquired or accumulated by the organization, the methods already put in place, but also the new knowledge that will be capitalized on. From this, we derive the so-called "applicable knowledge" which, like the requirements, will be used in the Design section at the center of the model.

Each artifact created undergoes a functional and iterative evaluation to ensure its relevance to the environment. Once it is, it becomes applicable and usable by the environment, and is considered as a new resource available in the organization's knowledge base. With regard to artifact creation, we plan to follow the same method proposed by Pef-fers et al. (2007). This method is divided into six activities (Fig. 4).

It's the iterative, multi-faceted aspect of this process that makes us the right choice for this study, as well as the fact that the organization, its members and the technologies in place are all taken into account, as it is on these points that gamification for knowledge management is based.

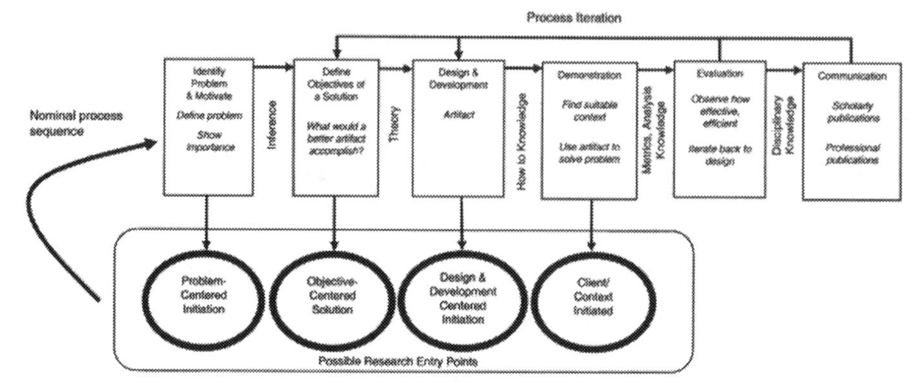

Fig. 4. Process for creating an artifact using the DSR method [11]

5 Conclusion

Gamification is thus seen as a tool to be used on already existing content. In this case, it's added to the content created by a pre-established knowledge management strategy and helps to structure and/or apply the chosen method. Knowledge is also to be considered as a valuable asset to share and manage for any company to be sustainable. In addition to performance, it is the human being who is at the center of the knowledge management process, just as he is at the center of gamification and its derivatives. Achieving performance goals and objectives successfully depends on their engagement and motivation, making it crucial to choose a solution that fits the specific context and needs of the member of those organization. Hence the importance of thinking iteratively about the integration of a gamified process, to ensure its relevance to the people to whom it is proposed. This is why the creation of artefact(s) following the DSR method was proposed and considered, in order to both meet an organization's knowledge management needs, and gradually integrate gamification into the process.

References

1. Swacha, J.: Gamification in knowledge management motivating for knowledge sharing, vol. 12, pp. 150–160 (2015)
2. Lam, A., Lambermont-Ford, J.-P.: Knowledge sharing in organisational contexts: a motivation-based perspective. J. Knowl. Manag. **14** (2010). https://doi.org/10.1108/13673271011015561
3. Besson, B.: Concepts KM. In Gestions des connaissances (s. d.)
4. Gardoni, M., Navarre, A. : Pratiques de gestion de l'innovation : guide sur les stratégies et les processus. Montréal: École de technologie supérieure (2017)
5. Nonaka, I., Takeuchi, H.: The knowledge-creating company: how Japanese companies create the dynamics of innovation. Oxford University Press (1995)
6. Kapp, K.M.: The Gamification of Learning and Instruction: Game-Based Methods and Strategies for Training and Education (Pfeiffer) (2012)
7. Ole, G.: Gamification Mindset. Springer Nature Switzerland (2019)

8. Kapp, K.M., Blair, L., Mesch, R.: The Gamification of Learning and Instruction Fieldbook: Ideas into Practice (Pfeiffer) (2014)
9. Chou, Y.: Octalysis: Complete Gamification Framework - Yu-kai Chou. Yu-Kai Chou:Gamification & Behavioral Design (2021). https://yukaichou.com/gamification-examples/octalysis-completegamification- framework/
10. vom Brocke, J., Hevner, A., Maedche, A.: Introduction to design science research. In: vom Brocke, J., Hevner, A., Maedche, A. (eds.) Design Science Research. Cases, pp. 1–13. Springer International Publishing, Cham (2020). https://doi.org/10.1007/978-3-030-46781-4_1
11. Peffers, K., Tuunanen, T., Rothenberger, M.A., Chatterjee, S.: A design science research methodology for information systems research. J. Manag. Inf. Syst. **24**(3), 45–77 (2007). https://doi.org/10.2753/MIS0742-1222240302

Drivers of Change Impacting Outcome-Based Business Models in Industrial Production Equipment

Olli Kuismanen(✉), Karan Menon, and Hannu Kärkkäinen

Tampere University, Tampere, Finland
olli.kuismanen@tuni.fi

Abstract. The objective of this paper is to explore the potential future changes on the value and feasibility of Outcome-Based Contracts (OBC) and Outcome Business Models (OBM) in the setting of industrial production equipment. As companies and industries are implementing these business models characterized by long value co-creation contracts, the impact of the world around changing is becoming both a risk as well as in other cases an opportunity for increased value creation. We conducted a futures research project with the emphasis in understanding the external changes impacting the value of the OBCs and OBMs for the contract parties. The paper contributes by highlighting the impact of Drivers of Change on OBCs for both parties as well as identifying a set of Drivers of Change for industrial equipment manufacturers and users.

Keywords: Outcome-based PSS · outcome-based contracting · Product-Service Systems · PSS · Servitization · Advanced Services · Pay-per-outcome · PPO · Pay-per-X · PPX · industrial production equipment · Drivers of Change

1 Introduction

Outcome-based business models have become a topic of great interest in academic literature in the recent years [1]. New ways of sharing risk and benefits when offering and sourcing equipment have been taking shape in industries since 1970's [2]. Product-Service Systems (PSS), Advanced Services, Outcome-based Contracts (OBC) and Pay-per-X (PPX) business models have been researched in recent few decades [3]. Outcome- or Performance-based contracting can be limited to bonus/penalty schemes "on top" of traditional equipment sales or include the asset with which the outcome is achieved, depending on the field of study. The above research streams discuss different business models for delivering products with a business and revenue model that is based on the value that is co-created between the supplier and the customer [4].

Little emphasis has been put on investigating the impact of potential business changes that take place during these industrial PSS relationships, during the evolution of OBC-types of business models and during the lifetime of related equipment. This is especially relevant in industrial production equipment, one of the areas where these business models

Published by Springer Nature Switzerland AG 2024
C. Danjou et al. (Eds.): PLM 2023, IFIP AICT 701, pp. 136–150, 2024.
https://doi.org/10.1007/978-3-031-62578-7_13

are utilized more and more, where there are large risks involved for both customers and suppliers, and where the lifetime of the assets are often measured in decades. The aim of this paper is to dive into the potential business related changes that may take place during the life cycle of these industrial production equipment and impact the feasibility of the Outcome-Based Contracts (OBC) [1] in a setting where the equipment to deliver the outcomes is part of the contract.

Industrial production equipment have a long useful life, typically decades and are often critical within their usage process. This results in long OBC contracts where the supplier provides the equipment and services to keep the equipment running optimally. Consequently, the Outcome Business Model (OBM) through which the suppliers operate these machines is a long game – it needs to survive not only a few years but decades. This makes them susceptible to changes in the business environment which can impact the feasibility and value of the OBC and on a larger extent the OBM [5].

As we have seen in recent years through COVID-19, supply chain disruptions and Russia-Ukraine war, the world is volatile and uncertain [6]. Therefore, we need to better understand the life cycle perspective of the OBCs and OBMs. In this study we study the changes through Drivers of Change (DoC) studying how DoCs impact the feasibility of the OBCs and OBMs.

As individual OBC's success, and consequently success of the OBM is dependent on the dyadic value the contract and equipment deliver, changes in the external world may have an impact [4, 7]. The changes can happen on multiple levels from the production line or equipment levels to the global megatrend level and can have a compounding effect on the parties [8].

Futures studies literature shows that different environmental changes impact different industries differently [8–11]. Recent history also shows us that the impact of individual changes can have a compounding effect in the global supply chains – the component shortage caused by Covid-19 pandemic is an example of entire industries shaking due to a single 'wild card' event [12, 13]. On the other hand, changes can present new opportunities. The current global energy crisis is an opportunity for companies in energy generation and energy efficient equipment to their customers [14].

There is a significant gap in literature related to the futures studies and life cycle dimensions in OBCs and OBMs, especially regarding the industrial production equipment. Although many studies have identified the long life time of OBCs and a few individual risks that this may impose on the supplier-customer relationship, none to our knowledge have researched the themes and potential changes that may impact the feasibility of the OBCs and OBMs in the context of production equipment.

To answer the research gap, we formulate our research question to:

RQ: *Which potential Drivers of Change impact the feasibility of the Outcome-Based Contracts and Outcome Business Models in industrial production equipment?*

The purpose of this study is to understand:

- Which Drivers of Change may impact the feasibility of the OBCs or OBMs for the supplier or the customer of industrial production equipment?
- Which Drivers of Change are most critical for the long term feasibility of the business models of industrial production equipment?

- Why the Drivers of Change are relevant, what are some of the ways in which the changes might impact the OBC parties?

By identifying potential changes, we contribute to the development of these business models to be resilient to change and help practitioners design better business and sourcing models and contracts.

2 Literature Review

To lay the foundation of our study we conducted a literature review to identify the potential changes as well as the potential ways to categorize these changes. We targeted our search to three distinct groups of extant literature: 1) Outcome-based business model-oriented literature (including Advanced Services, Servitization, outcome-based contracting, PSS, PPX and Product-as-a-Service literature), 2) generic business model and value delivery literature, 3) futures literature. Our target was to identify how literature deals with external changes [8] and to identify specific DoCs for these business models. We used here PESTLE model to categorize and describe the potential changes identified in literature, dividing them into political (P), economic (E), social (S), technological (T), legal (L) and environmental (E) changes.

Political and Industry Level Changes. We identified from literature some political and industry-level DoCs that potentially have important effects on OBCs and OBM's. They range from geopolitical changes in markets and their impacts on industries [15, 16] to timely sustainability requirements and their impact on production facilities' modernization [17–19], changes in the operational strategies of manufacturing companies [8, 20] to the ever increasing volatility driving business decisions [21–23] to governmental restrictions on resource usage that govern businesses [3]. All of these have an impact the OBCs in the way companies collaborate to deliver value and to improve their internal efficiency.

Economic Changes. Most of the relevant changes can ultimately be tracked down to the economical level, to impact either the cost or revenue of the party, but that is often a consequence of mitigating or exploiting the change, not a direct impact of it. From literature we found multiple DoCs that are purely operating on the economic level. Often these are a product of some other, often operational change, but were handled in our study as separate DoCs due to their immediate and clear impact on OBM's. Some of these were the changes in costs of materials, labor, taxes, [3, 24–26], financing [3, 17], transport and travel [15, 16] or new business opportunities increasing the revenues of either party [27].

Social Changes. Due to our focus on the external changes [7], we did not find many relevant DoCs that affect mainly on the social level. Practically all major changes, including Black Swans, Wild Card events and big global or regional events have a social aspects [15, 16], and e.g. social unrest can arise even without them, but these events are difficult for actors to manage other than through market selection and insurance policies. Additionally, we included in our list the risk of discontinuous relationships either due to key persons changing companies or taking on new responsibilities as it can have a significant impact on the long-term life cycle of the supplier-customer relationship [25].

Technological Changes. Our literature study found an abundance of technological DoCs that may impact industries in general or particularly industrial production equipment, their design or usage. Technological changes can occur either through overall technology advancements that are migrated to new industries, or through technical problem-solving within an industry. There are many technological shifts visible that have already changed how businesses are managed, some at the very core of the OBM's. Generic technological advancements that impact many if not all OBMs include connectivity, cloud computing, digitalization, Artificial Intelligence (AI), Internet of Things (IoT) and other IT technologies [3, 8, 21, 28–30], new materials [31]. Some very specialized technologies like blockchain have already been discussed in business model literature as a component of OBM's [32].

Any technological changes in the customer's production facilities is both an opportunity and a risk for the other equipment suppliers as the new technologies can either hinder or support the feasibility of the other equipment [25]. Additionally, as the global sustainability requirements are putting pressure on industrial production, creating a need for change that can be addressed at least partially through the OBC's where the value (energy saving) is shared between the supplier and the customer [33].

Legal Changes. Changes in the political and regulatory environment will naturally drive some changes in the legal frameworks in which businesses operate. We identified several DoCs that impact OBM's through the legislative or standardization routes. Due to the temporal nature of the OBCs these changes are not only possible, but even probable during the long life cycle of the equipment and contracts. These include more direct regulation, e.g. in the taxation, import/export regulation, financial accounting and e.g. depreciation [16, 24], industrial level legislation such as cyber security norms and legislation [20] and governmental actions to support or regulate certain types of businesses or certain types of business models [24, 33].

Environmental Changes. The environmental changes identified from the literature were on the other hand related to environmental restrictions caused by the currently pressing ecological crises and the supply of material and energy resources, already mentioned from the economic perspective and on the other hand through accelerating transformations in the consumer and therefore the industrial production spaces. Additionally, we identified several DoCs related to industrial and competitive environment and trends within industries. These include competitive hostility [3] or completely new entrants (e.g. generalist maintenance suppliers, system integrators) entering the market [34, 35], changes in overall firm production strategy, e.g. Manufacturing-as-a-Service [16, 21, 36], changes in company strategies, decisions on core vs. outsourced operation [10, 37, 38], that are constantly evaluated by manufacturing and industrial production firms. Environmental risks also have been identified by many scholars as a source of inefficiencies in production or in the continuity of the production [21, 39].

3 Methodology

The study combines literature research with a process of expert workshops to identify and prioritize the Drivers of Change. After identification of a list of potential DoCs from literature the DoCs were evaluated and prioritized by expert groups in multiple workshops.

3.1 Identification of Drivers of Change

As described in the previous section we first researched academic and business literature to identify a list of potential DoCs that have been identified in the business model, industrial production, and futures literature. These potential changes functioned as a starting point for our quest to identify a representative set of DoCs that would ultimately enable discovering how future changes affect the feasibility of the outcome-based business models.

From the list of identified potential changes we formulated a set of possible DoCs that can impact the OBCs and OBMs. As we wanted the DoCs to cover all of the main aspects of the OBMs we mapped them PESTLE [11] and in several business, strategy and management frameworks [8, 40, 41] to make sure that changes on different abstraction and operational level were included. We used a holistic model adapted from Bokrantz [8] to categorize the DoCs according to their origin and scope. Contrary to Bokrantz we wanted to include some DoCs also from the individual company level for those changes that have a clear impact on the other company in the dyadic relationship. These are usually not controllable by the impacted party, so they fit the description of 'external' from the impacted party's perspective.

Organizing the list of relevant DoCs was relatively straightforward by coding those sections of the articles. We also coded the accuracy or relevance of each DoC towards OBM's with "impact markers" (direct impact, indirect impact). This way the research team identified 57 potential DoC. After a few rounds of iterations (removing duplicates, reformulating the wording of the DoCs) the number was limited to 35 to avoid research fatigue [42].

3.2 Evaluation of the Initial Drivers of Change by Expert Panel

The list of DoCs was evaluated by a panel of 8 experts (industry and academia) in a 2-h online workshop. They were asked to review and compliment the list with possibly missing DoCs that they thought would be relevant for OBCs or OBMs. The timescale for analysis was set at 10 years. The experts prioritized the DoCs based on the probability and impact of the change and provided qualitative reasoning. In addition to validating the identified DoCs the panel also identified 7 new DoCs that were in their view missing from the list derived from literature.

The experts also provided insights on how to improve the evaluation process for the next round. There were suggestions e.g. on how to deal with the dualistic direction of change (e.g. energy costs can either rise or lower within a timeframe, and both of these changes might be significant for one or both contract parties). As we didn't want to have duplicates for all DoCs which have a negative as well as a positive possible change direction this helped to eliminate some of the redundancy from the list.

3.3 Prescreening the Drivers of Change by Researchers

Based on the expert panel discussion and initial evaluation of probability and impact the list of DoCs was restructured and prioritized. This helped reduce the list for the next, more comprehensive evaluation by a larger panel of experts. The number of DoCs after this round of analysis was 30.

3.4 Final Evaluation of the Drivers of Change by Expert Panel

Five industry and 7 academia experts on OBMs were then invited to a second 2-h workshop where the list of 30 DoCs were complimented and evaluated on probability and impact, much like the previous round. A summary of the panelists for both workshops is shown in Table 1. In this round only one new DoCs emerged.

Then the panelists evaluated each of the DoCs on probability and impact, similarly to the first evaluation round. Scale for evaluations was 0 to 10. After the evaluation the team discussed the highest scoring DoCs to identify some ways in which these changes impact the OBC parties.

Table 1. Workshop panelists

Category	Title	Description of background organization	Country
Industry	CEO	Globally operating company in industrial production equipment, rolling out OBC	Finland
Industry	CEO	Globally operating equipment manufacturer, working with multiple companies to enable OBMs	Finland
Industry	CCO	Globally operating company in industrial production equipment, 5+ years in OBM	Belgium
Industry	CEO	Leader of a financial institution providing financial services + capital for OBMs	Austria
Industry	Consultant	Several years of academic and industry experience on business models	Finland
Academia	Professor	Several years of experience in business models and knowledge management	Finland
Academia	Professor	Several years of experience in servitization, both in academia and industry	Switzerland
Academia	Professor	Several years of experience in research on smart manufacturing, business models	Mexico
Academia	Associate Professor	Several years of experience in research and industry on manufacturing, especially smart manufacturing, Internet of Things, product-service systems	USA
Academia	Associate Professor	Several years of experience in supply chain management research, manufacturing business model innovation	USA
Academia	Post Doctoral Fellow	Several years of experience working in industry projects in developing OBMs, special focus on connected equipment in industry	Finland
Academia	Post Doctoral Fellow	Several years of experience working in industry projects in developing OBMs, smart manufacturing	Finland

4 Results

Identified and Prioritized Drivers of Change. The validated and prioritized list of DoCs can be seen in Table 2. The list is organized according to the criticality of the DoC, which is calculated as the product of the probability and the impact of the DoC.

The evaluation of the panel of the DoCs was quite unanimous, the standard deviations in probability evaluations was between 0,07 and 0,31 (average 0,17) and in impact evaluations between 0,17 and 0,34 (average 0,24), even though the backgrounds of the

Table 2. List of prioritized Drivers of Change

Driver of Change	Probability Mean	Impact Mean	Criticality
Changes in costs	8,11	6,56	53,20
Product market price change	8,11	6,33	51,34
Technological disruption	7,00	7,22	50,54
Life cycle cost/value becoming customer decision criteria	7,56	6,67	50,43
Customer business strategy change	7,67	6,56	50,32
Changing customer expectations	7,67	6,44	49,39
Change in OBC related regulations and financial policies	6,78	6,33	42,92
Component supply problems	7,44	5,67	42,18
Sustainability goals driving modernization of equipment	6,89	6,11	42,10
Change in interest rates	8,22	5,11	42,00
Country/region incentives for local production and sourcing	7,33	5,67	41,56
Supplier's competitors adopting OBC business models	7,11	5,67	40,31
Scarcity of skilled labor	6,56	5,89	38,64
Wild card events (e.g. war)	5,89	6,56	38,64
Customer's production line related problems	6,67	5,56	37,09
Increase in OBM financing opportunities	7,11	5,00	35,55
Energy and material shortage	6,33	5,56	35,19
Shortening of product life cycles	5,33	6,00	31,98
Changes to Import/Export Rules	5,89	5,22	30,75
Restrictions on usage of natural resources	5,89	5,22	30,75
Digital platforms emergence	6,89	4,11	28,32
Changing customer IT needs	6,78	4,11	27,87
Political situation impacting the equipment and data flows	5,22	5,11	26,67
Digitalization leads to centralization of maintenance	6,44	4,00	25,76
Customers wanting to insource equipment competence	5,22	4,89	25,53
Cyber security affecting remote monitoring	5,11	4,78	24,43

(*continued*)

Table 2. (*continued*)

Driver of Change	Probability Mean	Impact Mean	Criticality
Changes in safe data sharing (e.g. blockchain)	5,33	4,33	23,08
Change of ownership of companies	5,78	3,89	22,48
Social unrest creating uncertainty	5,33	4,11	21,91
Change in union rules	4,22	4,89	20,64
Personal relationship change	5,44	3,11	16,92

panelists were quite varied and the future outlook at the time of the evaluation (July 2022) was particularly uncertain due to recent disruptions that were fresh on the minds of panelists (COVID-19 pandemic, supply chain disruptions, RUS-UKR war).

Most Critical Drivers of Change. Based on the evaluation of 12 experts the list of DoCs was prioritized based on probability and impact. The most critical DoCs and their evaluations were related to the cost and therefore the profitability of the OBC's in the long run, disruptions related to technologies, changes in the customer needs in the long life cycle of the equipment and OBC as well as more general disruptions that can shake the foundations of industries.

Discussion on Most Critical Drivers of Change. The most severe DoCs were then discussed by the panelists to discover why they felt that they are critical for OBCs or OBMs.

5 Discussion and Conclusions

The identified and evaluated list of potential DoCs impacting the feasibility of OBCs and OBMs shed light into the life cycle feasibility of these business models both from the perspective of suppliers and the customers of industrial production equipment by identifying external changes that may impact the feasibility and providing reasoning of how this impact might take place. The prioritized list provides a starting point for future research on OBC and OBMs, the design and implementation of these business models as well as shed light into the type of external changes that impact the resilience of these business models.

As the world is constantly changing and industries evolving to find better and more profitable ways of operating, we can estimate that the identified DoCs are only a tip of the iceberg in the wider scheme of things. There will be new Drivers of Change and consequently new ways to cope with the changes. The results of this study will surely provide insights into this process in the future.

This study selected and analyzed the DoCs from the perspective of OBCs and OBMs. However, the same changes likely also impact the traditional business models as well. The same impacts with different weights can be seen even wider than for the scope of this study.

Through this study, through the results and summarized commentary in Appendix 1 we have shown that understanding major external environment Drivers of Change is important for the implementation of outcome-based PSS and outcome-based business models, either from the suppliers' or the customers' point of view, or both.

5.1 Drivers of Change with the Highest Probability and Impact on Feasibility

The DoCs most probable to take place, based on the expert evaluations were 1) changes in interest rates, 2) changes in costs, 3) product market price change, 4) customer business strategy change and 5) changing customer expectations. The DoCs with the highest impact, based on the evaluations were 1) technological disruption, 2) life cycle cost/value becoming customer decision criteria, 3) changes in costs, 4) customer business strategy change, and 5) wild card events. Put together, the DoCs with the highest criticality, calculated as a product of the probability and impact, based on the evaluations were 1) changes in costs, 2) product market price change, 3) technological disruption, 4) life cycle cost/value becoming customer decision criteria, and 5) customer business strategy change.

We can note that many of the most critical DoCs are related to the future economic feasibility or profitability of the OBC. Most critical DoCs have to do with the costs, both from the perspective of cost of operating the OBC for the long life cycle profitably, as well as from the perspective of changing prices on the markets, both from the business model comparison perspective as well as from the operating cost perspective. This is logical, especially at the time of the study when a lot of changes have been recently seen in the cost and availability of different production inputs and volatility of markets, as can see in many of the highest criticality DoCs in Table 2. Some of the critical changes are stemming from the customers' interest to 'buy' flexibility for their operations, which in turn poses somewhat of a risk for the suppliers.

Interestingly, the impacts of truly external big changes (e.g., legislation changes, wild cards, social unrest, etc.) were not seen very high by the experts. In the discussion these were mentioned, but the experts saw them as less transformational than some of the 'closer' changes. Even with the abundance of recent historical data about these types of disruptions the experts saw them as not having an especially big impact, compared to the impact they have on the traditional business models.

Different DoCs and the underlying changes impact the OBC contract parties differently. Most of the DoCs were seen to impact the suppliers more, which is logical since the suppliers are taking a bigger role in the relationship compared to the traditional sell & service-business models. By becoming partners to their customers production outcome, they of course hope to gain benefits in many ways, but at the same time expose themselves to the effects of these type of external changes. In the traditional business model, the customers would have felt most of the impacts of the changes, but now they share that risk with the supplier. And at the same time the parties share the opportunities that changes introduce.

5.2 Theoretical Contribution

This paper contributes by firstly highlighting the importance of futures thinking and foresight when designing and operating OBCs and OBMs. For the first time we have practically attempted to forecast the changes that may happen during the life cycle of these contracts in the industrial production equipment setting, adding a systematic temporal dimension to the study of outcome-based business models in industrial production equipment PSS [3]. Although similar futures studies have previously been done in different arenas, for example the future of Industry 4.0 [20], additive manufacturing [28], future maintenance [8] and some other technical and operational fields, market and strategy-related studies, to our knowledge this study is the first from the outcome-based business model perspective. Therefore, it contributes by extending the view of OBC and OBM to a wider futures perspective [1]. This extension of the view expands the theory of PSS business models to be more extensively time-based instead of studying them primarily in one time instance and mostly noting the long life cycle dimension of the business [3]. We feel that this is an important contribution to the field.

Secondly, the study contributes by identifying specific current potential DoC that may have an impact on the feasibility of the OBCs and OBMs in industrial production equipment within a 10-year time frame. Thirdly, the study contributes by describing a process which can be used by researchers and practitioners to identify and prioritize the potential DoCs in their respective industries or settings.

5.3 Managerial Contribution

This study contributes to the practical application of the OBCs and OBMs by indicating an initial list of different DoCs impacting OBM's. These DoCs include a multitude of different sources of the possible changes that should help practitioners identify the changes that are most relevant to their industry and OBC type(s). The same identified DoCs also can be used to better understand the future changes within the wider realm of industrial production equipment, regardless of the business models used as the same changes may impact the operations and value. The list of prioritized DoCs should help practitioners dealing with industrial production equipment to better understand the future risks and opportunities, increasing their futures understanding.

5.4 Limitations and Future Research

Some of the most prevalent limitations of this study have to do with the limited geographical coverage of the experts (mainly Europe and North America), limited representation from the customer organization of the OBC as well as from the timing of the study in the time of turmoil (post Covid-19, during a global security crisis) that might influence the views of the participants (experts and researchers alike).

As for future research, there is clearly a need to better understand the life cycle feasibility and value of the OBCs and OBMs. Despite the growing trend of these contracting and business models there is still limited knowledge about the best practices to make these business models, contracts, and cooperation resilient to the changing world. Especially there is a need to understand in more detail the mechanisms through which these changes impact the parties, the direct and indirect, even systemic changes they invoke, prohibit or support. By understanding the temporal aspects of these business models better we would improve their value and feasibility for all parties involved and would enable their adoption at a faster pace. Furthermore, it would give us insights into the elements which are at the core of these business models – the risk and value and risk and value sharing as well as the decisions and alternatives necessary to design these business models to be resilient.

Appendix 1. Analysis of the Most Critical Drivers of Change

Driver of Change	Analysis on the significance of the Driver of Change	Impacted party
Changes in costs	These are probably happening in this time frame, especially considering current global inflation. Impact depends on business model design and contract. If cost escalations are built into contracts, then it is more fair for the parties. Where OBPSS outcome is savings (labor, material) then cost increase will increase the feasibility of OBPSS – in general OBPSS feasibility benefits from volatile markets. Energy cost increase leads to need for more efficient (modern) systems.	Either
Product market price change	If market price change is resulting from cost change, then this will be reflected on new OBPSSs pricing. Depending on the share of equipment cost vs. other life cycle costs the impact varies. On existing OBPSSs with no price modification or renegotiation clauses these might present an issue, depending on the relationship. Product price changes will impact the lifetime profitability and pricing power of OBBMs.	Either
Technological disruption	The risk of technological disruption seems to be lower than for traditional equipment sales. On the other hand, long-term contracts, higher profit of OBPSSs and supplier feeling the cost of spare parts will allow and push technological improvements in products.	Supplier

(*continued*)

(continued)

Driver of Change	Analysis on the significance of the Driver of Change	Impacted party
Life cycle cost/value becoming customer decision criteria	OBPSSs are a way to link life cycle costs to offering (price). The customer gets to benefit from the lower life cycle costs of better technologies. If this trend continues it will diminish the difference between transactional (sell and service) and outcome-based business models and lead to higher quality equipment prevailing. Also related to shortening product life cycles, which might impact OBBM feasibility, especially in areas where efficiency develops rapidly. OBBMs should have some extra value to share compared to traditional business model(s).	Supplier
Customer business strategy change	This can be catastrophic for the supplier in the case of plant closures and other very drastic changes. The opportunity for the customer that their potential future strategy changes might drive them towards OBPSSs as they are easier to terminate before equipment lifetime ends. If this happens then it might be a risk for the supplier that their equipment isn't needed any more. Even if equipment can be reused, transportation, de- and re-installation costs are lost.	Supplier
Changing customer expectations	Also related to technological disruptions, can also be a result of acquisitions. In most cases will be negative for the supplier unless the change is for more capacity that can then be sourced within the OBPSS. Supplier revenues are directly linked to customer production, both positively and negatively.	Supplier
Change in OBBM related regulations and financial policies	Sustainability related regulation is increasing, this might create a need for better technologies in production. Import and export legislation related to assets, IPR and intangibles will impact OBBMs	Supplier

(continued)

(*continued*)

Driver of Change	Analysis on the significance of the Driver of Change	Impacted party
Component supply problems	Very topical at the time of study. The difference is that the component supply problems for spare parts changes from customer to supplier in OBPSS. Supplier is also more capable of handling it than the customer who is mostly dependent on the supplier in any case. Modularity and interchangeability were seen as ways to counter this risk-	Supplier
Sustainability goals driving modernization of equipment	Aligned with increasing energy costs. Creating a market pull for more energy efficient technologies.	Both
Change in interest rates	Depends on who finances the equipment, also depends on contract (fixed/variable) for single OBPSS. Also interest rates will impact traditional and OBPSS alike (in the big picture) and can be included in contract pricing. Higher interest rates might steer customers towards OBPSS, depending on their cash situation and strategy.	Supplier
Country/region incentives for local production and sourcing	Less visible in complex equipment than in commoditized equipment.	Both

References

1. Korkeamäki, L.: Manufacturer perspectives on outcome-based service offerings: essays on the concept, financial consequences and customer relationships, p. 149 (2021)
2. Ng, I., Parry, G., Smith, L., Maull, R., Briscoe, G.: Transitioning from a goods-dominant to a service-dominant logic: visualising the value proposition of Rolls-Royce, p. 30 (2012)
3. Kohtamäki, M., Henneberg, S.C., Martinez, V., Kimita, K., Gebauer, H.: A configurational approach to servitization: review and research directions. Serv. Sci. **11**(3), 213–240 (2019)
4. Sjödin, D., Parida, V., Jovanovic, M., Visnjic, I.: Value creation and value capture alignment in business model innovation: a process view on outcome-based business models. J. Prod. Innov. Manag. **37**(2), 158–183 (2020)
5. Korkeamäki, L., Kohtamäki, M., Parida, V.: Worth the risk? The profit impact of outcome-based service offerings for manufacturing firms. J. Bus. Res. **131**, 92–102 (2021)
6. Bennett, N., Lemoine, G.J.: What a difference a word makes: understanding threats to performance in a VUCA world. Bus. Horiz. **57**(3), 311–317 (2014)
7. Stojkovski, I., Achleitner, A.-K., Lange, T.: Equipment as a service: the transition towards usage-based business models. SSRN J. (2021)

8. Bokrantz, J.: Maintenance in digitalised manufacturing: Delphi-based scenarios for 2030. Int. J. Prod. Econ., 16 (2017)
9. Fritschy, C., Spinler, S.: The impact of autonomous trucks on business models in the automotive and logistics industry–a Delphi-based scenario study. Technol. Forecast. Soc. Chang. **148**, 119736 (2019)
10. Glauner, F.: Future Viability, Business Models, and Values: Strategy, Business Management and Economy in Disruptive Markets. Springer, Cham (2016). https://doi.org/10.1007/978-3-319-34030-2
11. Schuckmann, S.W., Gnatzy, T., Darkow, I.-L., von der Gracht, H.A.: Analysis of factors influencing the development of transport infrastructure until the year 2030 — a Delphi based scenario study. Technol. Forecast. Soc. Chang. **79**(8), 1373–1387 (2012)
12. Dar, M.A., Gladysz, B., Buczacki, A.: Impact of COVID19 on operational activities of manufacturing organizations—a case study and industry 4.0-based Survive-Stabilise-Sustainability (3S) framework. Energies **14**(7), 1900 (2021)
13. Saritas, O., Smith, J.E.: The big picture – trends, drivers, wild cards, discontinuities and weak signals. Futures **43**(3), 292–312 (2011)
14. Wurlod, J.-D., Noailly, J.: The impact of green innovation on energy intensity: an empirical analysis for 14 industrial sectors in OECD countries. Energy Econ. **71**, 47–61 (2018)
15. "Tracking the Trends 2021: Closing the Trust Deficit," Deloitte Insights, p. 94 (2021)
16. von der Gracht, H.A., Darkow, I.-L.: Scenarios for the logistics services industry: a Delphi-based analysis for 2025. Int. J. Prod. Econ. **127**(1), 46–59 (2010)
17. Chicksand, D., Rehme, J.: Total value in business relationships: exploring the link between power and value appropriation. JBIM **33**(2), 174–182 (2018)
18. Holgado, M., Macchi, M.: A value-driven method for the design of performance-based services for manufacturing equipment. Prod. Plann. Control, 1–17 (2021)
19. Reim, W., Parida, V., Örtqvist, D.: Product–Service Systems (PSS) business models and tactics – a systematic literature review. J. Clean. Prod. **97**, 61–75 (2015)
20. Culot, G., Orzes, G., Sartor, M., Nassimbeni, G.: The future of manufacturing: a Delphi-based scenario analysis on industry 4.0. Technol. Forecast. Soc. Chang. **157**, 120092 (2020)
21. Bulut, S., Wende, M., Wagner, C., Anderl, R.: Impact of manufacturing-as-a-service: business model adaption for enterprises. Procedia CIRP **104**, 1286–1291 (2021)
22. Kowalkowski, C.: What does a service-dominant logic really mean for manufacturing firms? CIRP J. Manuf. Sci. Technol. **3**(4), 285–292 (2010)
23. Marcos Cuevas, J.: The transformation of professional selling: implications for leading the modern sales organization. Ind. Market. Manag. **69**, 198–208 (2018)
24. Christensen, C.M., Anthony, S.D., Roth, E.A.: Seeing What's Next: Using the Theories of Innovation to Predict Industry Change. Harvard Business School Press, Boston (2004)
25. Kowalkowski, C.: Dynamics of value propositions: insights from service-dominant logic. Eur. J. Mark. **45**(1/2), 277–294 (2011)
26. Töytäri, P., Rajala, R., Alejandro, T.B.: Organizational and institutional barriers to value-based pricing in industrial relationships. Ind. Mark. Manage. **47**, 53–64 (2015)
27. Raddats, C., Baines, T., Burton, J., Story, V.M., Zolkiewski, J.: Motivations for servitization: the impact of product complexity. Int. J. Oper. Prod. Manag. **36**(5) (2016)
28. Jiang, R., Kleer, R., Piller, F.T.: Predicting the future of additive manufacturing: a Delphi study on economic and societal implications of 3D printing for 2030. Technol. Forecast. Soc. Chang. **117**, 84–97 (2017)
29. Korkeamäki, L., Sjödin, D., Kohtamäki, M., Parida, V.: Coping with the relational paradoxes of outcome-based services. Ind. Mark. Manage. **104**, 14–27 (2022)
30. Parida, V., Sjödin, D., Reim, W.: Reviewing literature on digitalization, business model innovation, and sustainable industry: past achievements and future promises. Sustainability **11**(2), 391 (2019)

31. Plé, L., Chumpitaz Cáceres, R.: Not always co-creation: introducing interactional co-destruction of value in service-dominant logic. J. Serv. Market. **24**(6), 430–437 (2010)
32. Schlecht, L., Schneider, S., Buchwald, A.: The prospective value creation potential of blockchain in business models: a Delphi study. Technol. Forecast. Soc. Chang. **166**, 120601 (2021)
33. Schlaak, T., Franke, H., Brod, K.: Power Market Study 2030. Monitor Deloitte, April 2018. https://www2.deloitte.com/content/dam/Deloitte/de/Documents/energy-resources/Deloitte-Power-Market-Study-2030-EN.pdf. Accessed 12 July 2021
34. Baines, T.S., et al.: State-of-the-art in product-service systems. Proc. Inst. Mech. Eng. Part B: J. Eng. Manuf. **221**(10), 1543–1552 (2007)
35. Rabetino, R., Kohtamäki, M., Brax, S.A., Sihvonen, J.: The tribes in the field of servitization: discovering latent streams across 30 years of research. Ind. Mark. Manage. **95**, 70–84 (2021)
36. Cleary, K., Palmer, K.: Energy-as-a-service: a business model for expanding deployment of low-carbon technologies. Resour. Future, 6 (2019)
37. Eggert, A., Ulaga, W., Frow, P., Payne, A.: Conceptualizing and communicating value in business markets: from value in exchange to value in use. Ind. Mark. Manage. **69**, 80–90 (2018)
38. Töytäri, P., Rajala, R.: Value-based selling: an organizational capability perspective. Ind. Mark. Manage. **45**, 101–112 (2015)
39. Visnjic, I., Jovanovic, M., Neely, A., Engwall, M.: What brings the value to outcome-based contract providers? Value drivers in outcome business models. Int. J. Prod. Econ. **192**, 169–181 (2017)
40. Chesbrough, H., Lettl, C., Ritter, T.: Value creation and value capture in open innovation: value creation and value capture. J. Prod. Innov. Manag. **35**(6), 930–938 (2018)
41. Osterwalder, A., Pigneur, Y.: What is a Business Model? p. 16 (2005)
42. Mitchell, V.W.: The Delphi technique: an exposition and application. Technol. Anal. Strateg. Manag. **3**(4), 333–358 (1991)

Exploration of Multi-layers Networks to Elicit and Capture Product Changes

Vincent Cheutet[1(✉)], Aicha Sekhari[1], Chantal Cherifi[1], Justin Favre[2], Mohamed Douass[3], and Maxence Millescamps[3]

[1] University Lyon, INSA Lyon, Univ. Lyon 2, Univ. Claude Bernard Lyon 1, DISP UR4570, Villeurbanne, France
vincent.cheutet@insa-lyon.fr
[2] ENS Paris-Saclay, Gif-sur-Yvette, France
[3] INSA Lyon, Villeurbanne, France

Abstract. Change management is a key process in the Product Development Process. In a context of availability of large data on the process through Enterprise Information Systems like PDM (Product Data Management) or PLM (Product Lifecycle Management) systems, the structured data network can be analysed to identify key feature to improve the process performance.

In this work, we propose a multi-layer network to engineering change management that models product, organisation and change facets together. Once build, this network is analysed with some classical graph tools to identify some key features allowing to elicit some organisational behaviours. This approach is applied on two pedagogic examples.

Keywords: Engineering change · multi-layer network · product development process · PDM

1 Introduction

Changes are classical during the Product Development Process (PDP) and the importance of managing design change is accordingly well-recognised by practitioners and researchers [6]. In a context of strong digitalisation of the PDP with the support of recognised information systems like PDM (Product Data Management) and PLM (Product Lifecycle Management), change can be tracked at each stage through ECM (Engineering Change Management) process [13].

Nevertheless, ECM can be still improved and one aspect that has been less analysed is the relation between the product structure, the organisation and the change management. In particular, some can be interested to analyse the traces of changes through all data and information stored in PDM from previous PDP. By nature, all these data are structured and interconnected, and in this research work, we would like to explore the created data network to elicit and capture some key features that could explain the efficiency (or not) of the change management.

Published by Springer Nature Switzerland AG 2024
C. Danjou et al. (Eds.): PLM 2023, IFIP AICT 701, pp. 151–160, 2024.
https://doi.org/10.1007/978-3-031-62578-7_14

In this work, due to the large diversity of semantics existing in a PDM, we explore more precisely the concept of a multi-layer network [1,8,11]. Paraphrasing [11], a standard graph is often described by a tuple $G = (V, E)$ where V defines a set of vertices and E defines a set of edges (vertex pairs), such that $E \subseteq V \times V$. An intuitive definition of a multi-layer network first consists in specifying which layers nodes belong to. Because we allow a node $v \in V$ to be part of some layers and not to others, we may consider 'multi-layer graph' nodes as pairs $V_M \subseteq V \times L$ where L is the set of considered layers. Edges $E_M \subseteq V_M \times V_M$ then connect pairs $(v, l), (v', l')$. An edge is often said to be intra- or inter-layer depending on whether $l = l'$ or $l \neq l'$.

More precisely, we propose a multi-layer network for change management that models product, organisation and change facets together. Once build, this network is analysed with some classical graph tools to identify some key features.

The outline is as follows. Section 2 reviews the literature on change management with a network modelling approach. Section 3 presents the proposed multi-layered network approach and Sect. 4 presents an application on two pedagogic use cases. Section 5 concludes and presents some perspectives.

2 Literature Review

Authors of [2] provide a thematic analysis of state-of-the-art in design Change Propagation Analysis (CPA), and highlights opportunities for further work. In particular, they identifies seven use cases for CPA in the literature: CPA can support the generation of alternatives for implementing a change, the assessment of how a proposed change might impact a design, the assessment of how a proposed change might impact a product family, the assessment of a proposed change in terms of redesign cost, time and effort, the assessment of how a proposed change might impact production, the coordination of change activity and the improvement of designs with respect to potential future changes. The last two use cases cover our research questions.

Authors of [5] focus initially on the analysis of the causes of the change propagation and its management by the company: it then proposes design solutions in order to limit the impact of the propagation of changes. It also proposes a classification of the components of a product with respect to their character to transmit or not the changes during the propagation:

- "Absorbers": components that "stop" changes more than they involve other parts,
- "Carriers": components which transfer changes from one part to another (as many changes induced by the component as changes applied to the component),
- "Multipliers": components that pass on more changes than they receive from other parts.

If a large literature on engineering change management focus on the product itself only and specific approaches like DSM (Design Structure Matrix) [10,15], some explore this concept through a network vision [9]. Among the literature using a network approach, the CPM method (Change Propagation Model) [4] aims at predicting the propagation of changes within a complex product. This method is built using a network model and is based on 2 values evaluated by experts between the components directly in contact (mechanical, information or energy transfer) in the product in order to eventually calculate the image of these values for each pair of components of the product: the analysis, in the long run, thus allows to highlight the central components and those having a remote impact in the potential cascades of changes.

Based on these two root papers on the subject, others apply tools to analyse the network models defined through the CPM method, like [3] or [14].

Some authors explore multi-layered models for this context. For instance, in [7], the nodes are changes that have been proposed in the past on the product and the links are directed and unweighted to represent the relationship of parentage between the changes, i.e. a link is created from a source change to another target if this source change has propagated and generated this target change. The study is therefore based on real cascades of changes that have operated on the product. Attributes are also added to the nodes to identify which of the past changes were retained and actually made and which were rejected: some of the analysis tools used will focus only on the changes made. A study of the patterns of change propagation is also carried out in order to better understand how this phenomenon works. From this, several new tools could be developed:

- CPI (Change Propagation Index): is an index between -1 and 1 which allows to judge the category of a component according to [5] with the help of the number of changes undergone and transmitted by the component.
- The PDSM (Propagation Design Structure Matrix) and CPFM (Change Propagation Frequency Matrix): which refer respectively to the number and frequency of changes induced by one component on another. The frequency of changes is the ratio between the number of changes transmitted by the source component to the target component and the number of changes applied to the source component. Only completed changes are taken into account in this study.
- the CAI (Change Acceptance Index) and the CRI (Change Reflexion Index): 2 complementary indices which make it possible to evaluate the openness or obstinacy of a component to changes via the study of the attributes (retained or rejected) of the changes applied to the component.

Differently, the approach in [12] is based on 3 distinct layers: one relating to the product, one relating to the changes and one relating to the social organisation of the various actors gravitating around the design/re-design of the product. It should be noted, however, that in this study the links in the product layer are directed but not weighted as they were in the vast majority of previous studies. In this last so-called social layer, the nodes are people or groups of people and the links between them represent the existence of a communication link: these links are considered to be undirected and unweighted. This model has the advantage of being compatible with many of the tools presented previously but also allows new ones to be developed:

- PAR (Proposal Acceptance Rate): this is an index similar to the CAI; here we calculate the rate of acceptance of changes proposed by a node of the social layer, i.e. a person or a group of persons.
- The extension of the PDSM and CPFM matrices as well as the CPI index to the nodes of the social layer (as these tools, defined by Griffin (2009), were therefore only applied to the product network and using the information extracted from the change network).
- Propagation Directness (PD): this measure corresponds to the shortest path between 2 components of the product layer between which changes have propagated. As the links in the product layer are unweighted here, we can see whether the propagation is direct ($PD = 1$) or indirect ($PD > 1$) and judge the distance if the propagation is indirect.

In conclusion, the analysis of these different studies highlights several key points:

- Many efficient tools have already been developed and it seems interesting to try to build a network adapted to these tools in order to be able to re-use them, with some modifications if necessary.
- The construction of network models relies heavily on expert opinion in the choice of nodes and weights: it seems interesting to try to limit this component by proposing a construction protocol based on a medium containing the required data such as a PLM.
- The social dimension has been neglected in many articles: it seems interesting to keep this dimension in the study as it is one of the causes of the propagation of change.
- A network associated with past changes seems to be a very powerful tool if many changes are implemented in the network: it allows us to observe real cascades of changes in order to base the study on something other than a simple prediction model, and it therefore also allows us to question prediction tools by comparing the results of these tools with known data on past changes

Finally, it is a multi-layer network which will be retained in order to be able to preserve the study of the social organisation and past changes and this by trying to build a model making it possible to preserve the possibility of using the most powerful tools of analysis met in all these articles.

3 Proposition

In this section, the multi-layer network model is described, composed by several layers all linked together (Fig. 1):

- a set of layers associated to the hierarchical description of the product,
- two layers associated to the social organisation: "People" and "Teams".
- a last layer associated to changes.

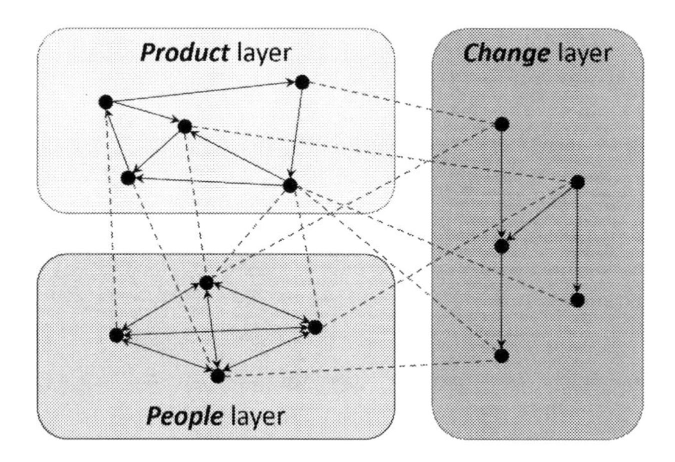

Fig. 1. A simplified representation of the multi-layer network, with only three layers. The dark edges are intra-layer edges, the blue dashed ones are the inter-layers ones. (Color figure online)

3.1 Definition of Layers and Intra-layer Edges

Product Layers. The product layers are based on the EBOM (Engineering Bill Of Material), closely to the product model structure proposed by [4]. More precisely, there are as many layers in the network as there are levels in the EBOM and the vertices in these layers correspond to articles in the EBOM (i.e. components of the product).

In a given layer, the edges represent the dependencies of change propagation that may exist (e.g. function/requirement sharing, kinematics joints, etc.). More precisely, an edge is created from vertex j to vertex i if, when component j is modified, this change may imply a significant change on component i. These edges are therefore oriented but also weighted by a number ranging from 0 to 1: the higher the weighting, the stronger the dependency between the components. The weighting, called risk of change and noted $r_{i,j}$, is defined as $r_{i,j} = l_{i,j} \times i_{i,j}$, where:

- $l_{i,j}$ represents the probability of propagation of the change from component j to component i,
- $i_{i,j}$ represents the impact rate of the change induced on i by j if the change has effectively propagated from j to i.

Organisation Layers. These layers are associated with the social organisation of those who design the product. It can be seen as an evolution of the "Social" layer of [12].

More precisely, two layers are defined: "People" where each vertex represents a person in that organisation, ad "Team" where each vertex represents a group of several people. These layers both exist since some changes are initiated by people and some by groups.

A link between 2 nodes represents the existence of communication between the 2 persons or teams represented. These links are unweighted and non-oriented because it is assumed that if communication exists between 2 people, then it exists in both directions: each person can provide and also receive information from a person with whom they are in contact. Generally speaking, this kind of links can be e-mail conversation, chat discussion or informal meetings but it is very difficult to capture these. So in this environment, we will first focus on empirical evaluation between teams (based on expert evaluation) and on requests between two people existing in a PLM framework.

Change Layer. This layer refers to the changes already made or proposed during the design/re-design of the product under study: each vertex represents one of these changes. This layer is inspired by the change network model of [7]. An attribute is associated with each node according to whether the change it represents has been retained or not (held, rejected or under consideration).

The edges in this layer represent the relationship between 2 changes: if a change 1 propagates from a component i (where it is applied) to a component j, then a change 2 associated with the propagated change on j is created in the "Change" layer and thus a link is created from vertex 1 to vertex 2. These links are oriented and unweighted.

Finally, we create a binary attribute which identifies whether the cascade of changes of which the node is a part is complete or not. It is automatically assumed that "Retained" changes have a complete cascade, but it may happen that a "Under consideration" change cascade is complete or unfinished, as a "Rejected" change cascade was after the change cascade was completed or not.

3.2 Definition of Inter-layers Edges

A first set of inter-layers edges represent inclusion:

- an edge between two vertices belonging to two different product layers represents a EBOM link,
- an edge between a vertex in "People" layer and another in "Teams" layer means that a person belongs to a team.

The edges between Product and Change layers mean that a change is applied to a given product component. It should be noted that not all changes are necessarily associated with a component: indeed, a system engineer (or respectively an engineer in charge of calculations) may propose a change following a modification of a requirement (or respectively following a calculation which calls into question the satisfaction of a requirement) which is not directly associated with a component but which will create child changes, most probably by other engineers, which will be relative to the product components.

Finally, an edge between a vertex of an Organisation layer and a vertex of the Change layer means that a people or a team has participated to this change at a phase of its lifecycle, whereas an edge between a vertex of an Organisation layer and a vertex of a Product layer means that a people (or a team has contributed on this component.

It should be noted that all inter-layers edges are non-oriented and non-weighted, due to their intrinsic nature.

3.3 Network Analysis Tools

Based on the literature review, we propose to explore several tools to analyse the network and its internal information (see [7,12] for more details):

- The degree, which is a measure of centrality for a node, applied on one type of link at a time,
- PDSM (Propagation Design Structure Matrix) and CPFM (Change Propagation Frequency Matrix), which analyse change propagation inside a layer (i.e. inter-layers edges between change layer and another one),
- Component categorisation indices (CPI - Change Propagation Index, DCPI - Degree Change Propagation Index, LCPI - L-Change Propagation Index), to classify a component in between "Absorber", "Carrier" or "Multiplier" [5],
- PAR (Proposal Acceptance Rate) that evaluate performance of a vertex from an organisation layer according to change proposition,
- CAI (Change Acceptance Index) and CRI (Change Reflexion Index) that evaluate performance of a vertex from a product layer according to change acceptation.

4 Application on Two Use Cases

In order to validate the completeness of the proposed approach and to estimate the quality of the analysis that can be made on it, the approach has been deployed on two pedagogic use cases: an office chair (composed of 10 components, 7 changes and 4 engineers from 2 teams, Fig. 2) and a bike (composed of 28 components, 7 changes and 6 engineers - no associated team, Fig. 3).

The office chair example, by its small number of edges, allow to manually explore the model and calculate all tools. For instance in Tab. 1, we can identify from DPCI that the seat is a "Multiplier", which is expected for a central part of the chair. Still with this example, the wheels are also a "Multiplier" due to the strong asymmetry in the relation with the leg extension: the weight of the (Wheels → Leg extension) edge is three times bigger than the (Leg extension → Wheels) one.

The bike example, with an increased complexity, allow us to perform some sensibility analysis on the different tools. For instance, by modifying the weight of the intra-product layer edges between the bike fork and the components in contact by 20% (increase for incoming edges and decrease for outgoing edges), this component change from a carrier behaviour to an absorber one. This behaviour change is easily identified with the degree. Moreover, the DCPI analysis highlights the impacts on the other components, with an increase of the change propagation inside the bike with this example.

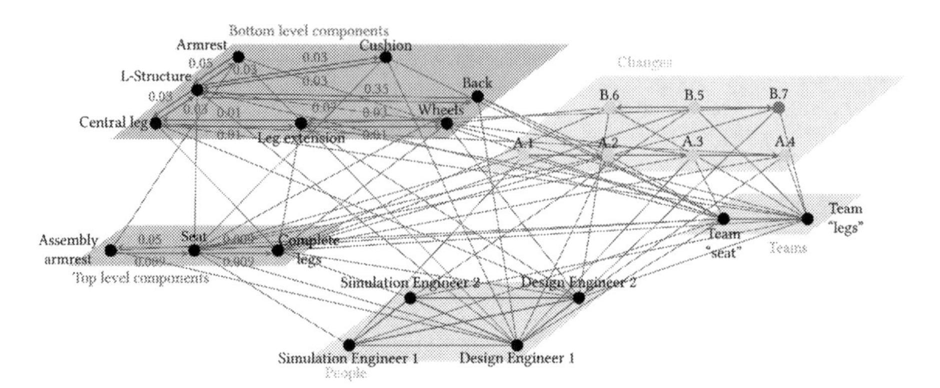

Fig. 2. Representation of the network associated to the office chair, with 2 product layers defined

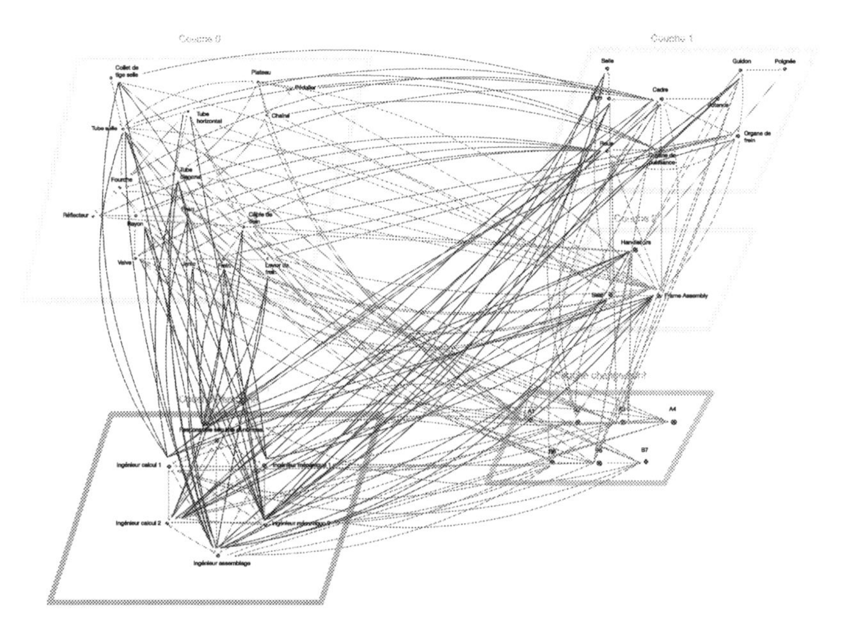

Fig. 3. Representation of the network associated to the bike, with 3 product layers and only the people layer defined

With these two examples, we can conclude on the relevance of this approach to explore the relations between the product and the organisation in the context of change management. Some nodes, either in the product or in the organisation, are central to some change propagation and so require a initial attention to ensure that the initial choices are consistent and will not provide too much loops in the future. The next step is now to directly implement this multi-layer network approach on data and information stored inside a PDM tool on existing prod-

Table 1. Determination of degrees and DPCI for edges from intra-product layers on the office chair example.

		Incoming Degree	Outgoing Degree	DPCI
Bottom level Comp.	L-Structure	0.46	0.16	−0.48
	Cushion	0.03	0.03	0
	Back	0.07	0.35	0.67
	Armrest	0.03	0.05	0.25
	Central leg	0.04	0.04	0
	Leg extension	0.04	0.02	−0.33
	Wheels	0.01	0.03	0.5
Top level Comp.	Seat	0.018	0.059	0.53
	Armrest Assembly	0.05	0.009	−0.69
	Complete legs	0.009	0.009	0

uct development, to better understand and navigate on these relations between organisation and product and so potentially find improvements for future product development projects.

5 Conclusions and Perspectives

Engineering change management is a strong research domain, with a large literature on tracking, visualisation and prediction of changes and their impacts. Nevertheless, the relations between the product structure, the organisation and the change management have been less analysed, in particular with the objective to improve the overall PDP. This issue can be more easily tackled nowadays with the strong digitalisation of the PDP and as a consequence the access to a large number of structured data and information in this context.

This article explores the use of a multi-layer network with graph analysis tools to elicit some structural and organisational behaviours. This multi-layer network models the product structure, the organisation and the changes together. The application of this approach on two pedagogic examples seems promising for future works.

The underlying graph that exists in any Enterprise Information Systems and which is developed during the PDP is an implicit representation of the organisation implied in the PDP and the analysis of this graph (completed with some external information to capture) can be a strong source of factual knowledge to help organisation behaviour improvements. It factually highlights the converging points of product and organisation, which should be monitored to ensure the fluidity of the PDP.

For the future, the application on more examples coming from real PDP and extracted from PDM systems is the first perspective, in order to really tackle the links with the organisational behaviour. On the other hand, changes are dynamic by nature and the evolution of these networks during the PDP life should be examined to better elicit some key features.

References

1. Aleta, A., Moreno, Y.: Multilayer networks in a nutshell. Ann. Rev. Condens. Matter Phys. **10**, 45–62 (2019)
2. Brahma, A., Wynn, D.C.: Concepts of change propagation analysis in engineering design. Res. Eng. Design, 1–35 (2022)
3. Cheng, H., Chu, X.: A network-based assessment approach for change impacts on complex product. J. Intell. Manuf. **23**(4), 1419–1431 (2012)
4. Clarkson, P.J., Simons, C., Eckert, C.: Predicting change propagation in complex design. J. Mech. Des. **126**(5), 788–797 (2004)
5. Eckert, C., Clarkson, P.J., Zanker, W.: Change and customisation in complex engineering domains. Res. Eng. Design **15**(1), 1–21 (2004)
6. Fei, G., Gao, J., Owodunni, O., Tang, X.: A method for engineering design change analysis using system modelling and knowledge management techniques. Int. J. Comput. Integr. Manuf. **24**(6), 535–551 (2011)
7. Giffin, M., de Weck, O., Bounova, G., Keller, R., Eckert, C., Clarkson, P.J.: Change propagation analysis in complex technical systems. J. Mech. Des. **131**(8) (2009)
8. Kivelä, M., Arenas, A., Barthelemy, M., Gleeson, J.P., Moreno, Y., Porter, M.A.: Multilayer networks. J. Complex Netw. **2**(3), 203–271 (2014)
9. Koh, E.C.: A study on the requirements to support the accurate prediction of engineering change propagation. Syst. Eng. **20**(2), 147–157 (2017)
10. Masmoudi, M., Zolghadri, M., Leclaire, P.: A posteriori identification of dependencies between continuous variables for the engineering change management. Res. Eng. Design **31**(3), 257–274 (2020)
11. Mcgee, F., Ghoniem, M., Melançon, G., Otjacques, B., Pinaud, B.: The state of the art in multilayer network visualization. Comput. Graph. Forum **38**, 125–149 (2019)
12. Pasqual, M.C., de Weck, O.L.: Multilayer network model for analysis and management of change propagation. Res. Eng. Design **23**(4), 305–328 (2012)
13. Storbjerg, S.H., Brunoe, T.D., Nielsen, K.: Towards an engineering change management maturity grid. J. Eng. Des. **27**(4–6), 361–389 (2016)
14. Wang, Y., Bi, L., Lin, S., Li, M., Shi, H.: A complex network-based importance measure for mechatronics systems. Phys. A **466**, 180–198 (2017)
15. Zheng, P., Chen, C.H., Shang, S.: Towards an automatic engineering change management in smart product-service systems-a DSM-based learning approach. Adv. Eng. Inform. **39**, 203–213 (2019)

A Model to Predict Span Time and Effort for Product Development Processes

Shourav Ahmed[1,2] and Vince Thomson[1(✉)]

[1] McGill University, Montreal, Canada
vincent.thomson@mcgill.ca
[2] Bangladesh University of Engineering and Technology, Dhaka, Bangladesh

Abstract. Companies are continuously trying to deliver products on time. However, traditional tools such as product lifecycle management, critical path method, and program evaluation and review technique cannot predict product development span time with reasonable accuracy. Product development involves large, multidisciplinary teams designing complex, interdependent systems; so, predicting span time is challenging. We have modelled product development microactivities to predict project span time and effort. The model uses a knowledge perspective where the difference between product knowledge requirements and designer knowledge capability drives product development speed. We simulate both process tasks and team behaviour, which are influenced by product complexity, development process complexity, and the difficulty of both technical and interface design. Our method predicts span time and effort at the start of a project or at any point during the development process as conditions change. It can also identify specific bottlenecks and poor levels of designer performance.

Keywords: Project span time · Design effort · Timeliness · Complexity · Product development · Agent-based

1 Background

Many academics have looked into product development (PD) methods in an effort to cut down on product develop time. They have stated that better methods, such as concurrent engineering, agile methods, lean, and systems thinking, have improved PD processes, and that information tools, such as computer aided drawing, computer aided engineering, and product life-cycle management, have improved design time (Wynn and Clarkson, 2018). These tools have significantly enhanced engineering-in-the-small since small teams can easily coordinate because they utilize the same design tools, communicate easily, and solve problems rapidly. However, engineering-in-the-large where numerous companies and individuals work together, such as in the creation of an aircraft, is still a challenge. The methods used in PD deal with novel creation, uncertainty due to incomplete information, and dynamics where technology changes and customer needs change. Additionally, a significant number of associated tasks frequently rely on data generated by other

Published by Springer Nature Switzerland AG 2024
C. Danjou et al. (Eds.): PLM 2023, IFIP AICT 701, pp. 161–170, 2024.
https://doi.org/10.1007/978-3-031-62578-7_15

processes (interdependence). When tasks are interdependent, exchange of information is crucial for timely task completion. It is almost impossible to document and track communication events using a traditional scheduler. Traditional scheduling typically uses a work breakdown structure to identify the effort of individual tasks, and then, optimizes the tasks on the critical path. According to a survey of 211 firms, 55% of PD projects are not delivered on time (Edgett, 2011). An organization suffers both financial and reputational harm when a product is delivered late. Therefore, timeliness is a vital performance metric for PD processes.

Numerous studies have been conducted to accelerate PD processes; however, faster PD does not guarantee timeliness. The two primary obstacles to achieving timeliness are: (i) methods to estimate span time, and (ii) methods to manage complex PD processes. Researchers have chosen two distinct approaches to resolve these issues: (a) improving scheduling, and (b) improving process flow through the use of suitable coordination mechanisms (organization structure, communication methods, ...). If the process is simple, scheduling techniques such as crashing are effective. However, most scheduling does not work well if the process is complex and has high task interdependency. The development of complex products requires highly specialised resources that are difficult to redeploy effectively, and if there is high task interdependency and if the critical path is dynamic, it is nearly impossible to do better scheduling.

Computer modelling and simulation is the best method for studying dynamic processes. Therefore, a computer-based PD process model was created that incorporated microlevel activities (technical work, interface development, design reviews, and communication) and measured the resulting span time and effort with span time used to estimate timeliness. In the past, lack of computer power hindered the simulation of highly complex processes involving billions of calculations. Recently, computers with high processing speed and the capacity to manage large data sets have been available. The objectives of the proposed research were to investigate the effects of various procedural practises on timeliness, including rapid, efficient, and frequent communication as well as to investigate the effects of macrolevel managerial decisions regarding process procedures and coordination mechanisms.

2 Critical Literature Review and Gap Analysis

Today, in order to stay ahead of the competition, it is not only necessary to provide superior products, but also to bring them to market on time and within budget. Timeliness is highly dependent on the precise estimation of span time, which is difficult to forecast due to task dynamics and process complexity. Traditional methods for estimating span time, such as the Gantt chart, critical path method (CPM), and program evaluation and review technique (PERT), are severely inaccurate (Ballesteros-Pérez et al., 2018). Gantt charts and CPM cannot account for the many iterations due to rework and are based on estimates of average task time, which underestimates the distribution of possible task times as well as the longest possible task time. In the case of many concurrent tasks, span time should be determined by the duration of the longest task, not the duration of the average task (Savage and Markowitz, 2009).

The complexity of PD tasks makes it difficult to manage and coordinate PD processes effectively. The most significant contributions to complexity are task interdependence, uncertainty, and novel knowledge requirements (Wynn and Clarkson, 2018). As the complexity of PD increases, so does the need for communication based on knowledge and coordination (Zhang and Thomson, 2019). McKinsey and Company (1994) found that exceeding the PD budget by 50% resulted in a 4% profit loss, whereas missing the target date by six months resulted in a 33% profit loss. Although researchers have recognized the need for faster PD and for developed methods to achieve it, few have stressed the need to achieve timeliness.

Simulation has proven to be robust and effective in predicting the span time of dynamic and complex PD processes. Among different simulation techniques, discrete event and time-stepping Monte-Carlo approaches have been used most frequently (Wynn et al., 2006; Suss and Thomson, 2012). One of the significant limitations is that they cannot model microlevel interactions between designers, such as communication, that plays a vital role in PD. A thorough understanding of the inner dynamics of the PD process is essential to developing a method that can predict span time accurately. We need a better understanding of how information exchange, PD process complexity, and organizational structure affect timeliness. Agent-based modelling has proven to be an ideal tool to capture microlevel communication and coordination. However, none of the existing agent-based PD models has focused solely on timely product delivery (Levitt et al., 1999; Zhang and Thomson, 2019). Most of the mentioned research emphasizes faster product development. However, accelerating a task that is not on the critical path does not affect timeliness. Since it is impossible to track and manage the activities on a dynamic critical path, it is time to try a different approach. We need to investigate what happens when we allow people to organically interact and solve problems that arise at the microlevel.

3 Knowledge Perspective

We wish to model PD as a problem-solving exercise since PD is a learning exercise where many issues have to be resolved. The time required to resolve any problem depends on two variables: the problem's complexity and the solver's expertise. The best characteristic to quantify these two elements is knowledge.

3.1 Knowledge Ability

We are aware that not all tasks demand the same amount of effort to complete. The amount of effort required to complete a task varies depending on a number of variables. How do we evaluate different tasks? Design reviews are used to appraise the quality of completed work. Depending on its quality, a task can either pass a review or be sent for rework. How do we determine a task's quality? These two questions can be answered by modelling the PD process from a knowledge perspective. The knowledge required to complete a task determines the inherent difficulty of that task. Multiple studies have identified the relationship between knowledge, design quality and PD performance (Kim et al., 2013; Markham and Lee, 2014). A research study compared two methods to estimate span time assessed against actual project durations. Using work breakdown structure resulted in a 27% error versus a 13% error when using a knowledge perspective (Zhang and Thomson, 2018).

Instead of capturing every possible aspect of knowledge, our model focuses on the technical knowledge needed for designers to complete designs. We categorize technical knowledge as either general or product specific. General knowledge (*GK*) is acquired through formal education or systematic training. For example, the ability to carry out stress analysis on a component can be considered as *GK* for a mechanical engineer. Product knowledge (*PK*) is defined as knowledge related to the development of a particular product that is acquired by past experience. A designer's *PK* anticipates issues regarding systems development and integration, and proactively suggests solutions. Individual tasks create product subsystems. Designers with good *PK* have a high probability of completing designs that can be successfully integrated with other designs to form a system. *PK* is defined on a scale of 0–100% to provide flexibility during simulation.

3.2 Knowledge Requirements (Complexity)

Complexity is the most significant factor that makes the PD process difficult. Complexity is the metric that is used to exhibit the inherent difficulty of a task and is composed of technical complexity and integration complexity. Generally, the amount of effort is proportional to task complexity. From a knowledge perspective designs that require more knowledge (items and levels) are more complex. Designers need *GK* for technical development and *PK* for the integration of functions (interfaces) (Fig. 1).

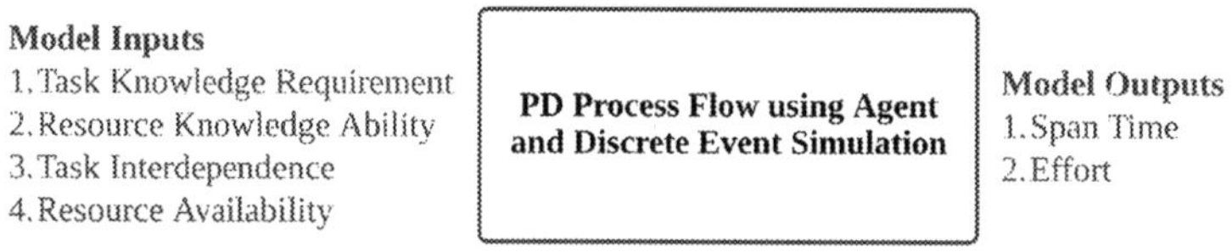

Fig. 1. Black box model of the PD process

4 Case Study: Aircraft Manufacturing

We partnered with an aircraft manufacturer that has a complex product and a very complex PD process due to the large number of subsystems and the large number of partners that have design authority. The aircraft manufacturer uses a stage-gate process for design and development. Our model focuses on a part of the PD process known as the joint definition phase (JDP). In this phase, the manufacturer and its suppliers develop a detailed definition of the aircraft and organize the ownership and development of the required interfaces into functional groups or work packages (WPs) such as structures, systems, power, etc. A single supplier, known as the prime supplier, is responsible for defining the functionality of a subsystem and managing the interface development. Interface management deals with managing communication, coordination, and responsibility distribution between subsystems. Interfaces are managed using interface control documents (ICDs).

We consider the development of an ICD as performing a task. Our PD process model determines how project organization, process characteristics, and workflow can improve timeliness.

The aircraft manufacturer uses eight different ICD categories that are required to define all aspects of aircraft specifications. The eight ICD categories are treated as eight process streams. At first, these ICDs are created by designers. Then, they are sent to reviewers by integrators, who manage the PD process, to review the technical aspects of the design. The integrators are experts within the design team who perform an initial quality check and decide the sequence in the case of multiple reviews. Since ICDs impact different aspects of the design, reviews and subsequent approvals take place across multiple WPs. If a design specification is not satisfactory from a technical or interface compatibility perspective, the design is sent for rework. After successful review, the interface design is sent for final approval to approvers. The process flow model of the manufacturer's JDP is shown in Fig. 2. The product development process model can be divided into four major tasks: create, review, approve, and rework. These events are discussed in detail in the following subsections.

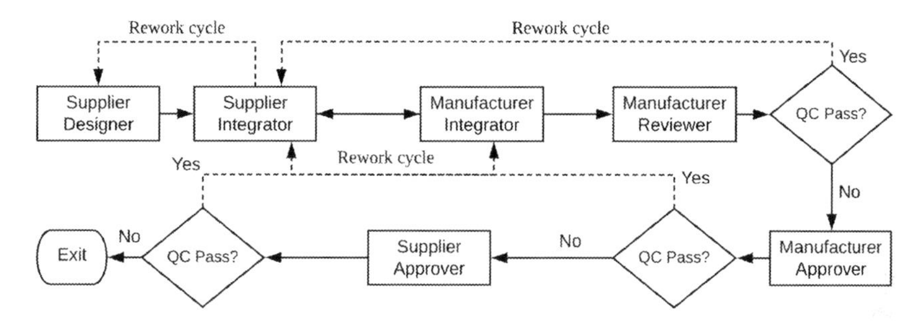

Fig. 2. PD Process flow model of activities for interface development during the JDP

4.1 "Create" Event

The JDP begins when ICDs are created by designers. There are a total of 394 designers spread across the 8 different ICDs. Designers mostly belong to supplier WPs. The creation of all 8 ICD types starts simultaneously as they use separate resource pools. It is important to note that designers not only create the ICDs, but also modify or correct them during rework. Therefore, there are instances when all designers are busy. If no designer is available, ICDs wait in a queue. We used ICDs as agents to store information related to knowledge requirements and capabilities. Whereas knowledge requirements are predetermined for each ICD, designer capability (knowledge level) is stored inside the ICD agent only after a designer is assigned to a task. This information is carried inside the agent for subsequent operations and is used in judging ICD quality. This information is stored as variables that can be changed dynamically during the simulation. For example, the knowledge capability of a designer increases through consultation with an expert. The ICD managed by a designer is updated automatically when a designer's new knowledge capability changes, as it is a major determinant of quality. A triangular distribution of effort in hours was obtained from the manufacturer and used to calculate effort. The effort to create an ICD is proportional to the knowledge difference between a task's knowledge requirement and a designer's capability. ICDs are then sent to reviewers via integrators who manage the process flow. Usually, there are integrators both on the supplier side and the manufacturer side. The supplier integrator first verifies the quality of the ICD before sending it to the manufacturer integrator. The manufacturer integrator decides the sequence of reviews by one or multiple reviewers.

4.2 "Review" Event

Reviewers check the quality of an ICD from both a technical and interface perspective. Due to the multidisciplinary nature of the ICDs, reviewers from multiple WPs need to verify the quality of an ICD from their perspectives. There are 206 reviewers in total. Some of the reviews are independent, while others are interdependent. A review is called independent when a reviewer from a WP is responsible for reviewing a single ICD type. The probability of passing a technical review is directly proportional to the general knowledge capability of a designer and inversely proportional to the technical complexity of the ICD. After a review is completed, a decision is made about whether an ICD should proceed forward or be sent for rework. In the case of interdependent reviews, reviewers often need to collaborate to ensure a consistent interface.

The need for collaboration depends on the integration complexity of the ICD. The higher the integration complexity, the higher the chance of requiring collaboration. Collaboration is done to ensure a consistent interface specification. There are generally two types of collaboration: vertical and horizontal. Vertical collaborations are done by reviewers from the same WP who work on different ICD types. Vertical collaboration meetings are held in pairs. Two reviewers from the same WP and working on two different ICDs meet and come to an agreement regarding the overlapping aspects of the design. Pairwise meetings continue until every reviewer is updated about the work of other reviewers in the same WP. In horizontal collaboration, all reviewers across multiple WPs and the designer in charge of the ICD meet to ensure all aspects of the design are satisfied. Reviewers from different WPs sometimes have conflicting objectives when making a design choice. For example, in choosing the material of a major component, the structural engineer would be more biased toward selecting a heavier material with high strength and rigidity. On the other hand, the engineer in charge of weights would be biased towards selecting a lighter material that would enable a higher payload. These collaboration meetings allow reviewers to discuss these differences and select the best design alternative. After collaboration, the reviewers decide if the ICD is consistent with other ICDs from a product knowledge perspective. The probability of passing this quality test is equal to the designer's product knowledge, which has a value between 0 and 1. The rationale behind this is that the greater the product knowledge of the designer who created an ICD, the better the chance of creating a consistent interface.

4.3 "Approve" Event

The final step in the JDP process is approval. During the approval process, an approver ensures that all the changes suggested in the previous steps are implemented properly.

No collaboration is required for this step. The approvers from different WPs work independently to ensure that the design specifications are consistent from their point of view. The amount of effort it takes to approve an ICD is calculated in a similar way to calculating review effort. The probability of passing a quality test is equal to the designer's level of product knowledge. The JDP is considered complete when ICDs from all 8 ICD types are approved.

4.4 "Rework" Event

Rework begins when an ICD fails a quality test at any point during the JDP. The designer with the highest level of technical knowledge is selected from the resource pool that created the ICD. Following rework, the ICD has a better chance of passing the quality test since the probability of passing is equal to the higher level of product knowledge of the new designer.

However, there are times when the designer from the pool with the most technical ability knows less than the designer who previously worked on an ICD. A consultation with an expert is needed in these cases. During a consultation event, the designer with lower knowledge meets with an expert who belongs to the same knowledge domain and has a significantly higher knowledge level. Since the number of experts is limited, the designer must wait for an expert before starting a consultation. The increase in knowledge after consultation is proportional to the knowledge gap between the expert and the designer as well as to the duration of the consultation. Missed consultation events result in significant wasted effort that increase span time and total effort. Reworked ICDs after consultation are sent to integrators to continue the JDP process.

5 Results

In our case study, we tested different scenarios and compared them with a baseline scenario. Each scenario was run 100 times with a different initial seed to account for the randomness in the probabilistic input data. For the baseline scenario, the span time was 1729 ± 213 h and the total effort was $345{,}000 \pm 5{,}500$ h. On average, the standard deviation of span time was 12% and the standard deviation of effort was 2%. All results are relative in nature. We investigated the effects on span and effort for several scenarios using data from the case study. A few of them are discussed below.

5.1 Varying Task Complexity

We investigated the impact of varying task complexity on span and effort. The use of new and improved technology is a good example of higher technical complexity. We tested 5 cases by lowering or increasing complexity in increments of 10% (Fig. 3). For each 10% rise in complexity, there was a 6% increase in span time and a 4% increase in total project effort.

5.2 Varying Designer Capability

We investigated the effects of a more mature work force on productivity, time, and effort. We increased designer knowledge level by 20% compared to the baseline knowledge level of 100% and we observed about a 2% reduction in span time and about a 6% reduction in total effort for each incremental change in knowledge (Fig. 4).

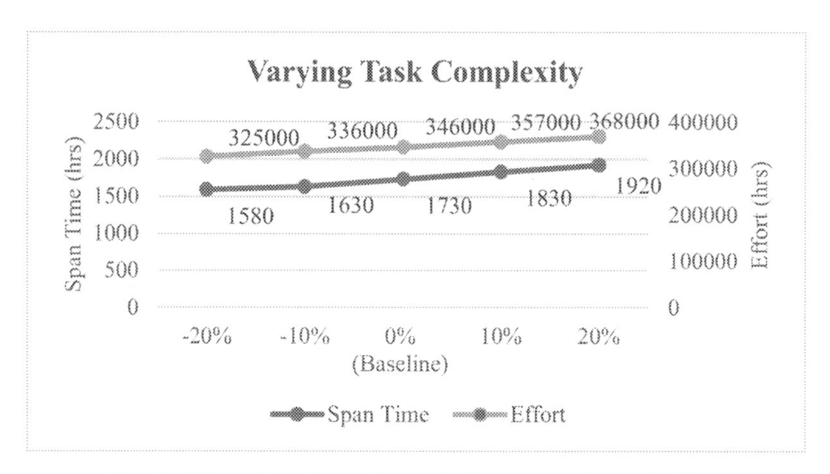

Fig. 3. Effect of varying task complexity on span time and effort.

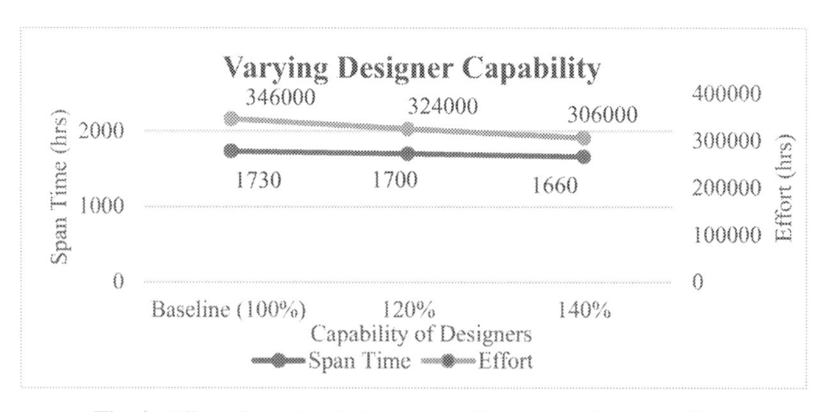

Fig. 4. Effect of varying designer capability on span time and effort.

5.3 Delayed Start of a Bottleneck Resource

Negotiations can often result in the delayed start of a resource. If it is a bottleneck resource, the effects on span time and effort are disastrous. We tested 2 delay points (3 and 6 months) and compared them with the baseline for span time. We observed that, for a 3-month delay, the span time increased by about 9.3% from 1730 h to 1890 h (Fig. 5). For a 6-month delay, the span time was 2140 h, an increase of 23% from the baseline. Total effort was unchanged as the delayed start did not add work.

5.4 Adding More Resources to Bottleneck Tasks

Finally, we wanted to study the effect of adding more resources to bottleneck tasks at different points, such as the beginning or midpoint of the PD process, compared to the baseline case. The span time decreased by about 24% when we added 20% more resources at the beginning (Fig. 6). When we did the same at the halfway point, the improvement dropped to 18%. There was no significant change in total effort because the amount of work in all three test cases remained constant.

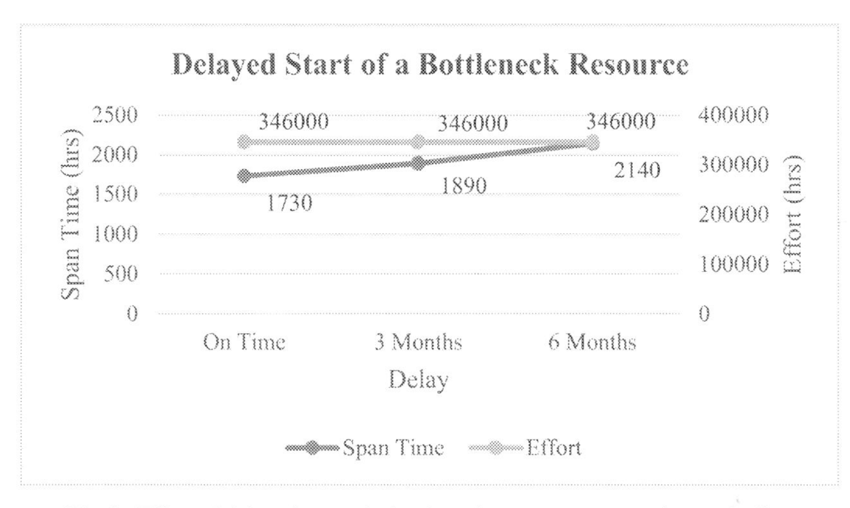

Fig. 5. Effect of delayed start of a bottleneck resource on span time and effort.

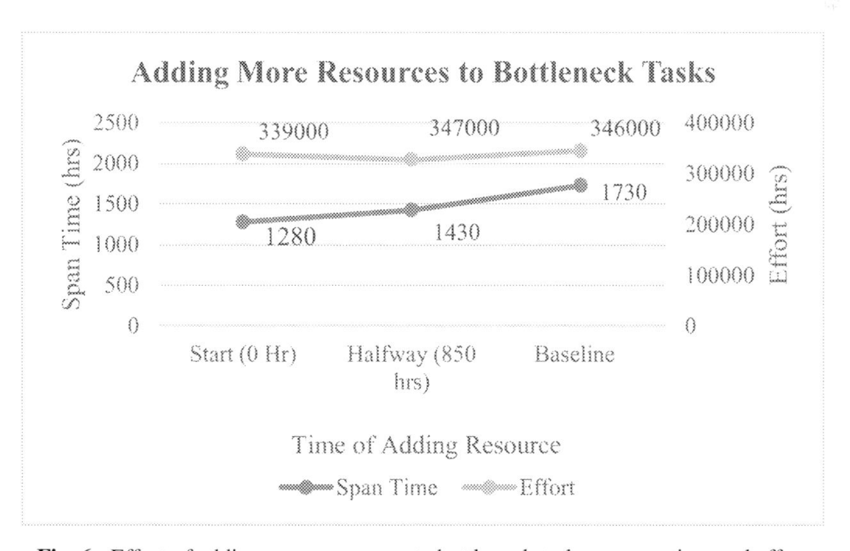

Fig. 6. Effect of adding more resources to bottleneck tasks on span time and effort.

6 Conclusion

Modelling PD processes using a knowledge perspective worked well. Simulations of an actual PD process gave results that were consistent in the estimation of span time and effort for variations of different process characteristics. We found that increasing task complexity increases both span time and effort, and increasing designer capability reduces both span time and effort. We also learned that delayed start of a bottleneck resource increases span time drastically.

The knowledge perspective model was able to estimate span time and effort, and thus, on time delivery of PD for complex, dynamic processes. The model can help managers to determine the impact of decisions on the timeliness of project outcomes. Managers can estimate the effect of technology (complexity) and resource maturity (capability) on future projects. The methodology will work on any type of project where knowledge is an appropriate characteristic to model the difference between product requirements and resource capability. In the future, we would like to extend the methodology to the entire PD process. This method can also be used for other coordination mechanisms such as agile methods.

References

Wynn, D.C., Clarkson, P.J.: Process models in design and development. Res. Eng. Design **29**(2), 161–202 (2018)

Edgett, S.J.: New Product Development: Process Benchmarks and Performance Metrics. Stage-Gate International (2011)

Ballesteros-Pérez, P., Larsen, G.D., González-Cruz, M.C.: Do projects really end late? On the shortcomings of the classical scheduling techniques. JOTSE J. Technol. Sci. Educ. **8**(1), 17–33 (2018)

Savage, S.L., Markowitz, H.M.: The Flaw of Averages: Why We Underestimate Risk in the Face of Uncertainty. Wiley (2009)

Zhang, X., Thomson, V.: Modelling the development of complex products using a knowledge perspective. Res. Eng. Design **30**(2), 203–226 (2019)

McKinsey and Company: Lead Local, Compete Global: Unlocking the Growth Potential of Australia's Regions. McKinsey (1994)

Wynn, D.C., Eckert, C.M., Clarkson, P.J.: Applied signposting: a modeling framework to support design process improvement. In: ASME 2006 International Design Engineering Technical Conferences and Computers and Information in Engineering Conference, pp. 553–562. American Society of Mechanical Engineers Digital Collection (2006)

Suss, S., Thomson, V.: Optimal design processes under uncertainty and reciprocal dependency. J. Eng. Des. **23**(10–11), 829–851 (2012)

Levitt, R.E., Thomsen, J., Christiansen, T.R., Kunz, J.C., Jin, Y., Nass, C.: Simulating project work processes and organizations: toward a micro-contingency theory of organizational design. Manage. Sci. **45**(11), 1479–1495 (1999)

Kim, T.T., Lee, G.: Hospitality employee knowledge-sharing behaviors in the relationship between goal orientations and service innovative behavior. Int. J. Hosp. Manag. **34**, 324–337 (2013)

Markham, S.K., Lee, H.: Marriage and family therapy in NPD teams: effects of We-ness on knowledge sharing and product performance. J. Prod. Innov. Manag. **31**(6), 1291–1311 (2014)

Zhang, X., Thomson, V.: A knowledge-based measure of product complexity. Comput. Ind. Eng. **115**, 80–87 (2018)

A Multicriteria Framework Proposition for Project Management Approaches

Márcio Leandro do Prado, João Felipe Capioto Seelent, Gilberto Reynoso-Meza, and Guilherme Brittes Benitez(✉)

Industrial and Systems Engineering Graduate Program, Polytechnic School, Pontifical Catholic University of Parana (PUCPR), Curitiba, Brazil
guilherme.benitez@pucpr.br

Abstract. One of the primary challenges in project management is to establish an effective methodology that can lead to both successful planning and stakeholder satisfaction. This is a critical decision that needs to be made at the beginning of the project, before the project's product, service, or object has been realized. In the case of technology-related projects, the complexity is even greater, as the evolution of systems and impacts on visibility may only become evident during the execution phase. To address this challenge, we aim to deliver and validate an MDCM (Multiple-Criteria Decision Analysis) framework that can support the selection of an appropriate project management methodology. To achieve this goal, we have extracted relevant literature on the influencing factors for project management. Our framework provides a comprehensive multicriteria approach that integrates the most critical factors identified in the literature to assist project managers in selecting the most suitable project management approach. In summary, this work proposes a quantitative method that combines the most influential factors in the current project management literature with a multicriteria framework to support project management approach selection.

Keywords: Project Management · Hybrid Approach · Technology

1 Introduction

The research on project management topics is mainly based on more waterfall literature, which does not always bring an adequate methodology or investigation of flow to solve or bring a workaround solution for the management problems. The PMBOK® itself is diffuse and multidisciplinary [1] and it is a fact that management is the sustainable basis of any project, always depending on a well-planned process [2]. The dynamic nature of projects involves many parallel streams of investigation, as Padalkar and Gopinath suggested [3]. Constant comparisons must be made to observe quick changes in how terms are used in the literature to identify emerging trends and avoid fads [1]. More waterfall methodologies have a standard framework, which needs to be adapted to each area to increase management efficiency and address the numerous complex problems that companies, and other organizations face daily [4]. In this sense, adaptable approaches

Published by Springer Nature Switzerland AG 2024
C. Danjou et al. (Eds.): PLM 2023, IFIP AICT 701, pp. 171–180, 2024.
https://doi.org/10.1007/978-3-031-62578-7_16

to project design and management in complex and turbulent operational environments suggest that projects conceived as experiments can contribute to decision-making [5]. The flexibility of the renegotiation must be maintained, where factors can assume different weight or relevance than initially planned, not only in the core disciplines but in the gaps created by more vertical project management. This helps to explore the intersection of project management and development to benefit from as-yet-untapped opportunities for project managers [6].

In this context, it is characterized as a "Problem Field" in decision-making, the choice of an approach and/or method for the most appropriate management for the project manager and her team, given the various influencing factors. Thus, the purpose of this study is to offer a framework based on multi-criteria methods that serve as a tool and guide for the project manager to select the most appropriate approach for a given project. To develop this framework, a literature review was carried out, where 27 articles were prioritized that included discussions on technological project management approaches and their influencing factors. The framework considers as input the project requirements, which comprises the market/customer demand, its product or service specifications, and essential factors such as scope, time, cost, and risk, always present in the projects. As an underlying basis, the framework considers the different approaches to project management, ranging from the waterfall method to the agile methodology, with all its intrinsic characteristics, such as rigidity or flexibility in management and documentation [7, 8]. It also considers each approach's influencing factors, such as the environment in which the project is inserted, and social, economic, legal, cultural, organizational, and maturity factors, to name a few.

The proposed framework then, through a multi-criteria decision-making support method, associates all input data, i.e., the project requirements, their influencing factors, and the characteristics of the management approaches, to order by degree of preference, the best approach to adopt and on which factor it should be adopted. Thus, it breaks the project into criteria and associates them individually with management approaches, providing a more granular view of the entire management of a specific project. Finally, the framework is validated from real applications in technology projects with different natures, the first being a technical platform evolution project, where information security modules and performance improvements are prioritized. In a second project, we have the development of a factory application, where the priority is the user experience and productivity improvements. Furthermore, in the last project, we have the expansion of a business unit, where service standards and operation synchronism are the main requirements of the project. Thus, the framework allows validating, from the selection of influencing factors in a literature review on project management, the selection of the most suitable approach for technological projects.

2 Theoretical Background

2.1 Critical Factors in the Waterfall (Traditional) Approach

The waterfall approach is applied to well-planned projects with defined content, executed according to pre-determined guidelines. These are predictable and linear projects, which allow for detailed planning and monitoring without many changes [2]. This brings security when we look at the main pillars of the waterfall approach, such as strict cost control and proposed deadline. However, in some technology projects, there is significant scope variability during their execution, either due to the need to change the process or even the frequent emergence of new technologies [8]. In addition, the product's very non-materialization in the project's initial phases causes frequent observations from stakeholders to cause changes in the course. This can generate customer dissatisfaction due to non-compliance with agreed items; after all, all rework generates increased costs and time.

2.2 Critical Factors in the Agile Approach

Project development is increasingly subject to the concept of agility, which still does not have a consensus on its definition, but which refers to the ability to change the configuration of a system, in the face of unforeseen changes, to gain competitiveness and improve the innovation capabilities of a project. This concept is related to factors of flexibility, speed, simplicity and readiness, and its management requires practices, tools, and techniques for project development and execution, as well as facilitators for its implementation [9–12]. In other words, agile approach is a way to manage a project by breaking it up into several short phases to accelerate its cycle.

In theory, these would be required features in any project, but this is not what happens in practice. For example, technology projects often require strict compliance with deadlines so as not to jeopardize an operation or even an information security guarantee. This can compromise the subsequent process in complex project chains, causing irreparable damage.

2.3 Critical Factors in the Hybrid Approach

While the waterfall methodology is focused on planning and validation, it is less effective in developing more unpredictable, uncertain projects needing rework. On the other hand, agile methodology is more innovative and focuses on organizational responsiveness and flexibility. The hybrid methodology combines the characteristics of the previous models. For this reason, it is not easy to adopt, as it requires a precise alignment between the project team, organizational objectives, and project implementation [13]. In technology projects, the hybrid approach presents itself as an alternative to optimize potentials and eliminate gaps from previous methodologies. However, the criteria for adopting it must be carefully observed. One should not make the mistake of adapting deficiencies in the methodology. If resources are lacking, this must be a gap mapped and resolved, regardless of the approach adopted. The selection of influencing factors can be a differential to achieve the desired results.

3 Methodology

The main objective of this article is to propose and validate a framework with a multicriteria approach to define the degree of priority, selection, and application of a technology project management method, depending on its influencing factors. To achieve this objective, qualitative-quantitative applied research was performed. The methodology for building the framework was a literature review on the conceptual foundations of project management approaches with the application of experimental cases in the technological area.

The framework was developed by applying the MDCM (Multiple-Criteria Decision Analysis) algorithm to support the selection and prioritization of the project management approach according to the factors that influence it, allowing these factors to be evaluated individually in terms of their method of management. Multi-criteria decision-making support methods are being increasingly adopted by managers in the political, organizational, and financial fields, being a critical support tool for solving increasingly complex problems where there are many risks and uncertainty criteria to be considered and possible alternatives to be adopted [14].

The TOPSIS method (Technique for Order of Preference by Similarity to Ideal Solution) was adopted for this framework because it is a multi-criteria decision analysis method that compares and selects alternatives based on criteria and their respective weights, normalizing their scores. Moreover, calculating the geometric distances between each alternative of the ideal solutions, seeking the approximation of the positive solution and the distance of the negative solution. This is an exciting solution for technology projects, as it mitigates the incidences of complexity and uncertainty.

4 Literature Review and Bibliometrics

This topic aims to explain the flow adopted for the research to determine the decision framework for choosing a methodology for application in technology projects. The research started by consulting keywords (Project cost OR Project time or Project scope OR Agile, hybrid AND traditional methodology OR Project Changes OR Project Success Factors OR External Factors in Projects) typically used in projects, forming a bank of 41,449 articles. We have noticed an evolution in the number of searches increasing significantly in recent years, as shown in Fig. 1.

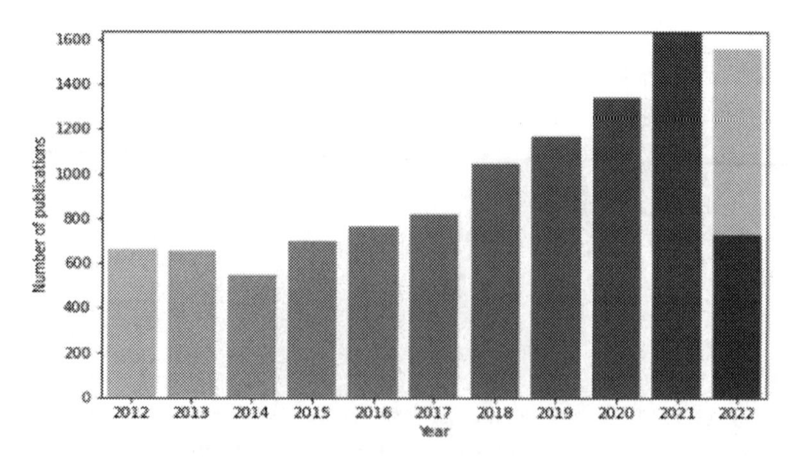

Fig. 1. Number of publications per year.

Figure 2 brings us a heat map, highlighting regions of the world with greater interest in the topic.

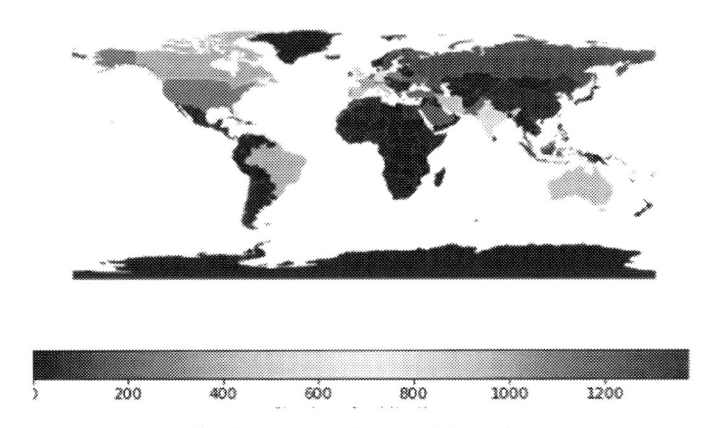

Fig. 2. Number of research by region.

The next step focused on applying inclusion and exclusion criteria, as shown in Table 1, to focus on the methodology approach theme, as well as segmenting for specific projects in the technology area:

Table 1. Exclusion and Inclusion Criteria.

I/E	CRITERIA	JUSTIFICATION
Exclusion	CRE-01	Papers with project management topics for specific products and services that do not address applied methodologies
	CRE-02	Papers not available in full
	CRE-03	Papers without interchange between project management chapters
Inclusion	CRI-01	Papers including uncertainties and complexities in project management
	CRI-02	Papers including project management methodologies in technology products or services
	CRI-03	Papers including changes and evolution of project management over the time
	CRI-04	Papers including practical experience in project management methodologies

After the exclusion and inclusion criteria, our final basket resulted in 27 prioritized articles, which served as the basis for the framework's application through the main influencing factors found in project management literature. The summary of these main prioritized articles can be found in the Supplemental File A of this research. In the next step, Wordclouds [15] was used to rank the importance of keywords according to the number of appearances, in addition to a grouping of similarities and translations, which shaped the influencing factors contained in the framework (Table 2).

5 Proposed Framework

Figure 3 shows the concept of the proposed framework, which is based on multi-criteria decision-making support methods for choosing the most appropriate approach to managing a technology project, considering all its influencing factors retrieved from our literature review on project management.

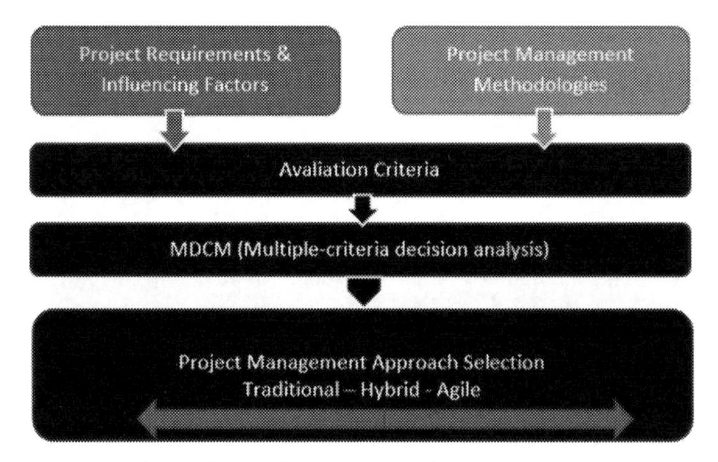

Fig. 3. Proposal Framework – Concept Diagram.

The influencing factors of the projects were previously raised by their frequency of occurrence in the studies found in our literature review, as shown in Table 2. They were related to the waterfall and agile approaches to technology project management according to their characteristics.

To develop such framework, we pondered the scales using TOPSIS method with the opinions of experts in the three projects. This supported our methodology to be validated in such a technological context. This is relevant since in a context where more technological and Industry 4.0 projects are demanded [17, 20, 21]. Thus, this can guarantee the sustainability [18] and innovativeness [19, 22] of projects (Fig. 4 and Fig. 5).

Table 2. Project Influencing Factors.

INFLUENCING FACTORS			
1	Scope	11	Complexity
2	Time	12	Technical
3	Cost	13	Behavior
4	Risks	14	Context
5	Stakeholders	15	Organization
6	Quality	16	Planning
7	Acquisitions	17	Control
8	Human Resources	18	Changes
9	Integration	19	Culture
10	Success	20	Monitoring

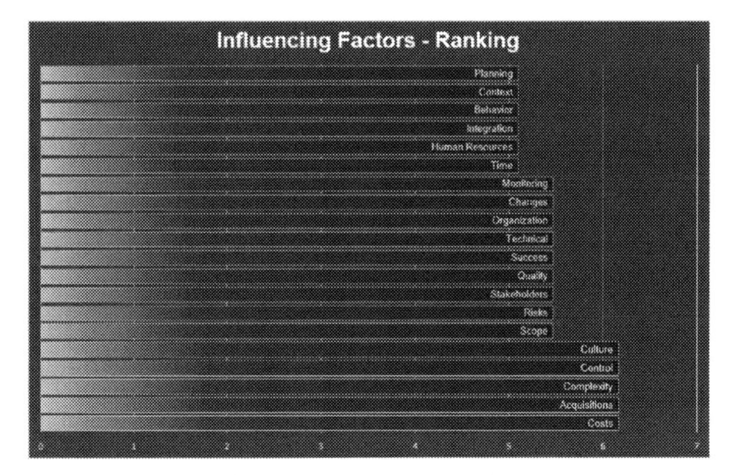

Fig. 4. Example of ranking of Influencing Factors in the project management: result of TOPSIS comparison.

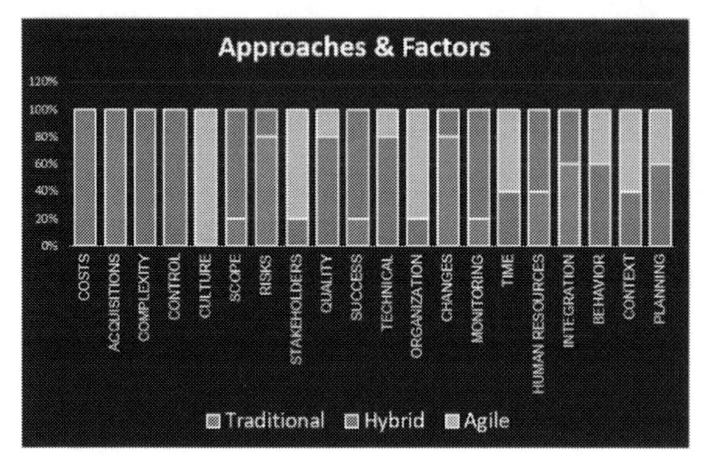

Fig. 5. Example of the individual relationship between project influencing factors and project management approaches.

6 Conclusion

The present study aimed to build and present a framework based on multi-criteria method, which serves as a tool and guide for the project manager to select the most appropriate approach to be applied in the development of a given project or even in a portfolio of projects. This is because considering the various influencing factors relevant to each management approach makes this choice difficult.

The literature review on project management approaches made it possible to identify their main influencing factors, which will be considered as criteria in the TOPSIS method for selecting and prioritizing the most appropriate approach to a given project. With the application of the proposed framework in cases of real projects, it was possible to verify the best approach to be adopted depending on the nature of the project, which facilitates the project manager's decision. Furthermore, the proposed framework, due to its versatility in the configuration and choice of influencing factors that serve as evaluation criteria, allows the manager to adapt new criteria and new approaches without specific limitations to a specific area of projects. In this sense, our main contribution is in the practical field for project managers, despite we considered the most relevant influencing factors on project management literature.

Thus, the framework allows the continuity of application for future studies by sharing knowledge with the scientific and academic community and with industry to bring both together in managing projects of any nature. Nevertheless, it is worth highlighting the limitations found in the development of this study about the lack of scientific studies that relate the factors of influence in the projects according to their management approach.

The scientific contribution of this work is in the tool to support the selection of the approach to be adopted in managing a project based on its influencing factors since the current models do not have this characteristic. The proposed framework combines decision-making support methods, project management approaches, and their influencing factors, which allows flexibility and versatility in choosing an approach depending on the specific characteristics of a project, enabling more fantastic alternatives for project managers. As a continuation proposal for future studies, the presented framework can be applied in other case studies to select the project management approach in other areas of development like artificial intelligence [23]. In addition, some multicriteria techniques could be integrated with such a framework for high technological context, as proposed by Almeida [16], to associate monetary values to each influencing factor and support decision-makers in the following steps.

References

1. Pollack, J., Adler, D.: Emergent trends and passing fads in project management research. A scientometric analysis of changes in the field. Proj. Manag. J. **33**(1), 236–248 (2015)
2. Lalmi, A., Fernandes, G., Souad, S.B.: A conceptual hybrid project management model for construction projects. Procedia Comput. Sci. **181**, 921–930 (2020)
3. Padalkar, M., Gopinath, S.: Are complexity and uncertainty distinct concepts in project management? A taxonomical examination from literature. Proj. Manag. J. **34**(4), 688–700 (2016)
4. Jovanovic, P., Beric, I.: Analysis of the available project management methodologies. Manag. J. Sustain. Bus. Manag. Solutions Emerging Econ. **23**(3), 1–13 (2018)
5. Picciotto, R.: Towards a new project management movement? An international development perspective. Proj. Manag. J. **38**(8), 474–485 (2020)
6. Ika, L.A., Hodgson, D.: Learning from international development projects: blending critical project studies and critical development studies. Int. J. Project Manage. **32**(7), 1182–1196 (2014)
7. Gemino, A., Horner Reich, B., Serrador, P.M.: Agile, waterfall, and hybrid approaches to project success: Is hybrid a poor second choice? Proj. Manag. J. **52**(2), 161–175 (2021)
8. Copola Azenha, F., Reis, D.A., Fleury, A.L.: The role and characteristics of hybrid approaches to project management in the development of technology-based products and services. Proj. Manag. J. **52**(1), 90–110 (2021)
9. Conforto, E.C., Amaral, D.C., Da Silva, S.L., Di Felippo, A., Kamikawachi, D.S.L.: The agility construct on project management theory. Proj. Manag. J. **34**(4), 660–674 (2016)
10. Serrador, P., Pinto, J.K., Pinto, J.K.: Does agile work? - A quantitative analysis of agile project success. Project Manag. J. **33**(5), 1040–1051 (2015)
11. Conforto, E.C., Salum, F., Amaral, D.C., Da Silva, S.L., De Almeida, L.F.M.: Can agile project management be adopted by industries other than software development? Proj. Manag. J. **45**(3), 21–34 (2014)
12. Hobbs, B., Petit, Y.: Agile methods on large projects in large organizations. Proj. Manag. J. **48**(3), 3–19 (2017)
13. Zasa, F.P., Patrucco, A., Pellizzoni, E.: Managing the hybrid organization: how can agile and waterfall project management coexist? Res. Technol. Manag. **64**(1), 54–63 (2020)
14. Wątróbski, J., Ziemba, P., Karczmarczyk, A., Zioło, M.: Generalised framework for multi-criteria method selection. Omega **86**, 107–124 (2019)
15. Wordclouds Homepage. https://www.wordclouds.com. Accessed 1 July 2022
16. Almeida, R.P., Ayala, N.F., Benitez, G.B., Kliemann Neto, F.J., Frank, A.G.: How to assess investments in industry 4.0 technologies? A multiple-criteria framework for economic, financial, and sociotechnical factors. Prod. Planning Control, 1–20 (2022)
17. Benitez, G.B., Lima, M.J.D.R.F., Lerman, L.V., Frank, A.G.: Understanding industry 4.0: definitions and insights from a cognitive map analysis. Br. J. Oper. Prod. Manag. [recurso eletrônico]. Rio de Janeiro, RJ **16**(2), 192–200 (2019)

18. Frank, A.G., Benitez, G.B., Ferreira Lima, M., Bernardi, J.A.B.: Effects of open innovation breadth on industrial innovation input–output relationships. Eur. J. Innov. Manag. **25**(4), 975–996 (2022)
19. Kai, D.A., Jesus, É.T.D., Pereira, E.A.R., Lima, E.P.D., Tortato, U.: Influence of organisational characteristics in sustainability corporate strategy. Int. J. Agile Syst. Manag. **10**(3–4), 231–249 (2017)
20. Kai, D.A., de Jesus, É.T., Pereira, E.A., de Lima, E.P., Tortato, U.: Influence of organizational characteristics in the sustainability strategy. In: ISPE TE, pp. 176–185, October 2016
21. Kai, D.A., de Lima, E.P., Cunico, M.W.M., da Costa, S.G.: Additive manufacturing: a new paradigm for manufacturing. In: Proceedings of the 2016 Industrial and Systems Engineering Research Conference, Availability, Development, vol. 14, p. 102 (2016)
22. Baierle, I.C., Siluk, J.C.M., Gerhardt, V.J., Michelin, C.D.F., Junior, Á.L.N., Nara, E.O.B.: Worldwide innovation and technology environments: research and future trends involving open innovation. J. Open Innov. Technol. Market Complex. **7**(4), 229 (2021)
23. Moraes, J.D., Schaefer, J.L., Schreiber, J.N.C., Thomas, J.D., Nara, E.O.B.: Algorithm applied: attracting MSEs to business associations. J. Bus. Ind. Market. **35**(1), 13–22 (2020)

Innovative Development in a University Environment Based on the Triple Helix Concepts: A Systematic Literature Review

Lucas Sydorak Lessa[1], Michele Marcos de Oliveira[2,3(✉)], and Osiris Canciglieri Junior[2]

[1] Industrial Engineering Undergraduate, Pontifical Catholic University of Parana, Curitiba, Brazil

[2] Industrial and Systems Engineering Graduate Program, Pontifical Catholic University of Parana, Curitiba, Brazil

michele.m@grupomarista.org.br

[3] PUCPRESS, Parana, Brazil

Abstract. Integrating university, industry, and government can represent innovation development in a university environment based on the concepts of the Triple Helix. Thus, there needs to be more research that considers university innovation ecosystems, indicating an opportunity to improve their understanding and functioning. This article proposes a Systematic Literature Review (SLR) to map the actions, actors, concepts and functioning of the innovation ecosystem that involves the university. The first step of the SLR was the search for articles from 2017 to 2022 containing the keywords: disruptive innovation, sustainability, and digital transformation, resulting in 27 selected articles. The research identified the roles and recommendations of each actor in the ecosystem and ways to develop it sustainably, focusing on actions to improve its current innovation strategy, contributing to the academic literature by providing initiatives in innovation ecosystems and synthesis of related concepts.

Keywords: Innovation Ecosystem · Triple Helix · University · Disruptive Innovation · Sustainability · Digital Transformation

1 Introduction

The demands for innovation have increased enormously in recent years. Due to the emergence of technologies, the focus on sustainability and cost reduction, improvements in the user experience and impact on society have produced initiatives to offer incremental or disruptive innovations in their different forms. With the growth of innovation systems, relationships between universities and companies have become an integral part of an ecosystem necessary to support the growth of these new initiatives. In this sense, the Triple Helix concept [1] contemplated the contribution of actors in representing three helices: university, industry, and government, to develop innovation. Posteriorly, this concept was revisited and expanded to the Quintuple Helix model [2], which adds the importance of involving civil society and the environment in the Triple Helix.

Published by Springer Nature Switzerland AG 2024
C. Danjou et al. (Eds.): PLM 2023, IFIP AICT 701, pp. 181–190, 2024.
https://doi.org/10.1007/978-3-031-62578-7_17

After conducting the initial phase of research on the subject, it was found that there is a need for an adequate conceptual structure that can fully address innovation ecosystems, as it is a current theme in constant evolution. So, through the definition of concepts and method of literature review, this study seeks to compile data about innovative development in the university's environment based on the Triple Helix concepts.

2 Theoretical Background

As previously mentioned, the Triple Helix concept considers the university, industry, and government as agents for innovation development [1]. In this sense, each agent has a respective role in this ecosystem to work and produce results. Moreover, "The interaction between university, industry and government is the key to innovation and growth in a knowledge-based economy" [3]. A literature review and analysis of the development of the triple helix ecosystem summarized five main aspects of the functioning of this ecosystem [4]: 1) The complex relationships between various agents in regional innovation are simplified according to the social geometry of triadic interactions; 2) The mechanism of Triple Helix interactions is "taking the role of the other"; 3) Its development is an evolutionary process that must be pre-structured and coordinated; 4)Triple helix interactions require integrating top-down coordination and bottom-up initiatives; 5) Certain tangible and intangible conditions make the triple helix model possible.

This article consists of understanding and analyzing the functioning of this ecosystem from different points of view and ways of analyzing the same ecosystem. However, all with the same theoretical basis based on the definition proposed by [1]. Although there are variations in how ecosystems operate due to geographic, cultural, and economic variations, there are common characteristics [4]. Studies suggest that trust in social relationships benefits innovation and interactive learning [5].

3 Method

An article that follows the Systematic Literature Review aims to identify the most relevant articles that direct the research on the theme [6], in this case, the innovation ecosystem. The articles' criteria for inclusion and filtering are described in the image below.

The last criteria for exclusion were a preliminary analysis of the articles and identifying aspects that were considered essential to be approached by the articles: University, Industry, Government, and Sustainability involving the social, environmental, and economic aspects. The articles that approach most or all these aspects were selected; with it, twenty-seven works were chosen for deeper analysis (Fig. 1).

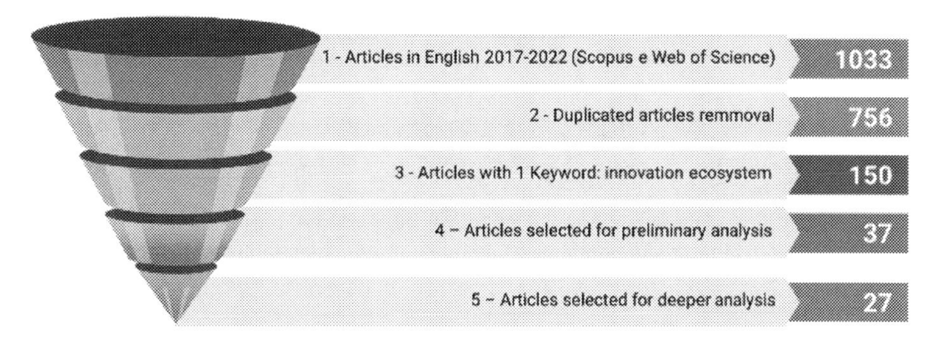

Fig. 1. Systematic Literature phases.

4 Results

Due to an analysis of the articles selected in the systematic literature review, it was possible to analyze them based on their contributions and applications.

4.1 Contribution

Most of the Contributions aim to describe and find a way to develop the interaction [7, 8, 35], relation [9] and cooperation [10] between the actors of the ecosystem, such as increasing the number of actions that aims to develop academic research [11], that may be applied in the industry. Also, promoting events to present research and projects to the market and with it stimulated its commercialization and application. Furthermore, most of the Contributions were analyses with several objectives, such as categorizing the interaction between universities and companies [8], comprehending the micro and macro dynamics of the open innovation in a sustainable model [12], a mechanism of analyses of the collaboration between research institutes and companies [13], the strategies implemented by an entrepreneurial university [14, 35], a structure to analyze and identify points to develop aiming to improve the ecosystem [15]. The Contributions were also related to identifying the roles of each actor in the ecosystem [16, 17] and the concept of sustainable innovation towards entrepreneurship [18, 19] to be maintained in the long term.

4.2 Application

Among the applications of the articles studied in this work, the most common topic was sustainable innovation. Works describing ways to develop it, such as digital application, sharing knowledge [13], co-working [25], and the impact of micro and macro actions for sustainable innovation [12], which reinforces the importance of the relationship between sustainability and innovation [23]. Studies about the innovation ecosystem helix [22] were found which aimed to describe and better understand the relation [8] and the importance of the actors in the triple and quadruple helix [7], and with it facilitates their actions and improve its operation [9, 16, 17].

Another topic was innovative entrepreneurship and its initiatives to develop and propitiate it [19, 28, 29], the importance of actions to encourage academics and in-tern politics to improve the way it is approached [11] and the student's point of view of the actions towards innovative entrepreneurship development [20]. The articles also studied the research and its relation as a step to development, as well as knowledge transfer, the teaching of innovation and social innovation [27], presenting ways to improve the development of emerging economies, such as the industry and university collaboration may be efficient [14] and to enlarge the range of engineering research and natural science creation [21]. The search for development in instability propitiates more sustainable creations [18].

5 Discussion and Conclusion

This research has discussed the content found in the twenty-seven articles selected, making it evident that the interaction and roles between the triple helix ecosystem actors are the central themes of the contributions and applications of the articles.

A better understanding of the innovation process for development allows us to develop responsible research and innovation methods. Research that will not only help a business make profits but also contribute to the common good of society and science. The development of innovative research requires transparency to implement and to build trust, which strengthens cooperation, and interdependencies between ecosystem stakeholders are stronger [7]. Since innovations emerge because of collaboration between all helices, there are desirable attitudes of all actors to responsible innovation and clear definitions of the types of responsibilities to be fulfilled by innovators [7]. Collaboration with big corporations is relevant, as they can become customers or partners, bridging the financing for innovation [16]. Each actor must manage resources, activities, value addition, and capture [15]. The university is a strategic actor recognised as a primary actor in the innovation ecosystem [8]. The universities with a relevant contribution to innovation ecosystems have as roles: support for start-up creation and growth, collaboration with police makers and firms, innovation sponsor, networking with other universities, stakeholder involvement and research, knowledge, and infrastructure share [29].

For universities which work in 4.0 domains, the authors suggested that universities also should increase or start actions to a) prioritise research and engineering projects; b) improve communication of related research results to the industry beyond the roles previously cited [29]. The authors suggest actions for university innovation which support sustainability in two categories of approaches: people-based approaches: a sustainability expert within a university innovation support unit; collaboration with a significant university sustainability coordinator/team; collaboration with a range of sustainability/cleantech experts in the university; collaboration with sustainability/cleantech-oriented organisations outside the university; sustainability objectives for the innovation support unit; environmental management system for the innovation support unit; sustainability reporting to university level; sustainability questions/criteria in project proposals/decisions; use of lifecycle analyses in projects [21].

Another perspective in the articles is that the entrepreneurial universities' innovation strategy calls for effective ways of integrating research, innovation, and application.

It also stresses the importance of international innovation cooperation [33]. However, in current higher education policies and practices, international research cooperation primarily aims for research excellence, while university ecosystems often have a local focus. Universities' engagement on a global scale should be emphasised in future policies since their engagement in innovation ecosystems crosses the boundaries of geographical locations [19]. There is a conceptualisation of two types of universities [19]: a) The entrepreneurial university: a knowledge producer for technology transfer from the academy to the industry as universities' reciprocal collaborations with industries and governments based on the triple helix model. Additionally, this university profile meets the societal needs of an entrepreneurial university concerning economic growth and innovation. b) The sustainable entrepreneurial university must be understood as an anchor organisation for knowledge exchange to help academics develop innovative research questions, conduct better research, and provide an improved understanding of research applications in industry and shape a better future society.

There are some components and conditions to entrepreneurship development: financing of entrepreneurship; state policy; state programs in entrepreneurship; entrepreneurial education; introduction of scientific and technical developments; commercial and legal infrastructure; market openness; physical infrastructure; cultural and social norms [24]. Moreover, some entrepreneurship and innovation indicators may evaluate the fostering factors: legislation; level of motivation for entrepreneurial activity; information accessibility; entrepreneurial culture and education; human capital; financial infrastructure, IT infrastructure and communication technologies, and market potential of the region [25].

To foster innovation in entrepreneurial universities and promote engagement actions is necessary to offer continued education programs on related topics such as frugal innovation, social inclusion, environmental challenges, and collaboration with external stakeholders. Additionally, it is suggested that curriculum design development and entrepreneurship education programs emphasise problem-based learning, STEM (Social, Technology, Engineering and Mathematics) and social disciplines [14]. As well as to prepare students to participate in the production process that generates income [30]. In this way, economic sustainability must be an essential pillar of entrepreneurship, which can be taught to help the student learn with practice, incorporate methods, network, and increase their interest in creating a new business [20].

In specific studies [20, 24], the training consists of students selecting a project from a technology portfolio and evaluating its marketing potential through the methodologies of the master's degree. Applying these methodologies has favored the emergence of sustainable initiatives within the students' projects, learning their business ideas, create a network with colleagues and teachers. This training allowed students to develop skills for innovation, technology transfer, the creation of new companies, the commercialization of innovative ideas, and entrepreneurial abilities since student becomes relevant in contributing to the solution of environmental issues and collaborates in economic development [20, 24]. Some challenges universities must face to succeed in innovation ecosystems are the reduction of favoritism that leads to unfair academic participation, the funding of technology acquisitions, the reduction of the inflexibility of university administration and restrictive regulations, the teaching overload of academics and inadequate industry links [29].

A challenge for university management and knowledge transfer is balancing the generation of technologies and their acquisition from the environment, how to transfer and commercialise research results and how to encourage university high-tech entrepreneurship and practical aspects. Other implications concern the difficulty creating relationships among firms, policymakers, and universities. There is a challenge when approaching the management of the University-Enterprise linkage due to the complex conditions of the Latin American ecosystem. Building interactive networks to develop specific programs and collaborative projects is a path to boost this new linkage [8].

5.1 Industry Roles for Innovation Ecosystems

Industrial ecosystems are "localised socioeconomic formations achieving sustainable development through the circulation of resources in the objective, environmental, process, and project subsystems" [31]. Furtherly, the ecosystem operates based on information, knowledge, technologies, or critical resources with distinct levels of exchange between business, industry, the scientific community, and government. These arrangements respond to digital challenges and ecological and industrial trends in innovation projects, products, digital platforms, and technologies [24, 34].

In the context of the new industrial revolution, the authors highlight the following principles for industrial ecosystems establishment: transboundary ecosystem processes; (self) organisation, regulation, and development; collaborative development, use of information, and intellectual resources; a continuous flow of projects; agility and flexibility to external challenges; project and client-orientation. Additionally, other principles may support the development of an industrial ecosystem: diversity of actors and network organisation design; collaboration based on partnership, trust, cooperation, and mutual help; balance between goals and objectives of actors; knowledge circulation; resources conservation priority; maintaining and development of each actor's potential; circularity principles enabling to extend the life cycle of resources and to regenerate them for use in other projects [25].

The university-industry connection enhances performance growth, and there are five main academic activities which contribute to the innovative process within firms take place: technological development carried out by academic research and linked to the industry; training and development of engineers and scientists able to deal with problems associated with the innovation process within companies; creation of new scientific instruments and techniques; creation of spin-offs by the academic community.

Recent studies demonstrate that different technologies have been applied to better understand and support ecosystem innovation development. The study shows that machine learning and artificial intelligence are used to understand and predict ecosystem innovations and collaborations to build constructive collaboration between international universities and industry cooperations. [10]. Additionally, digital platforms and applications are used to create innovation opportunities and strengthen sustainable innovation ecosystem alerts in heterogenous ecosystems, which is particularly interesting to entrepreneurs in potential as well as policymakers [29]. So, we define sustainable development of an industrial ecosystem as technological, innovative, and economic transformations fostering production and innovation potential development of all its actors and system through balancing digital transformation and circularity with human capital and technological development [25, 34].

The image below encapsulates the essence of the preceding text by serving as a concise summary that portrays the key elements and themes and the result of its application (Fig. 2).

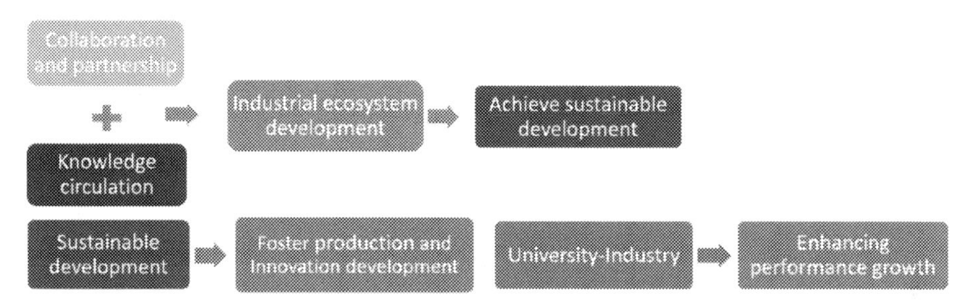

Fig. 2. Key elements and the result of its application.

5.2 Knowledge and Technology Transfer

The studies presented ways to contribute to knowledge transfer and regional and social development: universities acquiring and increasing their role in R&D projects, innovation performance and results. Additionally, universities should concern about high-tech entrepreneurship. University technology transfer is attracting greater attention from the high-tech segment [11]. Knowledge transfer is a common factor researched in the articles, with technology transfer being the specific link between universities and firms. Knowledge and technology transfer permits the exchange of different areas of knowledge by companies, and it improves their innovative capacity and market performance [32]; it can facilitate long-term growth, competitiveness, and transition towards sustainable development [28]. Especially regarding approaches for communicating corporate innovation capabilities outside the ecosystem [26] through university alliances. The Universities are acquiring and increasing their role in research and development projects, innovation performance and results to contribute to knowledge transfer and regional and social development.

5.3 Government Roles for Innovation Ecosystems

Government roles are related to encouraging knowledge exchange between the university and industry by promoting innovation as the natural path to a sustainable future [9]. Also, the government's role is to promote technical knowledge accumulation, develop internal learning processes in innovative firms, and widen the scientific and technological base essential to sustainable growth [9].

Policymakers should reinforce innovative practices, entrepreneurship, and presence in new markets, so they become more active rather than a simple regulator [9]. The state must have an "entrepreneurial" role, acting on allocating public resources to strategic areas where the private initiative has not yet invested, fulfilling uncertainty markets [9]. Policymakers must become more flexible and work with others outside their specialization to ensure the system's continued regional and national competitiveness [28]. The adaptability of policymakers towards continuous learning and interaction outside their direct field is necessary to aid knowledge-intensive entrepreneurial firms, which can be achieved by further policy experimentation, including dynamic monitoring practices, adaptation to recent problems, and working with adjacent policy fields to solve new challenges [28]. When approaching enterprises at distinct stages, they have different demand types (including product, applied, and basic generic technology) for collaborative research and development (R&D). In other words, there are differences in technologies to solve in R&D collaboration that were observed. The degree of such goal differences will harm knowledge transfer by affecting enterprises' learning willingness and absorptive capacity [13]. The demands from enterprises and R&D must be strictly linked to work together and achieve their respective goals. To accomplish goals towards development, universities must comprehend societal responsibility and the necessity of direct and continuous interchanges with firms and policymakers to develop a competitive knowledge-based society [11].

The university must have more flexible structures, a new action-oriented research approach, and social participation in networks to socialize the bidirectional flow of socially pertinent knowledge [8]. Sustained incentives to create bottom-up progress are necessary for existing start-ups and new entrants to the ecosystem. Given that, to move towards the future, the induction of existing actors to change their businesses to accomplish sustainable development [28]. With government facilitation, collaboration among companies can result in knowledge, product, and economic sustainability [12, 34].

Acknowledgment. The authors especially thank the financial support of Pontifical Catholic University of Parana (PUCPR) - Polytechnic School – Industrial and Systems Engineering Graduate Program (PPGEPS), The Brazilian National Council for Scientific and Technological Development (CNPq) and the Coordination for the Improvement of Higher Education Personnel in Brazil (CAPES).

References

1. Etzkowitz, H., Leydesdorff, L.: The dynamics of innovation: from National Systems and "'Mode2"' to a Triple Helix of university-industry-government relation. Res. Policy **29**, 109–123 (2000)
2. Carayannis, E.G., Barth, T.D., Campbell, D.F.: The quintuple helix innovation model: global warming as a challenge and driver for innovation. J. Innov. Entrep. **1**, 2 (2012). https://doi.org/10.1186/2192-5372-1-2
3. Etzkowitz, H., Zhou, C.: The Triple Helix: University–Industry–Government Innovation and Entrepreneurship. Routledge, London (2017). https://doi.org/10.4324/9781315620183
4. Cai, Y., Etzkowitz, H.: Theorising the triple helix model: past, present, and future. Triple Helix **7**, 189–226 (2020). https://doi.org/10.1163/21971927bja10003
5. Boschma, R.: Proximity and innovation: a critical assessment. Reg. Stud. **39**(1), 61–74 (2005). https://doi.org/10.1080/0034340052000320887

6. Fink, A.: Conducting Research Literature Reviews: From the Internet to Paper. Sage Publications, Thousand Oaks (2019)
7. Valackienė, A., Nagaj, R.: Shared taxonomy for the implementation of responsible innovation approach in industrial ecosystems. Sustainability (2021). https://doi.org/10.3390/su13179901
8. Álvarez-Castañón, L., Palacios-Bustamante, R.: Open innovation from the university to local enterprises: conditions, complexities, and challenges. Telos: revista de Estudios Interdisciplinarios en Ciencias Sociales. Venezuela, pp. 692–709 (2021). https://doi.org/10.36390/telos233.12
9. Costa, J., Neves, A.R., Reis, J.: Two sides of the same coin. University-industry collaboration and open innovation as enhancers of firm performance. Sustainability (2021). https://doi.org/10.3390/su13073866
10. Cai, Y., Ramis Ferrer, B., Luis Martinez Lastra, J.: Building university-industry co-innovation networks in transnational innovation ecosystems: towards a transdisciplinary approach of integrating social sciences and artificial intelligence. Sustainability (2019). https://doi.org/10.3390/su11174633
11. Centobelli, P., Cerchione, R., Esposito, E., Shashi, S.: The mediating role of knowledge exploration and exploitation for developing an entrepreneurial university. Manag. Decis. **57**, 3301–3320 (2019). https://doi.org/10.1108/MD-11-2018-1240
12. Yun, J.J., Liu, Z.: Micro- and macro-dynamics of open innovation with a quadruple-helix model. Sustainability (2019). https://doi.org/10.3390/su11123301
13. Li, Z., Zhu, G.: Knowledge transfer performance of industry-university-research institute collaboration in china: the moderating effect of partner difference. Sustainability (2021). https://doi.org/10.3390/su132313202
14. Fischer, B., Guerrero, M., Guimón, J., Schaeffer, P.R.: Knowledge transfer for frugal innovation: where do entrepreneurial universities stand? J. Knowl. Manag. **25**, 360–379 (2021). https://doi.org/10.1108/JKM-01-2020-0040
15. Helman, J.: Analysis of the local innovation and entrepreneurial system structure towards the 'wrocław innovation ecosystem' concept development. Sustainability (2020). https://doi.org/10.3390/su122310086
16. González Fernández, S., Kubus, R., Mascareñas Pérez-Iñigo, J.: Innovation ecosystems in the EU: policy evolution and horizon Europe proposal case study (the actors' perspective). Sustainability (2019). https://doi.org/10.3390/su11174735
17. Matt, D.T., Molinaro, M., Orzes, G., Pedrini, G.: The role of innovation ecosystems in Industry 4.0 adoption. J. Manuf. Technol. Manag. **32**, 369–395 (2021). https://doi.org/10.1108/JMTM-04-2021-0119
18. Dubina, I., Campbell, D., Carayannis, E., Chub, A., Grigoroudis, E., Kozhevina, O.: The balanced development of the spatial innovation and entrepreneurial ecosystem based on principles of the systems compromise: a conceptual framework. J. Knowl. Econ. (2017). https://doi.org/10.1007/s13132-016-0426-0
19. Cai, Y., Ahmad, I.: From an entrepreneurial university to a sustainable entrepreneurial university: conceptualisation and evidence in the contexts of european university reforms. High Educ. Pol. (2021). https://doi.org/10.1057/s41307-021-00243-z
20. Portuguez Castro, M., Ross Scheede, C., Gómez Zermeño, M.G.: The impact of higher education on entrepreneurship and the innovation ecosystem: a case study in México. Sustainability (2019). https://doi.org/10.3390/su11205597
21. Kivimaa, P., Boon, W., Antikainen, R.: Commercialising university inventions for sustainability—a case study of (non-)intermediating 'cleantech' at Aalto University. Sci. Public Policy **44**, 631–644 (2017). https://doi.org/10.1093/scipol/scw090
22. Yan, M.R., et al.: Evaluation of technological innovations and the industrial ecosystem of science parks in shanghai: an empirical study. Sci. Technol. Soc. 482–504 (2020). https://doi.org/10.1177/0971721820912906

23. Liu, Z., Stephens, V.: Exploring innovation ecosystem from the perspective of sustainability: towards a conceptual framework. J. Open Innov. Technol. Market Complex. https://doi.org/10.3390/joitmc5030048
24. Tolstykh, T., Gamidullaeva, L., Shmeleva, N., Woźniak, M., Vasin, S.: An assessment of regional sustainability via the maturity level of entrepreneurial ecosystems. J. Open Innov. Technol. Market Complex. (2020). https://doi.org/10.3390/joitmc7010005
25. Tolstykh, T., Shmeleva, N., Gamidullaeva, L.: Evaluation of circular and integration potentials of innovation ecosystems for industrial sustainability. Sustainability (2020). https://doi.org/10.3390/su12114574
26. Bem Machado, A., Secinaro, S., Calandra, D., Lanzalonga, F.: Knowledge management and digital transformation for Industry 4.0: a structured literature review. Knowl. Manag. Res. Pract. (2021). https://doi.org/10.1080/14778238.2021.2015261
27. Carayannis, E., Grigoroudis, E., Stamati, D., Valvi, T.: Social business model innovation: a quadruple/quintuple helix-based social innovation ecosystem. IEEE Trans. Eng. Manag. 1–14 (2019). https://doi.org/10.1109/TEM.2019.2914408
28. Gifford, E., Mckelvey, M., Saemundsson, R.J.: The evolution of knowledge-intensive innovation ecosystems: co-evolving entrepreneurial activity and innovation policy in the West Swedish maritime system. Ind. Innov. 651–676 (2020). https://doi.org/10.1080/13662716.2020.1856047
29. Yıldırım, N., Tunçalp, D.: A policy design framework on the roles of S&T universities in innovation ecosystems: integrating stakeholders voices for industry 4.0. IEEE Trans. Eng. Manag. https://doi.org/10.1109/TEM.2021.3106834
30. Edwards-Schachter, M., García-Granero, A., Sánchez-Barrioluengo, M., Quesada-Pineda, H., Amara, N.: Disentangling competences: interrelationships on creativity, innovation and entrepreneurship. Think. Sk. Creat. **16**, 27–39 (2015). https://doi.org/10.1016/j.tsc.2014.11.006
31. Kleyner, G.B.: Industrial ecosystems: a look into the future. J. Russ. Econ. Revival **3**, 53–62 (2018)
32. Chesbrough, H.: Open Innovation: The New Imperative for Creating and Profiting from Technology. Harvard Business School Press (2003)
33. Lamy, P., et al.: LAB – FAB – APP investing in the european future we want: report of the independent high-level group on maximising the impact of EU research & innovation 46 programmes. Directorate-General for Research and Innovation, European Commission, Brussels (2017)
34. Oliveira, M.M., Andreatta, I.G., Stjepandić, J., Canciglieri Junior, O.: Product lifecycle management and sustainable development in the context of industry 4.0: a systematic literature review. Transdisc. Eng. (2021)
35. Oliveira, M., Leite, B.R., Junior, O.C.: Universities Best Practices in Open Innovation and R&D. Proceedings of ICPR Americas, Editorial da Universidade Nacional del Sur, pp. 1237–1249 (2021)

Understanding Service Design in the Context of Servitization

Ana Maria Kaiser Cardoso, Pedro da Rocha Loures Robell, and Guilherme Brittes Benitez(✉)

Industrial and Systems Engineering Graduate Program, Polytechnic School, Pontifical Catholic University of Parana (PUCPR), Curitiba, Brazil
guilherme.benitez@pucpr.br

Abstract. In recent years, the manufacturing industry has sought to adapt its strategy by including services in its offerings, known as servitization. In this context, service design can be seen as a competitive advantage in the market, as it is considered an agent of transformation and innovation whose objective is to create efficient and effective experiences that meet the needs of users and organizations. However, the intangible nature of services can be a limitation when it comes to visualizing how design, which normally works with tangible aspects, can help in the development of services. Therefore, this article proposes to advance the understanding of service design and its pillars to show and discuss how this topic is currently treated in the servitization literature. Our findings point to a misalignment in service design, identifying a paradoxical understanding of its pillars, which limits its growth in the servitization literature. Finally, we propose questions that could guide future research to deepen the study of service design.

Keywords: Service Design · Servitization · Services

1 Introduction

As customer needs develop, manufacturing companies need to innovate, integrating services into their offerings in order to achieve greater competitive advantages in increasingly globalized markets (Baines et al., 2017; Gebauer, et al., 2011). As a result, service-oriented and product-oriented systems have undergone significant transformations (Ayala et al., 2017). This movement is known as "servitization", first introduced by Vandermerwe and Rada (1988); that is, this movement gives rise to customer-focused business models that consider individual customer behaviours and adapt to meet customer needs personalized needs. However, it is a major challenge to shift from a product-centric perspective to a service-oriented perspective (Bintner et al., 2008; Spring & Araujo, 2009; Baines et al., 2016), as services require capabilities, processes, responsibilities, culture that are different from what is used to deliver products (Story et al., 2016). Thus, as an alternative, we sought a holistic view of all processes and the environment that affect the development of services and their implications for the consumer (Lemon & Verhoef, 2016). Seeking this holistic view, we fall into service design research,

Published by Springer Nature Switzerland AG 2024
C. Danjou et al. (Eds.): PLM 2023, IFIP AICT 701, pp. 191–200, 2024.
https://doi.org/10.1007/978-3-031-62578-7_18

which has been adopted as a strategic and interdisciplinary approach that aims to plan and organize services to create a unique experience; in this way, companies that offer a certain service can benefit from service design. In other words, service design refers to the entire development and delivery process, not just the result (Stickdorn & Schneider, 2010). According to Stickdorn & Schneider (2010), service design is based on five pillars: user centered, co-creative, sequential, evident and holistic. Despite some definitions, service design and its pillars are little explored in the servitization literature, often being portrayed only as "services" or as a step in the service development process (Edvardsson et al., 2000). Thus, it is not clear which strategy or taxonomy the service design fits best, as the subject is approached in different ways and without much detail. Therefore, the aim of this article is to advance our understanding of how service design and its pillars are treated in the servitization literature.

2 Service Design Pillars

For a long time, the design of product was seen only as something related to aesthetics, but it is also related to functionality, usability, enabling competitiveness and sustainability (Baxter, 2018). Design is a creative problem-solving process that considers tangible and intangible values considering the user's needs (Evenson, 2006). Therefore, for Kotler & Rath (1984), using design in organizations is a strategic element, as it promotes cost reduction and innovation and is a company identity process leading customers to satisfaction. However, its management needs methodology and control, as it is a problem-solving activity with a systematic, logical and orderly process (Mozota, 2003). This means that awareness of the importance of design has grown, and many companies are discovering that design can also be used to improve the customer experience and the efficiency of internal processes, but what happens is that most of these studies are based on innovation through product design and hides the view that design can also be used as a strategic tool in offering innovative services (Gloppen, 2009).

Consequently, few studies portray the term service design due to the difficulty of answering the following question: "how to plan the design of something intangible like services?". Therefore, to try to help understand the fundamentals of service design and how it works, Stickdorn and Schneider (2010) raised five pillars on which service design is based. Such pillars are (i) user centered; (ii) co-creative; (iii) sequential; (iv) evident; and (v) holistic.

(i) User centered – Services are not standardizable and occur exactly from the provider's interaction with the user. Therefore, the experience of everyone involved and affected by the service must be considered, and it is necessary to understand the needs of users, their habits, and culture because only then will the development of something that is, in fact, useful be guaranteed.

(ii) Co-creative – It is not enough to place the user at the center of the process; it is necessary to consider all stakeholders. Each has different needs and realities, so this integration guarantees a good value proposition for a service.

(iii) Sequencing – A service is formed by user moments and interactions with the involved touch points. Therefore, each of these moments needs to be very well planned to be connected consistently and coherently in a sequence of interrelated actions.

(iv) Evidencing – Services, in general, are intangible, but in service design, it refers to the feeling that something is happening or has already happened. They can be used to convey tranquility to the user, emotional benefits, and make the experience much more prosperous.

(v) Holistic – Service design needs to keep a holistic look at the whole process; that is, it cannot just focus on the service itself but the overall context. Every environment and element of service must be considered. Physical artifacts, environments, settings, and procedures must be mapped.

3 Methodology

We carried out a literature review to better understand the proposed theme, through which we were able to identify gaps and recent developments in the study area, in addition to establishing the context for future research and formulating hypotheses and research questions. Briefly, we separate the review into two phases. In the first phase, we selected the keywords (service design and servitization) and the bases on which the searches would be carried out - Scopus, Science Direct and Web of Science. Already, in phase two, using search filters, the searches were carried out. We separated the inclusion and exclusion criteria by phases. The searches carried out in the three databases resulted in 1,445 articles using the terms "servitization" and "service design". We applied the primary exclude filter, where duplicate and unconnected articles were eliminated, resulting in 113 articles. To further narrow down these results, in these 85 articles, we applied one more primary inclusion criterion, where only articles that addressed services or servitization in the title were entered, resulting in 93 articles. From these 93 articles, secondary inclusion and exclusion criteria were applied, and articles that did not have access to the full text and that did not address services or servitization were excluded, and to be included, articles needed to address services, servitization, or service design in the abstract. With this filter, we go from 93 articles to 65. Of these 65, the tertiary inclusion and exclusion criteria were applied; that is, articles that talked about tools, processes, and approaches to service design were included, considering the five pillars in the introduction, conclusion, limitations, and contributions, and articles that do not explore the pillars from service design and focused only on servitization and conceptualization were excluded, resulting in 40 articles of which were analyzed in more detail. Using the exclusion criteria, we rejected a total of 1,405 articles.

4 Results and Discussion

Our results indicate that service design has been debated in a shallow and superficial way in the servitization literature, meaning that service design and its pillars are, most of the time, portrayed in an interpretive way. A general conclusion from the servitization literature is that companies need to be user-oriented and not product-oriented (Kowalkowski, Gebauer, Kamp, 2017; Kohtamaki, Hakala, Partanen, Parida & Wincent, 2015; Kindström, & Kowalkowski, 2014). Of the five pillars that shape the service design concept (see Stickdorn & Schneider, 2010), the most discussed in the servitization literature are user centered and co-creative (Fliess & Lexutt, 2019; Morgan, Anokhin & Wincent, 2019) as shown in Table 1.

Table 1. Service design pillars frequency of occurrence

Service Design Pillars related literature						
Authors	Year	User centered	Co-Creative	Sequencing	Evidencing	Holistic
Macdonald et al.	2011		x			
Kindstrom & Kowalkowski	2014	x	x	x		
Rabetino et al.	2015	x	x			
Tunisini & Sebastiani	2015	x	x			
Kohtamäki et al.	2015	x	x			
Baines et al.	2017	x		x		x
Green, Davies, & Ng	2017	x	x			x
Kowalkowski et al.	2017	x	x		x	
Hakanen, Helander, & Valkokari	2017	x	x			x
Story et al.	2017	x	x	x	x	x
Kuijken, Gemser, & Wijnberg	2017	x	x			
Mahut et al.	2017	x				
Batista et al.	2017	x	x			
Salonen, Saglam, & Hacklin	2017	x	x			
Song, W., & Sakao, T	2017	x		x		x
Martinez et al.	2017	x	x	x		
Lenka et al.	2018	x	x			
Oliveira et al.	2018	x	x			
Beltagui & Ahmad	2018	x	x			x
Costa et al.	2018	x	x	x	x	x
Iriarte et al.	2018	x	x	x	x	x

(*continued*)

Table 1. (*continued*)

Service Design Pillars related literature						
Authors	Year	User centered	Co-Creative	Sequencing	Evidencing	Holistic
Fliess & Lexutt	2019	x	x		x	x
Morgan, Anokhin, & Wincent	2019	x	x			
Raddats et al.	2019	x	x			x
Ruiz-Alba et al.	2019		x			
Hazée et al.	2020	x	x			
Kamal et al.	2020	x	x			
Magro & Pinar	2020	x	x			x
Sholihah et al.	2020	x	x		x	x
Jang, Bae, & Kim	2021	x	x			x
Struwe & Slepniov	2021	x	x	x	x	x
Rabetino et al.	2021	x	x	x		
Lievens & Blažević	2021	x	x		x	x
Kim & Lee	2021	x	x	x		x
Holgado & Macchi	2021	x	x	x		
Bigdeli et al.	2021	x				
Feng et al.	2021	x				x
Kreye & Van Donk	2021	x	x			
Frank et al.	2022	x	x			x
Kurpiela & Teuteberg	2022	x	x			

Another general conclusion is that success in servitization is achieved when customers are involved throughout the process; it is important to consider the customers' perspective to understand the situation and experience it throughout the service journey. This will ensure that customer needs and expectations are met satisfactorily and effectively, thereby increasing customer satisfaction and loyalty. (Jang, Bae & Kim, 2021; Alba et al., 2019; Batista et al., 2017;). Given the studies carried out, we found that the servitization literature often portrays service design in customization approaches, requiring a customization and flexibility strategy to meet the individual needs of the customer (Lenka et al., 2018). However, one of our main findings is that we found a transition problem; that is, some pillars are more related to the standardization of the service, while others are related to the customization of the service. While the servitization literature more emphatically discusses the user centered and co-creation pillars for the development of services, associating them with the customization strategy, it also mentions and links two other pillars (sequential and evident) to the standardization practices, thus causing a misalignment of service design concepts in the servitization literature. Consequently, this transition problem led companies to a service paradox problem, where companies should choose between customizing or standardizing services. We then created a framework (Fig. 1) to better visualize this conflict, which shows how service design, and its pillars are understood in the servitization literature, making it clear in which stages customization occurs and in which standardization occurs.

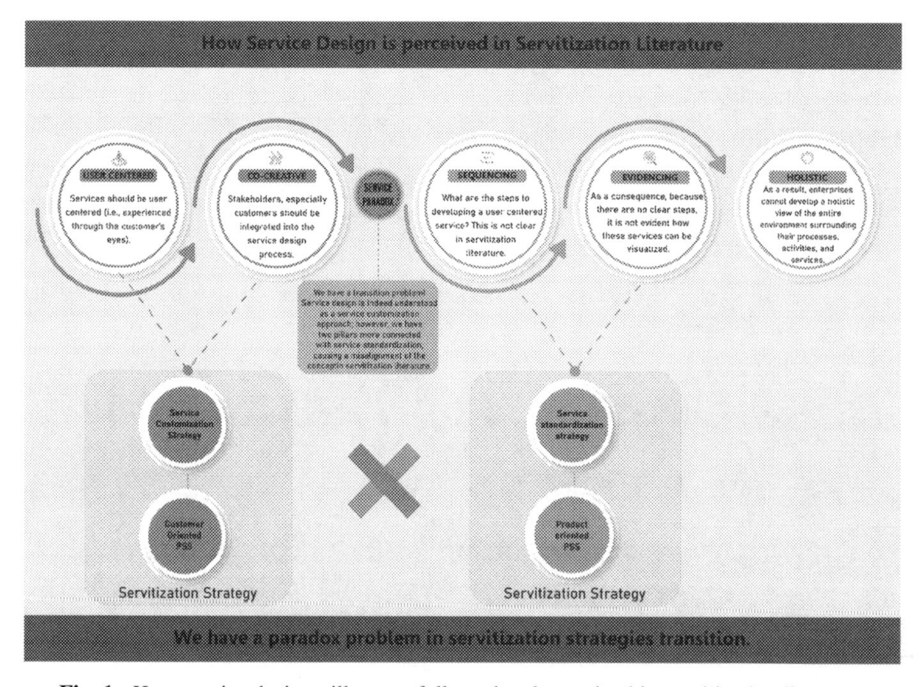

Fig. 1. How service design pillars are followed and perceived in servitization literature

5 Future Research Directions

Based on studies, and insights from quantitative, qualitative and conceptual work, we formulated seven research questions (RQs) about service design, its effects and the use of the five pillars. One of the objectives of this work was to identify how service design and its pillars were perceived in the servitization literature; thus, the questions and answers proposed below will support future research. Those questions were elaborated after an in-depth review of service design pillars in servitization literature.

RQ1: Why the servitization literature frequently portrays the service design approach's last step to be a holistic view of the environment?

Service design has become a challenge for companies that adopt servitization (Story et al., 2017), as it is seen as an area focused on creating experiences that are based on understanding the user's perspective. In other words, the lack of a line of research that guides servitized companies on how to design services makes some past and current authors (for example: Hanington, 2003; Holmlid, 2009; Meroni & Sangiorgi, 2016) state that service design is only an approach that focuses on human needs and concerns and understands that the result of this process is to provide a holistic service to the user. According to Stickdorn & Schneider (2010), having a systemic view is challenging, as considering all aspects of service is complex. However, it is essential for those involved to have this broad perspective of the process in which the service takes place to gain a complete understanding of the system and the different actors involved. Therefore, to achieve this holistic view, it is necessary to consider both the use and the context of the service, including emotional, environmental and behavioural aspects (Leonidou et al., 2018; Loureiro, Romero, & Bilro, 2020).

RQ2: Why does the servitization literature frequently connect only user centered and co-creative pillars in service design approaches?

We believe that these two pillars have a great impact on the development of the service design process, but developing only these two pillars does not guarantee the success of companies. Service design involves an interactive process and an interdisciplinary approach (Giacomin, 2014; Holmlid & Evenson, 2008). Everything needs to be integrated to enable the desired customer experience and orchestrate interactions between different socio-material elements (Pirinen, 2016; Yu & Sangiorgi, 2014).

RQ3: Is service paradox a key limitation that hamper service design approaches and studies?

Many studies have sought to provide an answer for companies to overcome this paradox and become successful service providers (e.g., Gebauer, Krempl, & Fleisch, 2008). However, due to the lack of clarity in the terminology used in service design studies, there is no consensus on the factors, and it is not clear when a service transition can be considered successful. Therefore, we believe that the paradox perspective can be a useful theoretical tool to understand the service design approach. We base this belief on Gebauer, Fleisch, & Friedli (2005), who claim that involving all departments and actors affected by the service transition can contribute to overcoming the service paradox.

RQ4: Why servitization literature relates user centered and co-creative pillars mainly to service customization strategy and sequencing and evidencing pillars to service standardization strategy?

Several authors argue that service personalization creates a deeper relationship, increasing loyalty and allowing a better understanding of users' needs (Cusumano, Kahl, & Suarez, 2008; Frank et al., 2022). However, like any strategy, customization can also come with risks. Therefore, we propose that these two pillars (user centered and co-creative) be used as a proxy for the respective servitization strategy and empirically tested to be validated, as is often mentioned in the servitization literature as pillars of customization.

6 Conclusions

This work sought to understand how service design and its pillars are perceived in the servitization literature and how this strategy can be developed to achieve success in the service journey. Our study contributes to the servitization literature by supporting an approach that is often simplified or little explored by scholars. We conclude that the service design approach is not rigid and immutable; it must be adapted to meet the diverse challenges of companies and their businesses. The issue here is that it must always be based on core values, driven by exploratory research; customer-centric; highly collaborative and creative; detailed and rigorous; focused on finding and creating measurable value; be strategic; designed for multichannel; and capable of simplifying complexity (Stickdorn & Schneider, 2010). One of our main contributions to the servitization literature is the proposal to study service design and its pillars as proxies for standardization, customization, and modularization strategies.

Furthermore, the lack of conceptualization and content in the literature on the pillars of service design is a great opportunity for scholars to develop servitization studies in the coming years. Until now, the service design tools, existing processes and even the types of services companies should offer (Spring & Araujo, 2009) remain uncertain (Sangiorgi & Junginger, 2015). The fact is that, according to bibliographical references, companies that manage to integrate and innovate with service design make their service offering relevant and have a great chance of being successful in their service journey (Gebauer & Friedli, 2005; Iriarte et al., 2018). However, the gap found is how a servitized company develops its service design approach, which can be paramount to success in servitization (Davies, Brady, & Hobday, 2007; Galbraith, 2002). Other contexts, such as Industry 4.0 (Benitez et al., 2019; Moraes et al., 2020), open innovation (Frank et al., 2022), sustainability (Kai et al., 2016; Kai et al., 2017), or even business models (Kai et al., 2016; Baierle et al., 2021) should be investigated in service design approach.

References

Baines, T., Bigdeli, A.Z., Bustinza, O.: Servitization: revisiting the state-of-the-art and research priorities. Int. J. Oper. Prod. Manag. **37**(2), 256–278 (2017)

Gebauer, H., Gustafsson, A., Witell, L.: Competitive advantage through service differentiation by manufacturing companies. J. Bus. Res. **64**(12), 1270–1280 (2011)

Vandermerwe, S., Rada, J.: Servitization of business: adding value by adding services. Eur. Manag. J. **6**(4), 314–324 (1988)

Spring, M., Araujo, L.: Service, services and products: rethinking operations strategy. Int. J. Oper. Prod. Manag. **29**(5), 444–467 (2009)

Bintner, M.J., Ostrom, A.L., Morgan, F.B.: Service blueprinting: a practical technique for service innovation. Calif. Manag. Rev. **50**(3), 66–94 (2008)

Baines, T., Bigdeli, A.Z., Bustinza, O.F., Shi, V.G., Baldwin, J., Ridgway, K.: Servitization: revisiting the state of the art and research priorities. Int. J. Oper. Prod. Manag. **37**(2), 256–278 (2016)

Ayala, N.F., Paslauski, C.A., Ghezzi, A., Frank, A.G.: Knowledge sharing dynamics in service suppliers' involvement for servitization of manufacturing companies. Int. J. Prod. Econ. **193**, 538–553 (2017)

Story, V.M., et al.: Capabilities for advanced services: a multi-actor perspective. Ind. Mark. Manage. **60**, 54–68 (2016)

Lemon, K.N., Verhoef, P.C.: Understanding customer experience throughout the customer journey. J. Mark. **80**(6), 69–96 (2016)

Stickdorn, M., Schneider, J.: This is Service Design Thinking. Amsterdam. BiS Publishers (2010)

Edvardsson, B., et al.: New service development and innovation in the new economy. Studenlitteratur (2000)

Baxter, M.: Product Design. CRC Press, Boca Raton (2018)

Evenson, S.: Theory and Method for Experience Centered Design (2006)

Kotler, P., Rath, G.A.: Design: a powerful but neglected strategic tool. J. Bus. Strateg. **5**(2), 16–21 (1984)

Kohtamäki, M., Hakala, H., Partanen, J., Parida, V., Wincent, J.: The performance impact of industrial services and service orientation on manufacturing companies. J. Serv. Theory Pract. **25**(4), 463–485 (2015)

Mozota, B.B.: Design Management: Using Design to Build Brand Value and Corporate Innovation, 256 p. Allworth Press, Nova Iorque (2003)

Gloppen, J.: Service design leadership. In: Conference Proceedings ServDes. DeThinking Service; ReThinking Design; Oslo Norway 24–26 November 2009, pp. 77–92. Linköping University Electronic Press (2009)

Kowalkowski, C., Gebauer, H., Kamp, B., Parry, G.: Servitization and deservitization: overview, concepts, and definitions. Ind. Mark. Manage. **60**, 4–10 (2017)

Kindström, D., Kowalkowski, C.: Service innovation in product-centric firms: a multidimensional business model perspective. J. Bus. Ind. Mark. **29**(2), 96–111 (2014)

Fliess, S., Lexutt, E.: How to be successful with servitization – guidelines for research and management. Ind. Mark. Manage. **78**, 58–75 (2019)

Morgan, T., Anokhin, S.A., Wincent, J.: New service development by manufacturing firms: effects of customer participation under environmental contingencies. J. Bus. Res. **104**, 497–505 (2019)

Jang, K.K., Bae, J., Kim, K.H.: Servitization experience measurement and the effect of servitization experience on brand resonance and customer retention. J. Bus. Res. **130**, 384–397 (2021)

Ruiz-Alba, J.L., Soares, A., Rodríguez-Molina, M.A., Frías-Jamilena, D.M.: Servitization strategies from customers' perspective: the moderating role of co-creation. J. Bus. Ind. Mark. **34**(3), 628–642 (2019)

Batista, L., Davis-Poynter, S., Ng, I., Maull, R.: Servitization through outcome-based contract – a systems perspective from the defence industry. Int. J. Prod. Econ. **192**, 133–143 (2017)

Lenka, S., Parida, V., Sjödin, D.R., Wincent, J.: Towards a multi-level servitization framework. Int. J. Oper. Prod. Manag. **38**(3), 810–827 (2018)

Gebauer, H., Fleisch, E., Friedli, T.: Overcoming the service paradox in manufacturing companies. Eur. Manag. J. **23**(1), 14–26 (2005)

Iriarte, I., Val Jauregi, E., Justel Lozano, D.: Service design visualization tools for supporting servitization in a machine tool manufacturer. Ind. Mark. Manage. **71**, 189–202 (2018)

Story, V.M., Raddats, C., Burton, J., Zolkiewski, J., Baines, T.: Capabilities for advanced services: a multi-actor perspective. Ind. Mark. Manage. **60**, 54–68 (2017)

Hanington, B.: Methods in the making: a perspective on the state of human research in design. Des. Issues **19**(4), 9–18 (2003)

Holmlid, S.: Participative, co-operative, emancipatory: from participatory design to service design. In: Proceedings of the 1st ServDes, Conference on Service Design and Service Innovation, pp. 105–118 (2009)

Meroni, A., Sangiorgi, D.: Design for Services. Routledge, London (2016)

Leonidou, E., Christofi, M., Vrontis, D., Thrassou, A.: An integrative framework of stakeholder engagement for innovation management and entrepreneurship development. J. Bus. Res. **119**, 245–258 (2018)

Loureiro, S.M.C., Romero, J., Bilro, R.G.: Stakeholder engagement in cocreation processes for innovation: a systematic literature review and case study. J. Bus. Res. **119**, 388–409 (2020)

Giacomin, J.: What is human centred design? Des. J. **17**(4), 606–623 (2014)

Holmlid, S., Evenson, S.: Bringing service design to service sciences, management and engineering. In: Hefley, B., Murphy, W. (eds.) Service Science, Management and Engineering Education for the 21st Century, pp. 341–345. Springer, Boston (2008). https://doi.org/10.1007/978-0-387-76578-5_50

Pirinen, A.: The barriers and enablers of co-design for services. Int. J. Des. **10**(3), 27–42 (2016)

Yu, E., Sangiorgi, D.: Service design as an approach to new service development: reflection and future studies. In: ServDes. Fourth Service Design and Innovation Conference "Service Futures". Lancaster, United Kingdom (2014)

Gebauer, H., Krempl, R., Fleisch, E.: Service development in traditional product manufacturing companies. Eur. J. Innov. Manag. **11**(2), 219–240 (2008)

Cusumano, M.A., Kahl, S.J., Suarez, F.F.: A theory of Services in Product Industries. MIT working paper, MIT Sloan School of Management, vol. 242, pp. 1–47 (2008)

Frank, A.G., de Souza Mendes, G.H., Benitez, G.B., Ayala, N.F.: Service customization in turbulent environments: service business models and knowledge integration to create capability-based switching costs. Ind. Mark. Manag. **100**, 1–18 (2022)

Sangiorgi, D., Junginger, S.: Emerging issues in service design. Des. J. **18**(2), 165–170 (2015)

Davies, A., Brady, T., Hobday, M.: Organizing for solutions: systems seller vs. systems integrator. Ind. Mark. Manag. **36**(2), 183–193 (2007)

Galbraith, J.R.: Organizing to deliver solutions. Organ. Dyn. **31**(2), 194 (2002)

Benitez, G.B., Lima, M.J.D.R.F., Lerman, L.V., Frank, A.G.: Understanding industry 4.0: definitions and insights from a cognitive map analysis. Brazilian J. Oper. Prod. Manag. **16**(2), 192–200 (2019)

Frank, A.G., Benitez, G.B., Ferreira Lima, M., Bernardi, J.A.B.: Effects of open innovation breadth on industrial innovation input–output relationships. Eur. J. Innov. Manag. **25**(4), 975–996 (2022)

Kai, D.A.; Jesus, É.T.D., Pereira, E.A.R., Lima, E.P.D., Tortato, U.: Influence of organisational characteristics in sustainability corporate strategy. Int. J. Agile Syst. Manag. **10**(3-4), 231–249 (2017)

Kai, D.A., de Jesus, É.T., Pereira, E.A., de Lima, E.P., Tortato, U.: Influence of organizational characteristics in the sustainability strategy. In: ISPE TE, pp. 176–185 (2016)

Kai, D.A., de Lima, E.P., Cunico, M.W.M., da Costa, S.G.: Additive manufacturing: a new paradigm for manufacturing. In: Proceedings of the 2016 Industrial and Systems Engineering Research Conference, Availability, Development, vol. 14, p. 102 (2016)

Baierle, I.C., Siluk, J.C.M., Gerhardt, V.J., Michelin, C.D.F., Junior, Á.L.N., Nara, E.O.B.: Worldwide innovation and technology environments: research and future trends involving open innovation. J. Open Innov. Technol. Market Complex. **7**(4), 229 (2021)

Moraes, J.D., Schaefer, J.L., Schreiber, J.N.C., Thomas, J.D., Nara, E.O.B.: Algorithm applied: attracting MSEs to business associations. J. Bus. Ind. Mark. **35**(1), 13–22 (2020)

Modelisation: CAD and Collaboration, Model-Based System Engineering and Building Information Modeling

SE Based Development Framework for Changeable Maritime Systems

Brendan Sullivan(✉) and Monica Rossi

Politecnico di Milano, Via Lambruschini 4B, 20156 Milan, Italy
Brendan.Sullivan@polimi.it

Abstract. Maritime vessels are complex systems that generate and require the utilization of large amounts of data for maximum efficiency. However, integrating different knowledge and data into the decision-making process during the design process remains a challenge. To address this problem, a development framework to support changeability for next generation vessels is proposed using Model Based Systems Engineering (MBSE) and associated SysML diagrams. This framework incorporates literature and industry interviews and can be integrated into the Systems Engineering development approach to improve decision-making through the inclusion of feedback loops. The contribution of this paper is the establishment of a development framework for incorporating change into the design, development, and deployment of next-generation vessels.

Keywords: Systems Engineering · Model Based Systems Engineering · SysML · System Development · Changeability · System Architecture · Extended Value · Maritime

1 Introduction

The maritime industry faces a variety of challenges related to sustainability (changing international environmental regulations), automation, and global economic conditions. So, while pursuing sustainability is important for the long-term health of our planet (reducing the carbon footprint of the industry through decreased reliance on fossil fuels), it often comes at a cost that must be carefully balanced with economic and social considerations. Designing for Changeability (DFC) is one approach that can become essential for meeting technological advances, changing regulations, and shifting customer demands [1–3]. By being agile and adaptable, maritime vessels can take advantage of opportunities to better serve the needs of the customers and stakeholders and thus be more capable of adapting to changing requirements and environments [4, 5]. The ability to meet such events proactively can extend the life of the vessel by delivering value irrespective of when/where changes occur [6, 7].

To address this challenge, this paper proposes a conceptual framework that leverages systems engineering to support the design of changeable maritime vessels. The guiding question for this research is what should be considered during the design process for

Published by Springer Nature Switzerland AG 2024
C. Danjou et al. (Eds.): PLM 2023, IFIP AICT 701, pp. 203–214, 2024.
https://doi.org/10.1007/978-3-031-62578-7_19

changeability to be realized in maritime vessels. The framework incorporates literature and industry interviews and can be integrated into the maritime development process to future proof and extended value of vessels.

1.1 Research Approach

A three-stage methodology was utilized for this study, consisting of a literature review, focus group, and a conceptual case. To identify the most relevant literature, SCOPUS, a database for academic literature, was used to search for literature pertaining to maritime vessel design, systems engineering, and changeability. The literature was collected using keywords (changeability, maritime vessel change, maritime vessel life cycle design, maritime ship design process) and filtered based on their field (Engineering), document type (paper & article), and language (English). Subsequent filtering was performed according to field of interest and the manuscript abstract.

A focus group was conducted with ten individuals (across 5 EU countries) involved in the design, construction, and testing of maritime vessels to identify and define changes faced by maritime vessels. The participants had an average of 6 years of work experience. Semi-structured interviews were conducted to discover precise insights of the actual vessel development process, particularly for aspects related to changeability, techniques to predict lifecycle operational contexts, etc. Each participant was asked to describe the most impactful changes that could be implemented to a current design to extend value according to modified stakeholder expectations. The combined expert opinions and literature findings were used to identify critical design considerations, and strategy's for implementing changeability into maritime vessels.

2 Approaches to Maritime Development

Maritime vessel design is a complex, iterative and multifaceted process, influenced by a number of internal and external factors [1, 8]. Depending on the vision or requirements set forth by the customer, designers are tasked with developing cost-efficient vessels capable of performing specific tasks, while adhering strictly to both international and national rules and regulations.

- **Concept Design** has the greatest impact on all subsequent stages, such as detailed design and construction. The aim is to define the ship's basic characteristics, such as type, deadweight, type of propulsion, and service speed, without requiring detailed calculations to be performed [8–10].
- **Preliminary Design** phase concerns the definition of the ship contract, as well as the completion of the maritime vessel's performance characteristics [9, 10].
- **Basic Design** phase involves a refinement process for the maritime vessel design, including the extension of the initial design to ensure ship performance characteristics, refinement of the general agreement (between the ship owner and the shipyard), basic design of the hull, and arrangement of ship systems (such as propulsion and electrical systems) concluding with a general production plan [9].

- **Detailed Engineering** begins with the creation of detailed material for maritime vessel hull production, as well as material procurement-related activities (such as ordering materials and equipment needed for the ship's construction) [9, 10].
- **Commissioning and Warranty** confirm the functionality of a technical system and obtain operational assurance. Technical assistance is provided during the production and warranty phases when the ship is sold. Feedback is collected during these phases to prevent possible system malfunctions and failures [11].

2.1 Design Spiral

The type of vessel design has a strong influence on the design choices and process, which is undertaken from conception to final customer delivery. This is due to both customer expectations and legal rules/regulations. The most common vessel design process is the spiral design process, which is often used in the shipbuilding industry [9, 12–14]. The 'Ship design spiral' is one of the most commonly used approaches to development and employs a sequential and iterative process [9, 15, 16]. The first step of the spiral design process is to establish requirements, which is a fundamental starting point before entering the concept design phase. This leads to preliminary power estimations, a propulsion system, hull shape, and preliminary cost estimations. Within each phase, solutions become more specific, and options are set, culminating in a design that is ready for authorization. The spiral does not involve exploration of potential solution variants, it relies on point-design making it well suited for the detailed phases but restrictive during the preliminary and conceptual phases.

2.2 Systems Engineering

Systems Engineering (SE) is "an interdisciplinary approach and means to enable the realization of successful systems by defining stakeholder needs, required functions, documenting requirements, then proceeding with design synthesis and system validation while considering the complete problem: operations, performance, testing, manufacturing, cost & schedules, training & support, as well as system disposal [17]." This approach has become increasingly relevant and important with the increasing complexity of modern vessels, as it provides engineers with a structured and formalized way to integrate new design elements for the creation of unprecedented systems. It also strengthens communication in the design process. By taking a systemic development approach, SE facilitates the decomposition of the system, which improves the ability for engineers to analyse technical and non-technical parameters. This effectively allows for requirements to be balanced and analysed in greater detail, which is critical for the development of unprecedented or complex systems. The emphasis on decomposition and analysis enables data-driven decisions to be incorporated into every stage of development and provides a means for testing and validating capabilities.

In the ship design process, systems engineering is a critical component that helps to ensure that the various systems and components of the vessel are integrated and work together seamlessly [2, 18]. By taking a holistic approach to ship design, systems engineering helps to identify potential conflicts and trade-offs between different systems and ensures that the vessel meets the specific needs of its intended use. This approach is particularly important for maritime vessels, which must meet a wide range of requirements, including speed, manoeuvrability, and emissions. Additionally, systems engineering helps to manage the complexity of modern vessels, which have become increasingly reliant on advanced technology. By using systems engineering principles, designers can develop vessels that are not only reliable and efficient, but also easier to operate and maintain. This can reduce costs and improve the overall effectiveness of the vessel. Overall, systems engineering plays a vital role in the design and construction of modern maritime vessels and has become increasingly important as vessels have become more complex and sophisticated.

2.2.1 Model Based Systems Engineering

Model-Based Systems Engineering (MBSE) is a formalized application of modelling that supports system requirements, design, analysis, verification, and validation activities [17]. Its use of digital models to simulate complex systems has revolutionized the way we approach engineering design and optimization.

In the maritime industry, MBSE has emerged as a powerful tool for designing and optimizing complex ship systems. By using modular computer-based design tools and model-based design, MBSE facilitates concurrent design, analysis, and optimization processes, resulting in cost-effective and shorter design cycles. MBSE has been particularly effective in the maritime industry is in the design and optimization of propulsion systems. By creating digital models of the propulsion systems, designers can simulate the behaviour of the system under different conditions and optimize its performance. This has led to significant improvements in fuel efficiency, reducing the environmental impact of shipping. MBSE has proven to be a valuable tool for designing and optimizing complex systems in the maritime industry. Its application has resulted in significant improvements in cost-effectiveness, efficiency, and environmental sustainability. As such, it is an area of research that warrants further exploration and development.

3 Changes Impacting System Value

Change is unavoidable in both reality and perception, as such when the life span of a system is extended the number of changes encountered increases. Consequently, the value of the system is a key consideration when any engineering design decision is made to accommodate or enable a potential change. Changeability seeks to enhance/sustain/maintain the value of a system throughout its lifecycle by either increasing the systems technical performance or reducing the cost of recursive changes that diminish system value. This requires the control and the management of mismatches between system offerings and stakeholder expectations, including responses for dynamic changes (initiated, emergent, or propagated) as shown in Table 1 [7, 25].

Table 1 presents a synthesis of changes identified by the US Defence Logistics Agency, and the Society of American Value Engineers (reviewed 415 implemented changes), as well as adjacent maritime literature on engineering changes to identify types of change that can reduce the value of an engineering system [19–22]. The changes are classified according to the change type, agent/initiator responsible [23];

Table 1. Classification of Value Diminishing Changes

Example of Change Reducing Value	Type	Initiator	Mech
Advances in technology: Incorporation of new materials, components, techniques, or processes (advances in the state-of-the-art) not available at the time of the initial design effort	Initiated Change	Ext. Tech. System	Sub
Excessive cost: Prior design proved technically adequate, but subsequent cost analysis revealed excessive cost	Emergent Change	Int. Tech. System	Sub
Questioning specifications: User's specifications are questioned, determined to be inappropriate, out-of-date, or over specified	Emergent Change	Int. Tech. System	Add. /Rem
Additional design effort: Application of additional skills, ideas, and information available but not utilized during design effort	Emergent Change	Int. Tech. System	Add. /Rem
Change in user's needs: User's modify or redefine of mission, function, or application of item	Propagated Change	Int. Tech. System	Mod
Feedback from test/use: Design modification based on user tests or field experience, parameters governing previous design	Propagated Change	Int. Tech. System	Mod
Design deficiencies: Prior design proved inadequate (e.g., characterized by inadequate performance, excessive failure rates, etc.)	Propagated Change	Int. Tech. System	Mod

activation/realization mechanisms used [24, 25]; Modification (change in components or interface), substitution, and addition/removal.

4 SE Based Changeable Framework for Maritime

The framework takes inspiration from the Win-Win and Theory-W approach to decision outcomes, which emphasizes that for an outcome to be value-creating (positive), everyone in the process needs to be a winner. This people-process centric concept allows the human and technological elements of engineering to encourage communication, integration, and knowledge sharing among its stakeholders to facilitate more effective vessel design. The framework serves as an adjustable complementary element to be built into the concept design phase of maritime vessels to facilitate the integration and consideration of value enhancing change.

This risk mitigating and cost-conscious approach to the integration of DFC increases and extends the value of a vessel. Allowing for change to be value positive, whereby the introduction of change to a vessel leverages the evolutionary nature of ship building and supports future technological adoption and digitalization. This encourages architects and ship designers to apply design rules, standards, and instructions to produce a design that ensures sufficient design margins so that problems of the past will not reoccur, while future problems can be mitigated.

4.1 Articulation Phase

The first step involves determining the scope and boundaries of the system, identifying the stakeholders and their needs, defining the system requirements and functions, evaluating the feasibility of the system design, establishing criteria for success, potential risks and mitigation strategies, strategy to support DFC and necessary design margins. The enables conceptual separation between elements within the system and its environment [25, 27] and can have overlap with Concept of Operations (ConOps) and Operational Concept (OpCons) documentation [26].

- **Usecase Diagram** is used to identify the different actors (users, systems, or external entities) that interact with the system and the different use cases (functions or services) that the vessel provides to each actor. For a changeable vessel, the use case diagram could identify different actors that interact with the vessel, as well as additional factors that are related to the use cases (navigation, communication, cargo handling, and passenger services).
- **Context Diagram**: The context diagram represents the vessel and relationships with external entities such as ports, other vessels, and weather systems as inputs and outputs. The diagram could also show different interfaces between the vessel and the external entities, such as the communication and navigation systems.

  - **Type of voyage** (intended use). Refers to the number of passengers, number of crew members, the number of onboard passenger vehicles, freight mass.
  - The **operating environment** areas in which the vessel will operate throughout its life cycle. Which includes environmental conditions such as: wind speed, wave height, wind force, fetch size, and water depth.
  - **Operating duration** that the vessel will be used for in the expected environment, which corresponds to the voyage and stakeholder requirements that include cruising speed, nautical coverage, fuel autonomy, and energy demand.
  - **Emissions regulations** for the operational context (geographical area + operating duration) to be addressed during vessel's lifecycle.

4.2 Prioritization Phase

Once the system has been identified, the next step is to develop and prioritize a set of requirements that the system must satisfy. The phase supports cognitive and team capabilities ensuring that all persons understand the needs and can effectively describe the value of the vessel in terms of capabilities, performance, function, and costs.

- **Requirements**: Functional, non-functional, technical, and mission requirements for the system to determine levels of correspondence based on clusters and the elements within the system boundary. Design margins and DFC can be utilized to manage uncertainty associated potential changes. Based on the heuristic ranking of elements the impact and bond between element and requirement demonstrate the understood priorities of the system designers [28].

  - **Requirements Diagram:** Is used to capture and organize the different requirements of the system. This can include both functional and non-functional requirements, as well as requirements related to performance, reliability, safety, and other factors.
  - **Sequence Diagram**: Models the interactions between different objects or components in the system. During the requirements process, designers should consider the potential for future changes or upgrades to the system and ensure that the requirements reflect this. This may involve identifying potential areas for modularity or standardization or considering how the system's layout can be made more flexible. This helps to identify the requirements related to system communication and data exchange.
  - **Traceability Matrix:** Is used to link the elicited vessel requirements to other model elements, such as system components, design elements, and test cases. This allows for the tracking of relationships between requirements and elements throughout the design process. The matrix can be leveraged to track changes made to alter the state of the system, including running tests to measure performance, functionality, or other key metrics. If the change has a negative impact, it may reduce the value of the system.

It's important to note that a poorly articulated system can cause the prioritization phase to fail, resulting in an ineffective architecture to the be designed. The stakeholder's involvement in the development process is crucial for the success of the framework. Even if the team assumes the system is well-defined, deviations from critical needs and ilities can cause significant problems. Therefore, the team should compare each relationship against one another to avoid loose understanding.

4.3 Evaluate

The evaluation phase considers the system elements and represents the general architecture of system and its components. This considers the relationships between the different elements/components and how they interact with each other.

- **Coupling within the system**: Coupling describes how closely related/connected elements and components of the system and how much they rely on each other to perform properly (tightly coupled, vs loosely coupled) [30]. Tight coupling is when components are highly dependent on one another while loose coupling is when there is little or no dependency between components. The differences between tight and loose coupling can also be described in terms of coordination and information flow. Within vessel design coupling can make the system more (loosely coupled systems) or less changeable (tightly coupled) due to the potential of propagated change, and difficulty predicting the full impact of changes.

- **Block Definition Diagram**: Enables the definition of the system architecture in terms of blocks (system components) and their relationships. For a changeable maritime vessel, the blocks might include the hull, propulsion system, navigation system, and/or communication system.
- **Internal Block Diagram**: Can be utilized to model the internal structure of each block and how its parts are interconnected. In this context, the model could be used to show the internal components of the propulsion system and how they are connected to the hull.

- **Solution viability and testing**: Determining viability involves assessing whether a solution, despite meeting stakeholder needs, is feasible given the complexity, changeability, and organizational/institutional factors. Equally important is testing to validate that the system can deliver the functions designed according to stakeholders' needs. The evaluation of system viability is based on determining whether a system is suitable for adopting or implementing DFC. Although all systems have the potential to be changeable, not all are well-suited, and not all changes or design solutions provide the most value to stakeholders. To determine suitability the change effect, cost, effort, and life cycle implications (extending/reducing the possibility for additional value enhancing changes).

  - **Activity Diagram**: The flow of activities within the vessel enables for the operations and maintenance actions of the vessel, and to understand how specific changes (parametric diagram) effect the system. This could include activities such as starting and stopping the engines, raising, and lowering the anchor, or adjusting the sails.
  - **State Machine Diagram**: The diagram is used to model the behaviour of the vessel in different states (the events that cause it to transition from one state to another), such as cruising, docking, or manoeuvring.
  - **Sequence Diagram**: Is used to model the interactions between different blocks or systems over time.
  - **Parametric Diagram**: Is used to model the relationships between system parameters, such as inputs, outputs, and constraints, as well as analyse the impact of changes, and the effect of the change on the overall system.

4.4 Reveal

In the concept design phase, the reveal process provides an opportunity for feedback and design review from all stakeholders. The process considers and leverages multiple models previously described to ensure the vessel meets the requirements and needs of the stakeholders. Using the system models developed, the current state of the system can be compared against various changes to demonstrate technical, performance or functional value increases. This model can be used to communicate the changes to stakeholders and get their feedback on whether the changes will provide value. The model can also be used to identify any potential issues or conflicts that may arise as a result of the changes.

- **System:** To ensure that a change made to the system provides value, the models can be used to simulate the behavior of the system before and after the change. This simulation can be used to test the impact of the change on the system and identify any potential issues or unintended consequences. By simulating the behavior of the system, you can ensure that the change will provide the desired value without negatively impacting the system.
- **Effect**: The evaluation of the change (current vs future state) allows for direct comparison of the action and helps to verify if the respective process can handle/manage the change according to the boundaries and considerations established. The updated model is used to analyse the impact of the change on the system. This can involve simulating the behaviour of the system before and after the change to identify any potential issues or unintended consequences.
- **Cost/Effort:** Through the comparison of the models and analysis of the impact of the change, an estimate can be made of the effort required to implement the change. This may include estimating the amount of time required to modify the system design, update documentation, and test the modified system. Once the effort is estimated, the cost can be calculated (labour, materials, equipment).

4.5 Update and Implement

Due to the inherent complexity of maritime systems once the changes are evaluated and compared, the system must return be revaluated using the models previously developed (Sect. 4.1–4.3).

- **Update**: The process for system update, takes the results generated from the reveal and evaluation process to improve the placement and allocation of within the system according to the evaluated elements and relationships. The update phase introduces a series of processed adjustments to the system before passing to the detailed design phase. This includes verification of the system, changeability, testing documents, design, and functionality. It includes activities such as inspection (measurement to verify the system elements conform to its specified requirements), analysis (the use of established technical or mathematical models or simulations, algorithms, or other scientific principles and procedures to provide evidence that the changeable system meets its stated requirements), and demonstration (actual operation of an item to provide evidence that it accomplishes the required functions under specific scenarios).
- **Implement**: The implementation phase serves as an authentication and engagement step according to feedback from the critical design review (reveal). This determines how the change should be incorporated into the system and must be performed in a concurrent manner. Implementation occurs through the following actions: (1) Define how the change will be integrated into the design based on the model diagrams, (2) Ensure the critical tests and outputs support value extension, and (3) Ensure the change in not in conflict with the system functions.

4.6 Execute

The execution phase is enacted according to the knowledge, value, and cost prioritization measures of the system to mitigate risk [32, 33]. Based on the outcomes derived during the concept development phase the requirements, interactions and functions must be reviewed and validated, reducing backlash.

5 Concluding Remarks

The presented conceptual framework cannot be applied without intense collaboration between shipbuilders and customers. This collaboration is crucial for the maritime industry to overcome its biggest challenge: customers sending their technical requirements directly to the company without allowing for a more profound collaboration. This approach leaves the customer less aware of the risks they may face without having a vessel designed for changeability, such as the inability to operate in the future due to emissions rules or the excessive cost of change. On the other hand, the shipbuilder does not have a deeper understanding of the customer's future needs or support in technology forecasting and related change-ability design choices. Therefore, a cultural change is necessary to allow design thinking principles to truly impact the development process, involving both customers and shipbuilders.

This study has contributed to the development of a specialized conceptual framework that can help maritime engineers and architects design changeable maritime vessels. The preliminary outcomes derived have helped validate the framework and have shown how a greater reliance on digital modelling tools could be used to improve the decision-making process.

References

1. Vossen, C., Kleppe, R., Randi, S.: Ship design and system integration. In: DMK 2013, no. September (2013)
2. Ricci, N., Rhodes, D.H., Ross, A.M., Fitzgerald, M.E.: Considering alternative strategies for value sustainment in systems-of-systems. In: SysCon 2013 - 7th Annual IEEE International Systems Conference Proceedings, pp. 725–730 (2013). https://doi.org/10.1109/SysCon.2013.6549963
3. Allaverdi, D., Herberg, A., Lindemann, U.: Lifecycle perspective on uncertainty and value robustness in the offshore drilling industry. In: SysCon 2013 - 7th Annual IEEE International Systems Conference Proceedings, pp. 886–893 (2013). https://doi.org/10.1109/SysCon.2013.6549989
4. Rehn, C.F., et al.: Quantification of changeability level for engineering systems. Syst. Eng. **22**(1), 80–94 (2019). https://doi.org/10.1002/sys.21472
5. Fricke, E., Gebhard, B., Negele, H., Igenbergs, E.: Coping with changes: causes, findings and strategies. Syst. Eng. **3**(4), 169–179 (2000). https://doi.org/10.1002/1520-6858(2000)3:4%3c169::AID-SYS1%3e3.0.CO;2-W
6. Mekdeci, B., Ross, A.M., Rhodes, D.H., Hastings, D.E.: Pliability and viable systems: maintaining value under changing conditions. IEEE Syst. J. **9**(4), 1173–1184 (2015). https://doi.org/10.1109/JSYST.2014.2314316

7. Sullivan, B.P., Rossi, M., Terzi, S.: A review of changeability in complex engineering systems. IFAC-PapersOnLine **51**(11), 1567–1572 (2018). https://doi.org/10.1016/j.ifacol.2018.08.273
8. Misra, S.C.: Design principles of ships and marine structures (2015)
9. Gale, P.: Ship design and construction. In: Lamb, T. (ed.) Ship Design and Construction. SNAME, NJ (2003)
10. Design, S.: Chapter 9 - Ship design, construction and operation. In: Molland, A.F. (ed.) The Maritime Engineering Reference Book, vol. 6, no. 1998, pp. 636–727 (2008)
11. A. Universtiy: Shipyard engineering Lecture 2–1: Design processes and systems (2016)
12. Mistree, F., Smith, W.F., Bras, B.A., Allen, J.K., Muster, D.: Decision-based design: a contemporary paradigm for ship design. In: Society of Naval Architects and Marine Engineers Annual Meeting, vol. seminar, p. seminar (1990). https://doi.org/10.1017/CBO9781107415324.004
13. Mikkelsen, J., Calisal, S.M.: The contribution of industry to ship design education. In: Proceedings of the Canadian Engineering Education Association (CEEA), 2011, no. February (2015). https://doi.org/10.24908/pceea.v0i0.3895
14. Tupper, E.C.: Ship Design, 5th edn. Elsevier Ltd (2013)
15. Vossen, C., Kleppe, R., Hjørungnes, S.R.: Ship Design and System Integration. In: DMK 2013, no. December (2013). https://www.researchgate.net/publication/273026917_Ship_Design_and_System_Integration
16. Harvey, E.J., Evans, J.: Basic design concepts. J. Am. Soc. Nav. Eng. **71**(4), 671–678 (1959). https://doi.org/10.1111/j.1559-3584.1959.tb01836.x
17. INCOSE, Systems engineering handbook: A guide for system life cycle processes and activities (2015)
18. Kossiakoff, A., Sweet, W.N., Seymour, S.J., Biemer, S.M.: Systems Engineering Principles and Practice, vol. 102, no. 3 (2011)
19. Giffin, M., De Weck, O.L., Bounova, G., Keller, R., Eckert, C.M., Clarkson, P.J.: Change propagation analysis in complex technical systems. J. Mech. Des. **131**(8), 0810011–08100114 (2009). https://doi.org/10.1115/1.3149847
20. Deubzer, F., Kreimeyer, M., Lindemann, U.: Exploring strategies in change management - current status and activity benchmark. In: DESIGN 2006, the 9th International Design Conference, Dubrovnik, Croatia, pp. 815–822 (2006)
21. US-DOD, Value Engineering Handbook. Value Engineering : A Guidebook of Best Practices and Tools (1986)
22. US-DOD, Value Engineering : A Guidebook of Best Practices and Tools. Office of Deputy Assistant Secretary of Defense Systems Engineering (2011)
23. Colombo, E.F., Cascini, G., De Weck, O.L.: Classification of change-related ilities based on a literature review of engineering changes. J. Integr. Des. Process. Sci. **20**, 1–21 (2016). https://doi.org/10.3233/jid-2016-0019
24. Ross, A.M.: Defining and Using the New 'ilities' (2008). http://seari.mit.edu/documents/working_papers/SEAri_WP-2008-4-1.pdf
25. Ross, A.M., Rhodes, D.H., Hastings, D.E.: Defining changeability : reconciling flexibility, adaptability, scalability, modifiability, and robustness for maintaining system lifecycle value. Syst. Eng. **11**(3), 246–262 (2008). https://doi.org/10.1002/sys
26. Ward, D., Rossi, M., Sullivan, B.P., Pichika, H.V.: The metamorphosis of systems engineering through the evolution of today's standards. In: 4th IEEE International Symposium Systems Engineering. ISSE 2018 - Proceedings (2018). https://doi.org/10.1109/SysEng.2018.8544426
27. Mekdeci, B.: Managing the impact of change through survivability and pliability to achieve variable systems of systems, no. 2002 (2013)
28. Ross, A.M., Rhodes, D.H., Hastings, D.E.: Using pareto trace to determine system passive value robustness. In: 2009 IEEE International Systems Conference Proceedings, pp. 285–290 (2009). https://doi.org/10.1109/SYSTEMS.2009.4815813

29. Ward, D., Sullivan, B.P., Pichika, H.V., Rossi, M.: Assessment and tailoring of technical processes: A practitioners experience. In: CEUR Workshop Proceedings, vol. 2248 (2018)
30. Valerdi, R., Sullivan, B.P.: Engineering systems integration, testing and validation. In: Handbook of Engineering Systems Design, p. 27. Springer, Cham (2020)
31. Bahill, A.T., Gissing, B.: Re-evaluating systems engineering concepts using systems thinking. IEEE Trans. Syst. Man Cybern. C Appl. Rev. **28**(4), 516–527 (1998). https://doi.org/10.1109/5326.725338
32. Lozano, A., Wermelinger, M., Nuseibeh, B.: Evaluating the relation between changeability decay and the characteristics of clones and methods. In: Aramis 2008 - 1st Int. Work. Autom. Eng. Auton. runtiMe Evol. Syst. ASE2008 23rd IEEE/ACM International Conference on Automated Software Engineering-Workshops, pp. 100–109 (2008). https://doi.org/10.1109/ASEW.2008.4686327
33. Moser, E., Stricker, N., Lanza, G.: Risk efficient migration strategies for global production networks. Procedia CIRP **00**, 104–109 (2016). https://doi.org/10.1016/j.procir.2016.11.019

Interface Modeling for Complex Systems Design: An MBSE and PLM System Integration Perspective

Yana Brovar(✉), Arkadii Kazanskii, Betania Tapia, and Clement Fortin

Skolkovo Institute of Science and Technology, Moscow, Russia
yana.brovar@skoltech.ru

Abstract. A continuous digital thread is not yet fully supported by the product lifecycle tools. A particular discontinuity appears between Model-Based Systems Engineering (MBSE) and Product Lifecycle Management (PLM) solutions and methodologies. Seamless data exchange is an essential element to achieve system components integration, that would support positive emergence and prevent negative ones. A significant integration challenge lies at the heart of MBSE and PLM solutions, since the former facilitates system conceptual design, while the latter focuses on detailed design. However, these two domains are missing common language elements and concepts. We foresee that this could be facilitated through better interface management definition, as one of the most critical integration processes which contains information about both behavior and form, thus, links conceptual and detailed design. In our paper, we propose to use a Design Structure Matrix as a supporting tool to achieve a continuous digital thread, necessary for Digital Engineering solutions. To verify this approach, an aircraft bleed-air system was used as a use-case. We demonstrate that this method can effectively enrich the interface management and control processes through the analysis of metaphors for interface representation proposed by the conceptual design language and product mock-up.

Keywords: PLM · MBSE · Digital Engineering · integration · interfaces · DSM

1 Introduction

During the product development process all systems are divided into subsystems that interact with one another. When the system is functioning and its components interact through interfaces, there is of emergence appearing, which refers to the functions of the product. Emergence is an important aspect during the design process, as it can lead to both desirable properties, for example, a new product with greater functionality and negative ones which can lead to accidents [1]. Thus, through the analysis of interfaces, we can support desirable functions and prevent negative ones. To support proper interface design, we need to ensure smooth exchange of interface design data between each phase of the product development process. However, the product lifecycle currently is

Published by Springer Nature Switzerland AG 2024
C. Danjou et al. (Eds.): PLM 2023, IFIP AICT 701, pp. 215–224, 2024.
https://doi.org/10.1007/978-3-031-62578-7_20

not supported by a continuous digital thread. A particular discontinuity appears between the conceptual design phase and the following stages of the design process. Thus, to accomplish upstream and downstream data flow throughout the product development process, Model-Based Systems Engineering (MBSE) and Product Lifecycle Management (PLM) tools need to be seamlessly integrated to implement Digital Engineering solutions. We thus propose to better represent explicit interfaces throughout the product lifecycle as a possible path to a better integration of these solutions.

1.1 Conceptual Design and Detailed Design

At the beginning of product design, the basic functionality and characteristics need to be defined. This part of the design phase is called the conceptual design, where multiple different product concepts are studied according to the product requirements [2]. Frequently, models are used to support the conceptual design process, and document the product concepts. Models of the product concept which encompass the composition of its architecture, as well as the product's interaction with the environment are at the core of Model-Based Systems Engineering. These models at the conceptual stage can be described with the help of various modeling and domain-specific languages, depending on the specific needs and context of the system being developed. In practice the Systems Modeling Language (SysML) [3] is the most used in industry. It is a general-purpose graphical modeling language with natural language descriptions for defining, analyzing, developing, and validating complex systems. SysML consists of nine diagrams, each focusing on a specific modeling aspect. For instance, the Requirements diagram captures requirements hierarchies and requirements derivation, and "satisfy" and "verify" relationships allow a modeler to relate a requirement to model elements that satisfy or verify the requirements, bridging requirements management tools and the system models. In turn, Product Lifecycle Management solutions are powerful tools for modeling complex products with an emphasis on spatial representations which are also coupled to powerful analysis tools, but lack explicit representations of product functions, which leads to discontinuities of the upstream and downstream data flow with conceptual design. Thus, to develop a continuous digital thread through these phases, MBSE and PLM methods and tools, that are currently applied for conceptual and detailed design phases respectively, need to be seamlessly integrated.

1.2 Context of MBSE and PLM Integration

To achieve effective integration of data between the conceptual and detailed design phases, MBSE and PLM should be properly integrated. Major PLM applications vendors have made very serious efforts to extend their solutions to the early and late phases of product development. However, in practice less than 30% of large-scale Digital Engineering projects which require the full integration of MBSE and PLM, are successful [4], as different product lifecycle phases need models with different purposes, which are not well integrated. Thus, to solve this issue, MBSE and PLM should be considered as complementary approaches rather than trying to reach a single solution for such diverse purposes. The need for a smooth data exchange between MBSE and PLM is a subject that is described a lot in the literature. For instance, a general future outlook on this problem [5] highlighted the need for MBSE and PLM integration around Digital Twins and their various applications. Unfortunately, the integration of MBSE with PLM solutions is still very limited in practice. The fundamental issue associated with the integration of MBSE and PLM is due to the fundamental essence of systems, which needs both explicit representations of time and space [6] to completely represent the system structural connections (that corresponds to form) and flow interconnections (that corresponds to behavior) throughout the product life cycle.

2 Literature Review

According to the NASA Systems Engineering Handbook [7] one of the processes applied to the early part of system design as well as to later stages is the Interface Management process. Interface management helps to ensure that different subsystems or components within a system that need to interact with one another, are compliant and work together effectively to ensure proper product behavior.

2.1 Interface Management

Parslov et al. [8] noted that interface management is a key product development activity for proper system integration. To store the key information during the Interface Management procedure, Interface Control Documents (ICDs) are used. ICDs record all interface information (such as drawings, diagrams, tables, and textual material) developed for a project. ICDs provide specifics and explain the interface or interfaces between subsystems. One of such standards regarding the ICDs, is the Space engineering Interface management normative (ECSS-E-ST-10-24C [9]). It states that the purpose of the ICDs is to define the design of the interfaces ensuring compatibility among involved interface ends by documenting form, fit, and function. However, this normative document contains minimal information on the development and realization of the ICDs. It just indicates what the ICD should contain. For example, ICDs could contain definitions of terms, descriptions of interfaces, product tree, etc. However, this standard does not include a clear description about the process and methodology in general.

2.2 Interfaces and Interface Relationships

As provided in the Expanded Guide for NASA Systems Engineering [10], one key aspect of interface management is the identification and definition of interfaces between different subsystems or components. This involves identifying the types of interactions that will take place between these components, as well as the specific requirements and constraints that must be considered to ensure proper communication and operation. System interfaces hold essential information about the interaction of system components with each other and with the environment. The essence of modeling, according to Zeigler [11], "lies in establishing relations between pairs of system description". According to Prof. Crawley's definition [1]: interfaces should be represented through form and function. There are two general types of relationships: functional and formal

[1]. Functional relationships emphasize the dynamic nature of interactions; in turn, formal relationships denote relationships that are stable for some period of time. To support interaction modeling, several taxonomies were proposed. In current design practice, the most widely-spread taxonomy of interactive systems modeling was proposed by Pimmler and Eppinger [12], where they consider four important types of interaction. The first type of interaction defines the need for adjacency or orientation between two components (structural connection). The other three types are associated with an exchange (flow) of material (or matter), energy, and information. In turn, Crawley et al. extended the four-dimensional taxonomy with an introduction of a 'spatial' type of interface relationship, that "capture absolute or relative location or orientation" [1].

2.3 Metaphors for Interfaces and Interface Relationships

Approaches that denote one kind of entity in place of another to suggest a likeness or analogy between them are defined as metaphors [13]. Metaphors are used to make abstract things more tangible, in order to get the proper level of understanding for the design team members [14]. A study by Jonathan H.G Hey and Alice M. Agonino [15] elaborated on the importance of metaphor use for design itself. They analyzed a variety of engineering design books that led them to discover various perspectives on design that need different metaphors for core design concepts which they defined as ideas, problems, and solutions. Metaphors are also very important for interface design. Referring again to the Expanded Guide for NASA Systems Engineering [10], the interface management procedure involves well defined documentation of the interfaces and analysis of interface compatibility. To realize this, a proper approach for interface modeling is needed. This could be done through formulating metaphors to achieve proper representations of interfaces to help design team members from different disciplines to collaborate and effectively highlight system interfaces. For instance, one the most obvious interface representation metaphor is arrow, which can metaphorically represent pipes, wires etc.

2.4 DSM

Another widely-used approach that proposes interface metaphor is the Design Structure Matrix [16]. It is a useful tool for representing the interactions between different elements, as it is an object-relationship model that sees value in both components and relationships. The notation utilized in DSMs is both clear and intuitive, and it offers as well the benefit of modularity and customization to create Multiple-Domain Matrices [17], enabling adaptation and extension according to the specific design phase or product context. DSM supports adaption to the domain-specific cases, as demonstrated in [18] with the formulation of new types of interface relationships for telecommunication systems. The metaphor to represent the interface consists of the designator (for example, *S* for structural connection, *Sp* for spatial orientation, *E* for energy flow, *M* for material flow, and *I* for information flow), and representation of the direction of the flow depending on the matrix convention. For instance, interactions below the diagonal indicate feed-forward interactions and above the diagonal interactions indicate feedback. Feedback and feed-forward are especially important for time or decisions-based sequences.

3 Methodology

Current tools implemented at the conceptual design phase are more focused on behavioral representation, which means that in terms of interfaces relationships, commonly available, only functional interfaces, such as material, energy and informational flow are modeled, ignoring spatial interposition of components and physical connections. However, these interface relationships are critical to ensure proper flows within the systems. Such lack of attention to these types of interface relationships leads to the gap between MBSE and PLM integration. MBSE tools are more focused on flow representations, while PLM lacks specific flow representation, but can represent spatial and structural interface relationships. To overcome this gap, we propose proper metaphors for all interface relationships types both at the conceptual and the detailed design phases. At the conceptual design phase, we foresee DSM as a tool that supports the interface modeling as well as the analysis of system integration. To support interface relationships representations at the detailed design stage, we propose to use 3D representations as interface metaphors. We foresee such representations as concrete implementation tools to increase the clarity of products and systems to support the reduction of negative emergence. Thus, DSM and 3D metaphors of interface relationships allow a transition between the early phases of the product development process and CAD modeling and analysis.

4 Bleed-Air System Use-Case

To demonstrate the proposed method, an aircraft bleed air system within the aircraft pylon was chosen as a use-case (see Fig. 1).

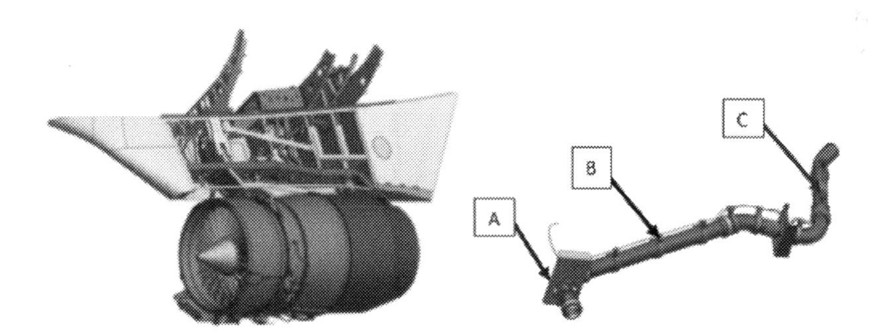

Fig. 1. Aircraft pylon (left hand-side) and Bleed-air system (right hand-side) example

The bleed-air system main role is to transfer a pressurized airflow from the engine compressor to the aircraft fuselage for the heating and cooling of the cabin. One of the most critical emergent properties of the bleed air system is connected to the potential risk of a gas leak, which could lead to a fire. In order to reduce potential risks, leak detection wires are attached to the bleed air system and are used to monitor the temperature at the bleed air pipe wall. Thus, bleed air systems must be designed carefully and must meet a large number of requirements coming from stringent certification rules to prevent the negative emergence of the aircraft system [19].

4.1 Conceptual Design

The conceptual design was conducted in the MagicDraw [20] software that utilizes the SysML modeling language containing numerous diagrams which define the system from different perspectives. For the purpose of the paper, we discuss the Requirements diagram and the Internal Block Diagram, as they offer specific metaphors for the representation of the interfaces. The requirements diagram proposes a combination of a graphical modeling and natural language approaches to represent interfaces; it has a specific entity called "interfaceRequirement" to emphasize some constraints dedicated to the system interface through the formulation of the sentence using a natural language. For instance, in Fig. 2 the main function of the bleed-air system is represented as a performance requirement that is strongly interconnected with positive emergence of the climate control function that could be reached through the proper integration of the main compressor, the bleed-air system and the airframe.

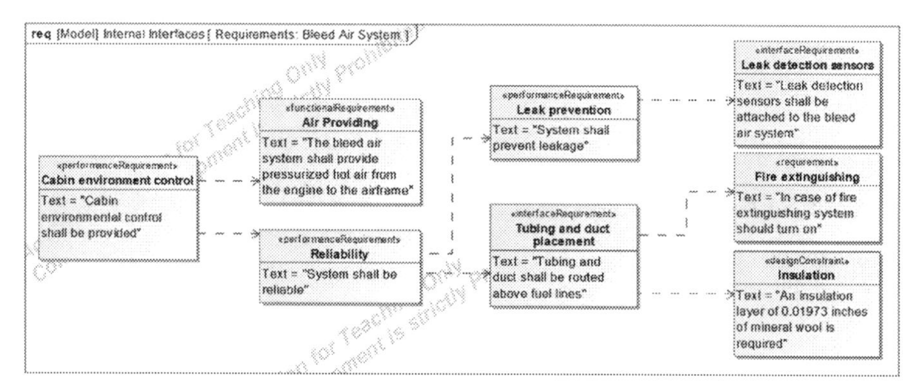

Fig. 2. Requirement diagram for the system of interest

Spatial interfaces are defined through a natural language in the interface requirement of the duct placement, that states that the bleed-air system duct shall be routed above the fuel lines, to avoid a potential fire in the pylon. Thus, in order to prevent a negative emergence of a fire, a structural interface between the bleed-air system and the leak detection sensors is needed to track the thermal flow (energy flow). These interfaces are extremely important, as the error with their design could lead to a high possibility of a system failure. However, the verbal form of the interface representation could lead to a discrepancy, and it is a challenge to manage and analyze textual representations.

The next step of interface design during the conceptual design phase of the Bleed Air System is represented through the SysML Internal Block Diagram. A critical point of the system is demonstrated in a separate diagram for the best understanding of the interfaces between subsystems. As could be seen in Fig. 3, numerous interfaces are represented as several flow modeling constructs. Metaphors for these interfaces are represented as colored lines with directions of flow between subsystems and port identifiers. Also, a connection to the requirement on the leak detection sensors was provided as a rationale for the information flow from "Leak Detection Lines block to Fire Extinguishing System block". In actual SysML representations, only information, energy and material flows are present. The flow metaphors are listed in the legend of Fig. 3. However, it would be necessary to include structural and spatial interface representations in order to connect all of the requirements to the 3D detailed model of the system.

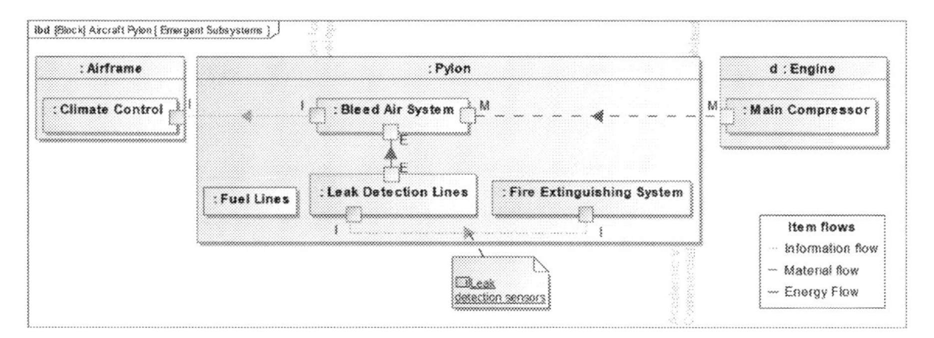

Fig. 3. Internal Block Diagram for the critical elements

4.2 Detailed Design

A solid model representation of the pylon and bleed air system are shown in Fig. 1 (right-hand side). The bleed air system is divided into three segments as shown in Fig. 1. These three segments represent the main components of the bleed air system. Each component is connected through main interfaces. There are many secondary interfaces in the bleed air system which include bellows to consider the expansion of the pipe due to large temperature differences and also connectors which ensures the sealing of the insulated pipe segments. The letters in the box represent the segments of the bleed air system. Segment A has a structural component, which connects it to the firewall. Segment B includes the middle section with bellows at both ends. Segment C includes structural components which connect the system to the aircraft fuselage. The leak detection lines cover all three segments of the bleed air system. This is a critical subsystem, especially as it prevents negative emergence (leakage) that can occur.

4.3 DSM as Integration Tool

As discussed above, functional interfaces at the conceptual design stage are represented through SysML notation with verbal description in the Requirement Diagram and complemented with Internal Block Diagrams, which allows them to represent flows that exist between system components. In turn, structural interfaces are represented through assembly models in 3D models at a detailed design stage. And there is almost no interface data exchange between these two phases. At the same time, spatial interfaces are not explicitly represented within any of these tools, as they are usually just noted in the Requirement Diagram or some specific positioning of components within a Product Mock-up. Thus, none of these metaphors cover Crawley's definition [13] of an interface, as the conceptual phase is focused mostly on the function representation, and detailed design is focused mainly on the representation of the form. We foresee DSM as a proper tool for the interface representation, as it contains system components (as form) and types of interaction at the intersection of columns and rows (as function). Furthermore, the DSM-based interfaces management model is an appropriate tool to reduce complexity, as it permits to simplify the visualization of the system on one page and supports quantitative analysis by adding values at the intersection of corresponding columns and rows.

			1	2	3	4	5	6
Airframe	Climate control system	1	I	S				
Pylon	Bleed air system	2	S, I	M	Sp	S, Sp	Sp	S,M
	Fuel lines	3		Sp	M			
	Leak detection lines	4		S, Sp, E		I	S	
	Fire extinguishing system	5		Sp		S, I	M	
Engine	Main compressor	6		S				M

Fig. 4. Design Structure Matrix of interface relationships for the pylon bleed-air system

The higher interconnectivity between the bleed air system with the airframe and the engine requires accurate management of interfaces to prevent system failures. The bleed air system functional relationships include the flow of energy (*E*), material (*M*) and information (*I*) and represent the dynamic aspects of the system. All of these types of interactions are inherently unidirectional, in that they have a specific flow direction and must be read from column to row in the DSM (Fig. 4). The off-diagonal element of the matrix is marked with *S* when two given components have a structural relationship between each other. The matrix element is marked with *Sp* when the relationship between two components is Spatial. Both structural and spatial relationships are bi-directional. For example, fuel lines have a critical requirement to a spatial relationship to the bleed air system to prevent an overheat. A matrix element is marked with *I*, *M* or *E* when there is an Informational, Material or Energy flow, respectively. Informational flow represents signals from the bleed air system to the leak detection lines that need to check whether there are any leaks from the ducts of the bleed air system to prevent overheating and fire. The flow of energy in this DSM represents the thermal exchange between the bleed-air system and the leak detection sensors, data from which is converted to information about the bleed-air system state. Material flows represent fluid flows that are contained within the fuel lines. So, the DSM contains information that is relevant for the concept development as it contains the information about functional interactions (flows), and at the same time it represents the formal interfaces through the structural and spatial interfaces in the detail design. Thus, the DSM is a link between function to form and vice versa and is thus a powerful integration tool within the Digital Engineering chain.

5 Discussion and Conclusion

Figure 5 shows the proposed interface data flow thread through the product development process with the identification of the tools applied for each phase (green rectangles) of the product development process, already existing metaphor (black font), and metaphors that we propose to add (blue font) to fulfill the gaps between MBSE and PLM models. The process begins with an iterative process of requirements and system structure identification. Interface requirements are commonly represented in a SysML Requirement Diagram through a natural language metaphor using the verbs "shall' or "must". In turn, at the stage of the system structure identification, the system interactions are indicated on the Internal Block Diagram through the SysML ontology elements, such as connectors, ports, item flow etc. However, in this type of metaphor, there is no representation of structural and spatial interfaces which are missing in SysML. This creates an important gap in the interface management chain, which results in a broken digital thread that leads to inefficiencies and errors.

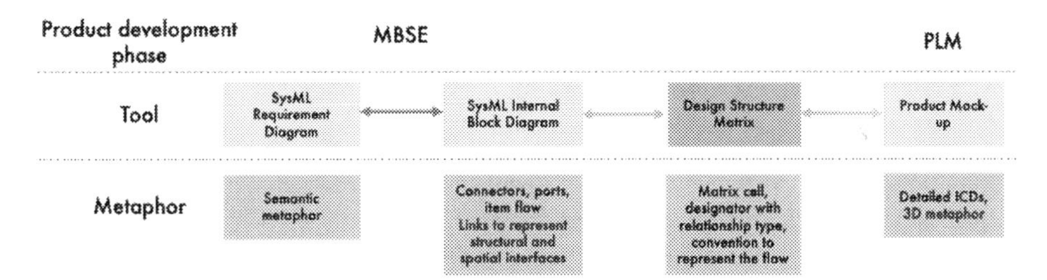

Fig. 5. Representation of interface metaphors interoperability and continuity

Also, interfaces are only implicitly modeled in PLM tools. Currently, there are almost no specific methods to explicitly represent interfaces in the Product Mock-up other than in the 2D drawings of the ICDs. To solve these problems within the digital thread, we propose to complement solid modeling ontologies with three-dimensional metaphors for representing interfaces of various types. In order to ensure a proper level of continuity and interoperability between the early stages of the design process with PLM solutions that could support Digital Engineering solutions, we foresee the Design Structure Matrix (orange rectangle in Fig. 5) as one appearance of mixed interface metaphors, which could integrate information about all types of interfaces in an easy to use and clear representation. Moreover, metaphors offered by the DSM-based approach could be supplemented or replaced with crosses or dots, numbers or the color identification at the intersection of the corresponding row and column. As future work, from a conceptual design perspective there exists a strong need to properly define structural and spatial interface representations, in order to provide a complete set of the interface information and better integrate upstream and downstream processes to ensure a continuous digital thread which will help to prevent negative emergence. As well, it is planned to extend the DSM analysis with the addition of quantitative characteristics to add quantifications to evaluate and analyze both positive and negative emergence. We also emphasize the need for proper 3D interface metaphors to be developed for each interface type as defined in a DSM, in order to increase clarity and properly support the representation of product functionality in a 3D model and analysis. As a way for the validation we plan to provide comparison of proposed approach with the existing methods implemented in industry, for instance, "RFLP" (Requirements engineering, Functional design, Logical design and Physical 3D CAD design).

Acknowledgments. The authors acknowledge the contributions of Eagle Star Aviation that was staffed by students from different Quebec universities affiliated with the "Comité sectorial de Main-d'oeuvre en Aérospatiale" (CAMAQ) and their industrial partners.

References

1. Crawley, E., Cameron, B., Selva, D.: System Architecture: Strategy and Product Development for Complex Systems. Prentice Hall Press, Boston (2016)
2. Horvath, I.: Conceptual design - inside and outside. In: 2nd Seminar and Workshop EDIPro, p. 10 (2000)
3. Object Management Group: OMG Systems Modeling Language: The official OMG SysML site. OMG. See Object Management Group (2008). http://www.omgsysml.org/
4. McKinsey&Company: Unlocking Success in Digital Transformations, Organization, 14 p, October 2018
5. Gerhard, D., Cordero, S.S., Vingerhoeds, R.A., Sullivan, B., Rossi, M., Brovar, Y., et al.: MBSE-PLM integration: initiatives and future outlook. In: Proceedings of 19th IFIP International Conference on Product Lifecycle, pp. 1–10, July 2022
6. Menshenin, Y., Knoll, D., Brovar, Y., Fortin, C.: Analysis of MBSE/PLM integration: from conceptual design to detailed design. In: Nyffenegger, F., Ríos, J., Rivest, L., Bouras, A. (eds.) PLM 2020. IAICT, vol. 594, pp. 593–603. Springer, Cham (2020). https://doi.org/10.1007/978-3-030-62807-9_47
7. NASA. Systems Engineering Handbook (2016)
8. Parslov, J.F.: Defining Interactions and Interfaces in Engineering Design. Technical University of Denmark (2016)
9. Space Engineering Interface Management, ECSS-E-ST-10-24C (2015)
10. Hirshorn, S.R.: Expanded guidance for nasa systems engineering. Volume 1: Systems engineering practices. Technical report (2016)
11. Zeigler, B., Muzy, A., Yilmaz, L.: Artificial intelligence in modeling and simulation. In: Meyers, R. (eds.) Encyclopedia of Complexity and Systems Science. Springer, New York (2009). https://doi.org/10.1007/978-0-387-30440-3_24
12. Pimmler, T.U., Eppinger, S.D.: Integration analysis of product decompositions. In: ASME Design Theory and Methodology Conference (1994)
13. https://www.merriam-webster.com/dictionary/metaphor#:~:text=%3A%20a%20figure%20of%20speech%20in,as%20in%20drowning%20in%20money). Accessed 13 Feb 2023
14. Erickson, T.D.: Working with interface metaphors. Readings in Human–Computer Interaction, pp. 147–151. Morgan Kaufmann (1995)
15. Hey, J.H., Agogino, A.M.: Metaphors In Conceptual Design, 10 June 2016
16. Steward, D.V.: The design structure system: a method for managing the design of complex systems. IEEE Trans. Eng. Manag. (3), 71–74 (1981)
17. Lindemann, U., Maurer, M., Braun, T.: Structural Complexity Management: An Approach for the Field of Product Design. Springer, Heidelberg (2008). https://doi.org/10.1007/978-3-540-87889-6
18. Meskoob, B., Brovar, Y., Menshenin, Y., Blanchard, F., Fortin, C.: Interface management for telecommunicayion system design through OPM and DSM-based approaches. In: The 17th Annual IEEE International Systems Conference (2023)
19. Brovar, Y., Menshenin, Y., Fortin, C.: Study of system interfaces through the notion of complementarity. Proc. Des. Soc. **1**, 2751–2760 (2021). https://doi.org/10.1017/pds.2021.536
20. https://www.magicdraw.com/. Accessed 13 Feb 2023

An Implementation of Integrated Approach in Product Life-Cycle Management Tool to Ensure Requirements-In-Loop During Complex Product Development: A Cubesat Case Study

Mubeen Ur Rehman(✉) and Clement Fortin

Skolkovo Institute of Science and Technology, 121205 Moscow, Russia
{Mubeenur.Rehman,C.Fortin}@skoltech.ru

Abstract. The complexity of innovative products is increasing due to implementation of emerging technologies and rise in system complexity because of greater system and software functionality and inter connectivity, to address the challenge, multiple engineering disciplines must be well-coordinated in product lifecycle management tools. Complexity of novel products have also demanded the establishment of life cycle spanning development from the conceptualization till product realization. The application of MBSE inside PLM tools to keep requirements-in-loop becomes paramount for timely decision support. Unfortunately, keeping a requirements traceability within product development is still a challenge. In aerospace domain, due to innate complexity results in huge number of mission critical requirements which need to be verified and validated to make well informed design decisions throughout the product development process. In this paper, we have proposed a method to implement MBSE using RFLP approach in PLM tool using a Cubesat case study, to keep requirements-in-loop during initial phase of product development.

Keywords: Model Based Systems Engineering · RFLP · Requirements-in-Loop · Product Life-cycle Management · Decision Support

1 Introduction and Problem Description

The purpose of this paper is to see the capability of the modern PLM tools to allow the requirements traceability during the development of complex products especially in the conceptual phase of the product development. As the expectations of the stakeholders is on the rise due to advancements in the emerging technologies, the requirements which are getting stringent and complexer too. Requirement traceability through out the development of complex products is seen to be paramount in the successful development of the product.

Published by Springer Nature Switzerland AG 2024
C. Danjou et al. (Eds.): PLM 2023, IFIP AICT 701, pp. 225–234, 2024.
https://doi.org/10.1007/978-3-031-62578-7_21

Due to the potential for trade offs and conflicts in technical specifications, it is crucial to integrate requirement management tools and methodologies into PLM platforms. This integration can lead to better efficacy and more successful product development, resulting in greater customer satisfaction [1].

The requirement traceability refers to the ability to describe and follow the life of a requirement in a both forward and backward direction [2]. The need for traceability is there because numerous changes are made by product development teams to various subsystems during the system life-cycle.

Altough, optimal usage of PLM tools in a relatively less complex product may lead to successful execution of the project. But especially in space domain, where the products are getting extremely complex needs an integrated approach instead of stand along tools for requirement traceability. This integrated approach can be incorporated by applying MBSE. As described by the [3], that the integration of MBSE and PLM are not aligned. Yet it in its current form it provides solution to some industries especially automotive industry [4] but not to all especially space industry. Due to intrinsic nature of the space domain, still domain specific tools are used which may not be able to fully integrate with current versions of PLM tools.

Application of MBSE in PLM tool can be performed using an approach called RFLP. This approach is called on the four modelling pillars of MBSE [5]. RFLP approach aligns itself well with the modified V-model. In this paper an innovation method especially for space sector is defined to keep requirements-in-loop during the product development cycle. It is done by verifying each requirement during each phase by assigning a numerical parameter to each requirement. The functional, logical and physical simulations are performed with Design of experimentation intend and checked whether the design conforms to the requirements. The product being design is optimized using design of experiments tools which allows to built a web of different configurations and test each of them out to provide decision support [6] during the product developmental cycle. It allows to timely manage changes and reduce the complexity of configuration management all along the way.

The RFLP approach is implemented using Dassault Systemes 3DExperience platform to simulate a single viewpoint of a Cubesat, focusing on the electrical power view. The model is simulated and results are obtained, with issues faced during the exercise discussed in a later section of the paper.

2 Research Methodology

This research follows the design research methodology framework presented by Blessing and Chakrabarti [7]. The prescriptive study is based on the actual implementation of the approach to evaluate how to keep requirements in loop during complex product development using the modified V-model of product development.

According to Kleiner and S Kramer [8] RFLP approach is derived from the V-Model. Original V-model lacks the functional and logical architectures framework between the requirements and the physical 3D aspect of the product. RFLP approach tends to combine them.

The incorporation of RFLP approach in PLM tools also helps to apply MBSE. Although there are alot of integration issues but it provides the basic framework. PLM in automobile sector is particularly very well suited but aerospace products presents different set of issues. This research papers shows the implementation of RFLP approach to a aerospace related product to evaluate the domain specific nature of RFLP along with the integration of MBSE in it.

The requirements are associated with parameters that measure whether the current system configuration satisfies them. A simulation scenario is created to assess the system's performance against these criteria. The RFLP approach consists of four methodological stages, represented by its acronym: requirement specification, functional, logical, and physical. These stages are followed sequentially to ensure that the requirements are effectively incorporated into the system design and development process.

Here we want to discuss the model structuring strategies. Purpose is to understand the usefulness of RFLP for detail design of the product.

The level of abstraction decreases and the applied models in RFLP tends to represent the model systems in more details as the product development progresses. This also depends upon the decisions made upon the information from the more abstracted models. Another model structuring strategy is based on systems views. There can be multiple views of each system. Logical, functional an physical, all are different viewpoints each modeled to represent the system in different form. These are defined in the Fig. 1. RFLP allows to view the system from various perspectives or viewpoints [9]. This approach allows systems engineers to perceive the product in high level of abstraction and then gradually adding more details and refining the model as needed. This allows deeper understanding of the system and its behavior, as well as the identification of potential issues and design trade-offs [10].

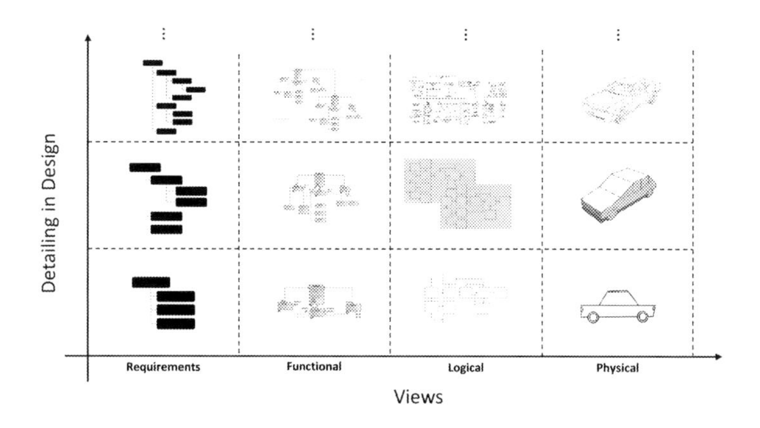

Fig. 1. Level of abstraction in RFLP in the form of viewpoints

Dassault 3DExperience allows the implementation of RFLP approach. The tool in the study is extensively used in the industry and remains the preferred choice to develop complex product portfolio,

3DExperience is a product life-cycle management tool which supports the model based development processes. It uses functional and logical design engineering tools to support RFLP approach. It aims to provide the ability to link requirements with functional and logical architectures to provide consistency. Functional and logical viewpoints are in SysML language and physical aspect of the product developments uses CAD-models.

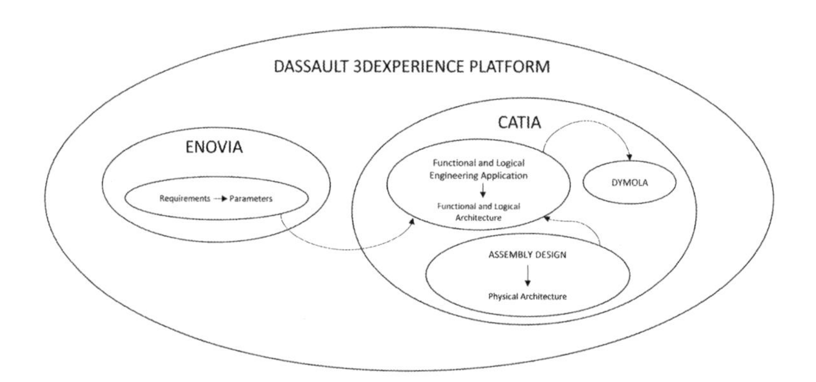

Fig. 2. Applications to support RFLP in 3DExperience platform

The Fig. 2 shows that the 3DExperience platform uses different applications to support RFLP approach. Requirements structure is defined in ENOVIA environment and numerical parameters are appended to the requirements. Functional and logical engineering application in CATIA environment is made to read the requirement parameters which it then uses in behavioral modelling in Dymola. Using Assembly/part design application, physical architectural is connected with functional and logical architectures using implement relations.

3 Case Study: Cubesat

RFLP approach has been used in a case study of Cubesat where the aim is to study the power curves of the onboard battery and optimize the sequence of operations to make sure the batteries does replenish after eclipse. It is particularly important as irradiation is the only source of energy harvesting in space and optimal usage of power is crucial for the success of the mission.

In the study we have investigated the classical subsystems of the Cubesat mission: mission design, systems engineering, propulsion, OBC, ADCS, EPS, thermal, communication and structural sub systems. We did not considered on launch segment since we have concentrated on the spacecraft part.

Concurrent product development lies at the heart of RFLP approach. Therefore a simulation scenario is created to keep track of the requirements during design of experiments. The scenario was built in which the Cubesat after releasing into designated orbit had already detumbled and stabilized. The simulation scenario illustrates the digital test to confirm that the performance requirements have been satisfied.

3.1 Requirements Definition Modelling

Requirements are gathered from stakeholder analysis and integrated into the PLM environment to ensure traceability throughout the product development phase. Numerically measurable requirements are used to verify that the current system configuration satisfies the requirements. To this end, variables such as voltage value and depth of discharge/state-of-charge (SoC) have been introduced into the requirements. The primary energy storage requirement specifies that the battery's SoC should not fall below 60% during mission operation, even under worst-case power requirements. This is crucial to ensure that the battery retains enough energy reserve to power the system during the mission without running out of power.

One way to meet this requirement is to oversize the energy storage capacity to handle the worst-case power requirement scenario without dropping below the 60% SoC threshold. However, this approach is constrained by the system's mass limit, which prohibits the use of a larger battery. Therefore, alternative approaches, such as optimizing the battery management system to maximize the battery's efficiency and ensuring that the system's power requirements are well within the battery's capacity, need to be considered.

To ensure that the energy storage system could meet the requirement, the design process took into account the expected usage patterns and mission duration through simulation.

3.2 Systems Functional and Logical Architecture

The RFLP approach was further implemented to formulate the logical and functional architecture that provides a comprehensive overview of the system, with the requirement specification being an integral part of it. This architecture helps to ensure that the requirements are accounted for throughout the system design process.

The functional architecture of the system was defined using SysML, which captures the system's behavior, structure, and functionality. This approach facilitated the identification and analysis of relationships between various components and interactions among different sub-systems of the Cubesat.

The Cubesat is composed of various subsystems, each with unique functionalities that contribute to the mission objectives. The architectural design of these subsystems is defined based on the SysML standard, which captures their behavior, structure, and functionality. However, SysML alone cannot compute the behavior of the Cubesat. To address the limitation of SysML in computing the behavior of the Cubesat, a Dymola model was created for each logical component within the system. The Dymola model uses numerical methods to simulate and evaluate the behavior of the component. The model compares its computational results against the numerical parameters of the specified requirements to determine if the current configuration satisfies the requirements. This integration of SysML and Dymola enables a comprehensive evaluation of the Cubesat's behavior and helps ensure that the requirements are met.

The RFLP approach relies on the use of implementation functions to enable the flow of information between the requirements defined in ENOVIA and the Dymola model. These implementation functions act as the foundation for the seamless transfer of information between the two systems throughout the entire product development life cycle.

The schematic representation of the Cubesat focuses on the electrical and power aspects of the spacecraft, providing a specific viewpoint for analyzing its behavior. This representation is just one of many possible perspectives, as multiple viewpoints at varying levels of abstraction can be employed to gain a deeper understanding of the Cubesat's behavior.

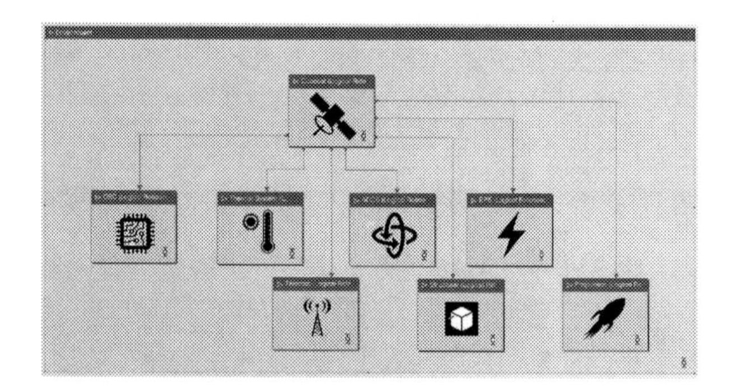

Fig. 3. Logical System Model of the Cubesat

In the schematic shown in Fig. 3, each subsystem of the Cubesat is represented by a Dymola behavior model, enabling simulation of the system. The interfaces between each subsystem are defined to transfer data in a specified manner, allowing for seamless communication between subsystems.

To perform the simulation, a global environment is created with parameters that represent the dynamics of the Low Earth Orbit (LEO). The Cubesat is comprised of seven subsystems, each connected to exchange information and quantities with one another. These subsystems include communication, attitude control, power, thermal control, command and data handling, payload, and structure. Through the exchange of information and quantities, the Dymola behavior models of each subsystem work together to simulate the behavior of the entire Cubesat.

Following a simulation, the parameters obtained are compared to the required parameters to determine if they meet the necessary requirements. If the requirements are not satisfied, the next experiment with different configuration is conducted as the earlier configuration is deemed infeasible. The Design of Experiments (DOE) technique is employed to modify the battery parameters associated with the EPS and power requirement parameters. An experiment matrix consisting of hundreds of potential configurations is created to evaluate and identify the optimal battery parameters that fulfill the desired power requirement. This technique guarantees that all requirements are met and is an effective tool for tracing the requirements.

3.3 Web of Connections

"Web of Connections" refers to the interconnected relationships between various modeling domains, such as requirements, behavior, structure, and verification, to cover all aspects of a system design.

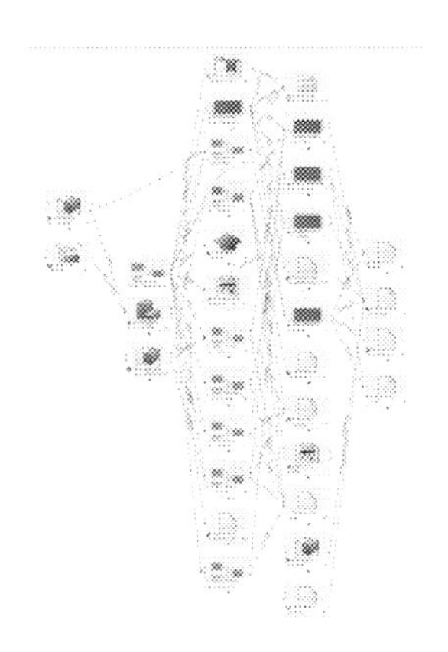

Fig. 4. Connections between different modelling domains

The Fig. 4 depicts the web of connections. It shows how entities in different modelling domains are connected. These connections enable the sharing of data and information between different domains, facilitating the analysis, validation, and verification of system requirements.

3.4 Results

The utilization of DOE to generate a pool of configurations and the integration of a mission planning tool specifically designed to predict the behavior of Cubesats in orbit has facilitated the development of an effective behavioral model for the Cubesat.

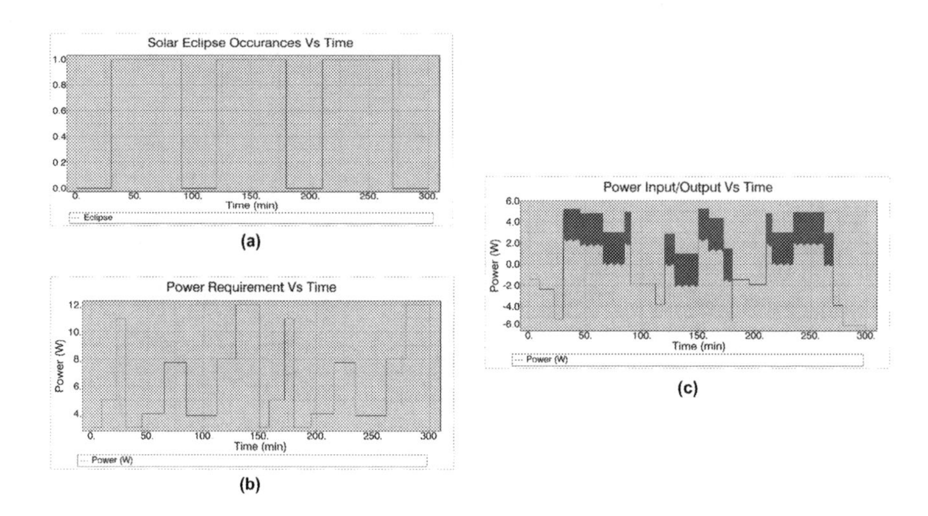

Fig. 5. Solar eclipse occurrences, Power production, Net power

In particular, the behavioral model was supplied with essential mission planning data, such as solar eclipse data shown in Fig. 5(a), and the desired power profile depicted in Fig. 5(b), to calculate the power production from the solar panels. The surplus power, as seen in Fig. 5(c), was used to charge the batteries for use during solar eclipse. A power management and distribution unit manages the power mixture of the Cubesat.

The voltage of a battery is one of the key indicators of its SoC. Generally, the higher the SoC of a battery, the higher its voltage. Conversely, as the SoC decreases, the voltage of the battery will also decrease. Here, the relationships is linear. The Voltage of the battery initially is approx. 8.226V which is about 95% state of charge. In Fig. 6(a) it can be seen that the voltage varies and dips down when the Cubesat is in eclipse as all of the power requirement is fulfilled by the battery. Similarly, when the power is in excess the battery is charged using surplus power. The important information that is visually depicted is that after 300 min the battery has replenished to its initial value. It shows that the Cubesat will have no trouble during its operation. It also verifies that the battery capacity is sufficient and power management unit is working as intended.

One of the key requirements for the battery operation in a particular application is that its SoC shall remain above 60% even during the worst-case scenario with respect to power requirements. In order to ensure compliance with this requirement, it is necessary for the battery to remain charged above the specified threshold at all times during operation. Analysis of the SoC data in Fig. 6(b) reveals that the SoC of the battery remained above 90% for a period of 300 min, which indicates that the margin is well-suited to satisfy the SoC requirement. It is important to note, however, that during the period of operation, both the battery and solar panels will degrade, which may have implications for long-term battery performance. Therefore, it is important to consider the long-term effects

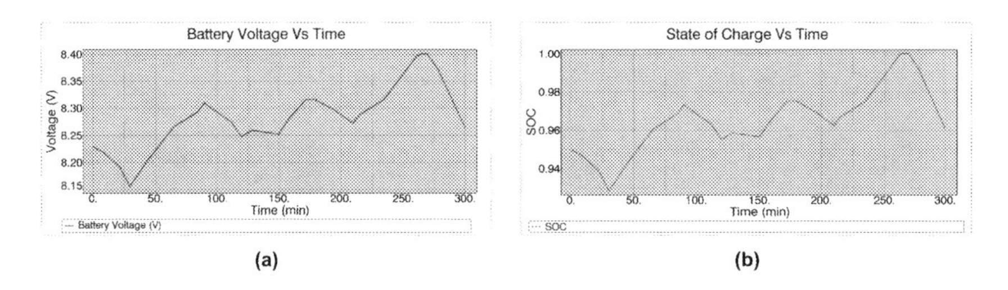

Fig. 6. Graph showing Voltage and State of Charge of Battery

of battery and solar panel degradation when evaluating the suitability of the margin for meeting the SoC requirement.

4 Conclusion

In conclusion, the implementation of the RFLP approach in Dassault Systemes 3DExperience tool has proven to be highly effective in ensuring requirements-in-loop during complex product development. By following the RFLP approach, all requirements are captured and addressed across various stages of development, ensuring that the final product meets all necessary requirements and functions as intended.

This approach has the potential to revolutionize the way complex product development, ensuring that all requirements are accounted for at each level of development, including the functional, logical, and physical levels during the early phase of product development. The success of this implementation highlights the importance of comprehensive requirements management in ensuring the success of a product or system.

Additionally, while the implementation of the RFLP approach in PLM tools is an effective way to ensure requirements-in-loop during complex product development, there are some challenges that need to be addressed. Modelling the behavior of systems in logical architectures can be challenging, and although Dymola is a useful tool, many domain-specific tools are widely used in industries. The closed nature of PLM tools can make it difficult to integrate these domain-specific tools, leading to increased complexity and costs.

To address these challenges, it is important to encourage collaboration and standardization across different tools and systems. Open standards and protocols can help to enable the seamless integration of domain-specific tools into PLM systems, reducing complexity and improving interoperability. Furthermore, the development of more flexible and customizable PLM systems can also help to address these challenges, allowing organizations to tailor their PLM tools to their specific needs and integrate different domain-specific tools.

Overall, while there are challenges associated with implementing the RFLP approach in PLM tools for complex product development, these challenges can be addressed through collaboration, standardization, and the development of more flexible and customizable PLM systems. With these solutions in place, the benefits of the RFLP approach, such as improved requirements traceability, can be fully realized in multiple engineering domains.

References

1. Violante, M.G., Vezzetti, E., Alemanni, M.: An integrated approach to support the Requirement Management (RM) tool customization for a collaborative scenario. Int. J. Interact. Des. Manuf. **11**, 191–204 (2017). https://doi.org/10.1007/s12008-015-0266-3
2. Gotel, O., Finkelstein, A.C.W.: An analysis of the requirements traceability problem. In: Proceedings of IEEE International Conference on Requirements Engineering, pp. 94–101 (1994). https://doi.org/10.1109/ICRE.1994.292398
3. Menshenin, Y., Moreno, C., Brovar, Y., Fortin, C.: Integration of MBSE and PLM: complexity and uncertainty. Int. J. Prod. Lifecycle Manag. **13**, 66–88 (2021). https://doi.org/10.1504/IJPLM.2021.115701
4. Krog, J., Suden, A.T., Schneider, D., Vietor, T.: Definition and use of logical and physical architecture within vehicle concept development. In: IEEE International Symposium on Systems Engineering (ISSE), Vienna, Austria, pp. 1–8 (2022). https://doi.org/10.1109/ISSE54508.2022.10005374
5. Viapiana, D., Riggio, G., Barbieri, L., Bruno, F.: An integrated approach to ensure requirements traceability during the product development process. In: Rizzi, C., Campana, F., Bici, M., Gherardini, F., Ingrassia, T., Cicconi, P. (eds.) ADM 2021. LNME, pp. 328–335. Springer, Cham (2022). https://doi.org/10.1007/978-3-030-91234-5_33
6. Fourgeau, E., Gomez, E., Adli, H., Fernandes, C., Hagege, M.: System engineering workbench for multi-views systems methodology with 3DEXPERIENCE platform. The aircraft RADAR use case. In: Cardin, M.-A., Fong, S.H., Krob, D., Lui, P.C., Tan, Y.H. (eds.) Complex Systems Design & Management Asia. AISC, vol. 426, pp. 269–270. Springer, Cham (2016). https://doi.org/10.1007/978-3-319-29643-2_21
7. Blessing, L.T.M., Chakrabarti , A.: DRM: a design research methodology. In: DRM, a Design Research Methodology. Springer, London (2009). https://doi.org/10.1007/978-1-84882-587-1_2
8. Kleiner, S., Kramer, C.: Model based design with systems engineering based on RFLP using V6. In: Abramovici, M., Stark, R. (eds.) Smart Product Engineering, pp. 93–102. Springer, Heidelberg (2013). https://doi.org/10.1007/978-3-642-30817-8_10
9. Rashmi, J., Anithashree, C., George, E., Robert, C.: Exploring the impact of systems architecture and systems requirements on systems integration complexity. IEEE Syst. J. **2**, 209–223 (2008). https://doi.org/10.1109/JSYST.2008.924130
10. Price, M., Raghunathan, S., Curran, R.: An integrated systems engineering approach to aircraft design. Prog. Aerosp. Sci. **42**(4), 331–376 (2006). https://doi.org/10.1016/j.paerosci.2006.11.002
11. Promyoo, R., Alai, S., El-Mounayri, H.: Innovative digital manufacturing curriculum for industry 4.0. Procedia Manuf. **34**, 1043–1050 (2019). https://doi.org/10.1016/j.promfg.2019.06.092
12. Donoghue, I., Hannola, L., Papinniemi, J., Mikkola, A.: The benefits and impact of digital twins in product development phase of PLM. In: Chiabert, P., Bouras, A., Noël, F., Ríos, J. (eds.) PLM 2018. IAICT, vol. 540, pp. 432–441. Springer, Cham (2018). https://doi.org/10.1007/978-3-030-01614-2_40

Holistic Perspective to the Drug-Device Combination Product Development Challenges

Yaroslav Menshenin(✉), Romain Pinquié, and Pierre Chevrier

Univ. Grenoble Alpes, CNRS, Grenoble INP, G-SCOP, 46 Av. Félix Viallet, 38000 Grenoble, France
yaroslav.menshenin@grenoble-inp.fr

Abstract. The steady growth of life expectancy calls for a new view on the importance of MedTech combination product development. One can envision that to keep a higher quality of life, more people would require some form of therapy depending on their individual state of health. This would require more medical devices of different types and complexity to be designed for specific drugs. However, from a holistic perspective, several challenges emerge in the development of drug-device combination products. An additional layer of complexity appears due to the increasing complexity of the MedTech devices, as they interact with the other systems and products in the design environment. This paper discusses those potential challenges and proposes ways to mitigate them. The primary focus of this work is on the drug delivery systems, such as autoinjectors.

Keywords: Combination Product · MedTech · PLM · Systems Engineering · MBSE · Digital Engineering

1 Introduction

From 1970 to 2021 life expectancy has been showing a steady growth globally: in the US, life expectancy reached 77,2 years in 2021 compared to 70,7 years in 1970 (9,2% increase). In France, it was 82,5 years in 2021 – 16% growth from 1970 when it was 71,2. Considering the other regions, one can observe a similar trend: in Brazil, the life expectancy has increased from 57,2 years (1970) to 72,8 years (2021) – 27% growth; in Nigeria – from 39,7 years (1970) to 52,7 years (2021) – growth by 33%; in China – from 56,6 years (1970) to 78,2 years (2021) – 38% increase [1]. Although such global crises as pandemics could potentially influence those numbers, nevertheless, if this positive trend continues, one can expect a life expectancy of close to 100 years by 2070 in the most developed countries.

Such improvement in the general population's health is not happening without a significant improvement in the healthcare sector itself. The reason for the metrics improvement is not that people became magically healthier, but because of the ability to receive the treatment earlier and deliver the drug to the patient with a specific disease ensuring he or she can receive it throughout their entire lifetime. Pharmaceutical companies are

C. Danjou et al. (Eds.): PLM 2023, IFIP AICT 701, pp. 235–242, 2024.
https://doi.org/10.1007/978-3-031-62578-7_22

primarily responsible for drug development, while MedTech device companies – for the drug delivery systems. In this paper the focus is made on the autoinjectors as one of the types of such systems.

This context is setting an ambitious high-level goal for the Healthcare and MedTech industries, which should act in close cooperation. Reaching such goal requires the integration of the lifecycle processes of how drugs and MedTech devices are developed – both processes meet well established regulatory landscape, managed by the Food and Drug Administration (FDA) in the U.S., the European Medicines Agency (EMA) in Europe, and the other respective regulatory agencies in the other parts of the world.

MedTech product development field deals with uncertainty appearing because of an increasing number of interactions, such as: (1) between components within the system (device) and outside it – with external products and systems; and (2) interactions with external stakeholders, from the MedTech company perspective. For this study, the only second type of interactions is explored with an emphasis on such stakeholders as drug development companies (Pharmaceutical companies), and regulatory bodies (such as FDA or EMA). This paper discusses the challenges associated with the combination product development process considering it from a holistic perspective, combining the systems engineering (SE) and product lifecycle management (PLM) approaches. The Digital Engineering (DE) tool is used to reflect the complexity of the combination product development.

This paper is structured as follows. Section 1 is the Introduction. In Sect. 2 the literature review is primarily focusing on systems engineering and product lifecycle management as the means for a holistic view of the product (Sub-sect. 2.1), and a combination product through the lenses of such a holistic view (Sub-sect. 2.2). The research method is discussed in Sect. 3. Section 4 presents the drug-device combination product development challenges – drug-device design processes alignment (Sub-sect. 4.1) and stakeholders capturing (Sub-sect. 4.2). The discussion and conclusion are made in Sect. 5, where the pathway for future research is also outlined.

2 Literature Review

2.1 Holistic View on the Product Through the Systems Engineering

A holistic and systemic approach to the product/system and its boundaries is needed to establish a proper view of the system from different stakeholders' perspectives. Systems engineering has grown as the discipline to reduce uncertainty and to manage the complexity in very large acquisition programs, such as in aerospace [2, 3] and defense [4, 5] industries. Over time, systems engineering was applied to other industries, such as automotive [6, 7], oil and gaz [8], and healthcare [9]. Therefore, it is not by coincidence the SE/DE methods and tools are applicable to MedTech and healthcare industries.

There were the efforts to use the SE and Model-Based Systems Engineering (MBSE) approaches for MedTech. The attempts to develop good design practices for MedTech development can be traced to two decades ago [10], however digitalization and advanced modeling capabilities could potentially move the design on a new level engaging a more agile [11] and model-based approach. However, systems engineering methods and tools have been used for the development of combination products in a limited way. For example, in [12] the authors have used SysML as an MBSE tool to represent a high-level system architecture for the drug delivery device, constructing the system model for the system itself and only barely mentioning the issues associated with the combination product consideration. SysML has also been applied to risk and safety management of the medical devices [13]. Simulink model implementation to create software for MedTech is presented in [14]. Requirements capture for the medical device development has been presented as the workbook in [15], focusing on the device only; and in [16] as the Master's Thesis. The model-based representation of the dialysis machine has been presented in [17], and to a wide variety of other applications in the healthcare domain – in [18]. From the value delivery perspective, the MedTech device is only creating a benefit to society when it is combined with the drug, and ultimately, when both entities - device and drug - function as a system to deliver a drug to the patient. Therefore, the combination product should be considered holistically, taking into account the context – packaging, instruction for use, etc.

Systems engineering value increases when its principles and methods are applied across the product lifecycle. The integration of systems engineering and Product Lifecycle Management (PLM) has been studied in the literature and applied to different domains, as both approaches are complementary to each other to facilitate the design process across the product lifecycle [19–21] establishing a common glossary [22].

2.2 Combination Product Through the Lenses of Holistic View

A combination product is "a product comprised of two or more regulated components, i.e., drug/device [...] that are physically, chemically, or otherwise combined or mixed and produced as a single entity" [23]. The combination product development involves two or more industries (for example, MedTech company, biotechnology company, and pharmaceutical company) with their own design processes and established procedures, which later should comply with each other and satisfy the regulatory landscape. The opportunity associated with the development of a combination product lies in the ability to combine the best knowledge and practice from all industries involved, revealing a product with greater functionality and, ultimately, becoming a pioneer up the industry standards [24]. Therefore, proper integration of the design processes between all the actors involved in the combination product development is highly needed. Regulatory pathways of combination products development in the USA and in the EU presented in [25] "...could result in a complex scenario for companies marketing the same product in both jurisdictions" [25].

To create new generations of MedTech innovative products, such as drug delivery systems, the product development teams should possess new skills, such as design thinking "to understand user needs" and systems engineering "to manage complexity and ensure interoperability" [26]. Such a setting is calling up for a holistic view of the product under development. The definition of a product and its boundaries vary depending on the perspective: either this is a MedTech device designer, or the Pharma company designer, or the end user, or even within the MedTech company – e.g. regulatory affairs vs. R&D.

The system view is needed to capture the combination product development complexity from different perspectives and to track stakeholders involved. We hypothesize that when the digital design thread is properly managed in DE environment, the links to all stakeholders can be established and the design knowledge can be managed across the lifecycle.

3 Research Method

The first step of the research method is digging into the state-of-the-art of the MedTech combination product development, holistic view on the product, and stakeholders consideration for such a complex product. This is achieved through the literature review of the related topics described in Sect. 2.

The second step of the research method (Sect. 4) is capturing the challenges associated with the device development process and the drug development process, according to the guidance provided by FDA. These challenges are discussed in Sub-sect. 4.1 (misalignment of the drug-device development processes) and in Sub-sect. 4.2 (tracing stakeholders for combination product). To demonstrate these challenges, the MBSE tool is used and applied to the stakeholders consideration.

4 Combination Product Development Challenges

4.1 Drug-Device Design Processes Alignment

Figure 1 presents the FDA-regulated processes for drug development (upper part of the Figure) and device development (lower part of the Figure). In the current practice, those processes are conducted in parallel with some level of uncertainty of drug properties for the MedTech company. For the combination product development, the FDA is only describing the general development considerations pointing out that "…because of the breadth, innovation and complexity of combination products, there is no single developmental paradigm appropriate for all combination products" [27]. The combination product development is also regulated by the CFR Combination Product 21 CFR Part 4 [28]. The challenge is related to the misalignment of the processes for the combination product (drug-device) development which would support MedTech with the new products initiation, conceptualization, and realization.

FDA acknowledges that even though the drug constituent of the combination product and the device constituent of the combination product might be approved separately, "new scientific and technical issues may emerge when the drug and device are combined or used together" [27].

Figure 1 demonstrates that the drug development company might need the device prototype from the device development company after Phase I of the clinical research. It implies that the device company has been working on such prototypes (most likely, a series of Minimum Viable Products MVPs) before this (see device discovery and concept; and preclinical research – prototype phases at the bottom of Fig. 1), which imposes uncertainty for MedTech company on early stages of its device design. Such a work is based on the internal assumptions of drug properties (volume, viscosity, etc.) under development. Figure 1 also captures the iterative nature of the communication between the drug developers and the MedTech device company.

To mitigate the risk of misalignment in the combination product development process, the MedTech design team should have an established communication with the device team at the Pharmaceutical company, shifting and facilitating the discussion on drug properties to the early stages of the product development and advocating that DE/MBSE tools would support this process allowing to traces the design knowledge to a later stages in a systemic way.

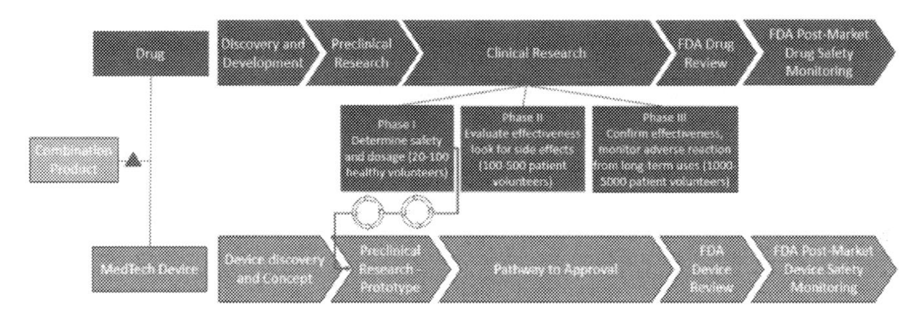

Fig. 1. Combination Product and FDA's drug/device development processes.

4.2 Tracing Stakeholders of the Combination Product

Another core challenge is reasoning about the stakeholders of the combination product. As discussed earlier, the combination product is comprised of the device developed by the MedTech company, and the drug, for which the Pharmaceutical company is responsible. Therefore, from the MedTech device company perspective, the Pharma partner is one of the stakeholders, alongside the patients, healthcare practitioners, and regulatory bodies. However, the primary customer for MedTech device company is Pharma. Healthcare practitioner and patients (final users of both – drug and device) are operating the Combination Product. This makes it difficult for the MedTech device company to directly negotiate with the final user. Rather, from the business-to-business perspective, in practice it is working directly with the Pharma partners trying to investigate their requirements, such as drug viscosity, to be able to reduce uncertainty and predict the design space for the prototypes to be developed.

This complexity is reflected in Fig. 2, which uses the DE/MBSE tool to capture stakeholders of the combination product development. The Object-Process Diagram (OPD) for the Combination Product is built in the OPCloud environment [29], following the Object-Process Methodology (OPM) [30]. This Figure reflects the complexity for the MedTech company: although it can negotiate with final users, its primary customers are the Pharma partners.

Figure 3 is the Object-Process Language for the combination product in the OPCloud environment, which is automatically generated from OPD and supports the stakeholders-processes allocation in a natural language, clearly describing who is responsible for what, and which processes are undertaken.

To mitigate the risk of losing an especially important link to specific stakeholder, the DE tools should be used, such as the one presented in Fig. 2.

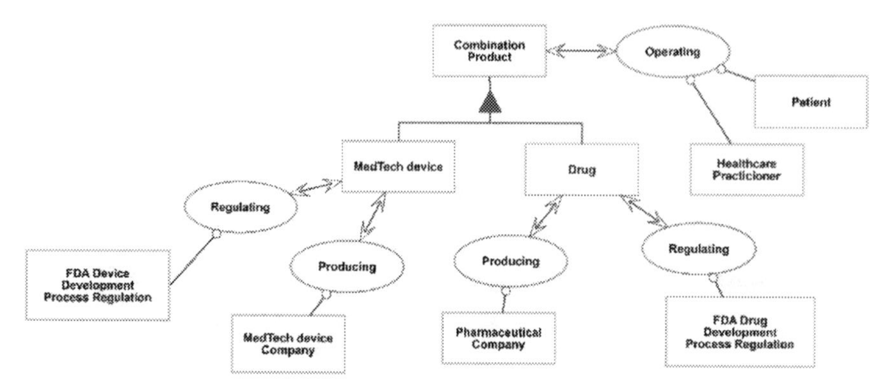

Fig. 2. Object-Process Diagram (OPD) for the Combination Product in OPCloud

MedTech device is an informatical and systemic object.
Drug is an informatical and systemic object.
Combination Product is an informatical and systemic object.
MedTech device Company is an informatical and systemic object.
Pharmaceutical Company is an informatical and systemic object.
Patient is an informatical and systemic object.
Healthcare Practicioner is an informatical and systemic object.
FDA Device Development Process Regulation is an informatical and systemic object.
FDA Drug Development Process Regulation is an informatical and systemic object.
Combination Product consists of Drug and MedTech device.
Operating is an informatical and systemic process.
Operating requires Healthcare Practicioner and Patient.
Operating affects Combination Product.
Producing is an informatical and systemic process.
Producing requires Pharmaceutical Company.
Producing affects Drug.
Regulating is an informatical and systemic process.
Regulating requires FDA Drug Development Process Regulation.
Regulating affects Drug.
Producing is an informatical and systemic process.
Producing requires MedTech device Company.
Producing affects MedTech device.
Regulating is an informatical and systemic process.
Regulating requires FDA Device Development Process Regulation.
Regulating affects MedTech device.

Fig. 3. Object-Process Language (OPL) for the Combination Product in OPCloud

5 Discussion and Conclusion

This paper discusses the challenges associated with development of drug-device combination products, focusing on the drug delivery systems as an example. These challenges are identified in two areas. The first is related to the misalignment of the drug-device design processes. A potential mitigation of this risk for MedTech design team is to establish a communication link with the device team in the Pharmaceutical company focusing on the importance of capturing the core design information (such as drug properties) at the early stages of product development, and advocating that MBSE/DE would support this process. For this, a larger study is required. Such a study should involve a large group of design team members to be interviewed on the design process. These design interviews should capture not only the R&D team, responsible for the MVPs development but also broader representatives - marketing, regulatory, and quality, to name a few.

The second challenge is related to the complexity of stakeholders representation. To mitigate the risk of losing a core stakeholder and the data from such stakeholder, the DE tool should be used. For the drug delivery system developer, the primary customer is the Pharma company, while the final user is the patient/healthcare practitioner. To capture this properly, MBSE-based solutions could track that information. This future work would require building the system model within the product boundary (decomposing the drug delivery system on its own subsystems, functions definition and requirements management, and assigning the core team members to those subsystems); and outside the system model – outlining the interfaces with the other stakeholders and systems.

References

1. Our World in Data Homepage. https://ourworldindata.org/life-expectancy. Accessed 23 Jan 2023
2. NASA: System Engineering Handbook, SP-2016-61 (2016)
3. Fortescue, P., Swinerd, G., Stark, J.: Spacecraft Systems Engineering. Wiley, Hoboken (2011)
4. Dahmann, J.S., Baldwin, K.J.: Understanding the current state of US defense systems of systems and the implications for systems engineering. In: 2nd Annual IEEE Systems Conference, pp. 1–7 (2008). https://doi.org/10.1109/SYSTEMS.2008.4518994
5. Piaszczyk, C.: Model based systems engineering with department of defense architectural framework. Syst. Eng. **14**(3), 305–326 (2011)
6. D'Ambrosio, J., Soremekun, G.: Systems engineering challenges and MBSE opportunities for automotive system design. In: 2017 IEEE International Conference on Systems, Man, and Cybernetics (SMC), pp. 2075–2080 (2017)
7. Loureiro, G., Leaney, P.G., Hodgson, M.: A systems engineering framework for integrated automotive development. Syst. Eng. **7**(2), 153–166 (2004)
8. Golkar, A.A., Keller, R., Robinson, B., de Weck, O., Crawley, E.F.: A methodology for system architecting of offshore oil production systems. In: DSM 2009: Proceedings of the 11th International DSM Conference, pp. 343–356 (2009)
9. Griffin, P.M., Nembhard, H.B., DeFlitch, C.J., Bastian, N.D., Kang, H., Muñoz, D.A.: Healthcare Systems Engineering. Wiley, New Jersey (2016)
10. Alexander, K., Clarkson, P.J., Bishop, D., Fox, S.: Good Design Practice for Medical Devices and Equipment: A Framework. University of Cambridge, Institute for Manufacturing (2001)
11. Glazkova, N., Menshenin, Y., Vasilev, D., Fortin, C.: MedTech product development framework for post-pandemic era. Proc. Des. Soc. **2**, 1273–1282 (2022)
12. Corns, S., Gibson, C.: A model-based reference architecture for medical device development. In: INCOSE International Symposium, vol. 22, no. 1, pp. 2066–2075 (2012)

13. Malins, R.J., Stein, J., Thukral, A., Waterplas, C.: SysML activity models for applying ISO 14971 medical device risk and safety management across the system lifecycle. In: INCOSE International Symposium, vol. 25, no. 1, pp. 489–507 (2015)
14. Hoadley, D.: Using model-based design in an IEC 62304-compliant software development process. In: MBEES, pp. 129–131 (2010)
15. Ward, J., Shefelbine, S., Clarkson, P.J.: Requirements capture for medical device design. In: Proceedings of ICED 2003, the 14th International Conference on Engineering Design, pp. 65–66 (2003)
16. Wang, H.: Multi-level requirement model and its implementation for medical device. Master's thesis. Purdue University (2018)
17. INCOSE: The Medical Device Digital Engineering Thread. https://www.incose.org/docs/default-source/working-groups/healthcare/public-library/2018-se-in-healthcare-conference/the-medical-device-digital-engineering-thread.pdf?sfvrsn=e90897c6_2. Accessed 13 Feb 2023
18. Zwemer, D., Intercax, L.L.C.: Technote: Applications of MBE for healthcare. Intercax LLC, pp.1–15 (2016)
19. Messaadia, M., Farouk, B., Eynard, B.: Systems Engineering and PLM as an integrated approach for industry collaboration management. IFAC Proc. Vol. **45**(6), 1135–1140 (2012)
20. Sharon, A., de Weck, O.L., Dori, D.: Improving project-product lifecycle management with model-based design structure matrix: a joint project management and systems engineering approach. Syst. Eng. **16**(4), 413–426 (2013)
21. Gerhard, D., et al.: MBSE-PLM integration: initiatives and future outlook. In: Noël, F., Nyffenegger, F., Rivest, L., Bouras, A. (eds.) PLM 2022. IFIP, vol. 667, pp. 1–7. Springer, Cham (2022). https://doi.org/10.1007/978-3-031-25182-5_17
22. Pinquié, R., Rivest, L., Segonds, F., Véron, P.: An illustrated glossary of ambiguous PLM terms used in discrete manufacturing. Int. J. Prod. Lifecycle Manag. **8**(2), 142–171 (2015)
23. FDA Homepage: Combination Product Definition Combination Product Types (2018). https://www.fda.gov/combination-products/about-combination-products/combination-product-definition-combination-product-types. Accessed 09 Jan 2023
24. Couto, D.S., Perez-Breva, L., Saraiva, P., Cooney, C.L.: Lessons from innovation in drug-device combination products. Adv. Drug Deliv. Rev. **64**(1), 69–77 (2012). https://doi.org/10.1016/j.addr.2011.10.008
25. Rocco, P., Musazzi, U.M., Minghetti, P.: Medicinal products meet medical devices: classification and nomenclature issues arising from their combined use. Drug Discov. Today (2022). https://doi.org/10.1016/j.drudis.2022.07.009
26. Kühler, T.C., et al.: Development and regulation of connected combined products: reflections from the Medtech & Pharma Platform Association. Clin. Therapeut., 768–782 (2022). https://doi.org/10.1016/j.clinthera.2022.03.009
27. US Food and Drug Administration, Guidance for industry and FDA staff: early development considerations for innovative combination products (2006). https://www.fda.gov/media/75273/download. Accessed 01 Feb 2023
28. Code of Federal Regulations Title 21 Chapter 1 Part 4, Regulation of Combination Products (2023). https://www.ecfr.gov/current/title-21/chapter-I/subchapter-A/part-4?toc=1. Accessed 08 Feb 2023
29. Dori, D., et al.: OPCloud: an OPM integrated conceptual-executable modeling environment for industry 4.0. In: Systems Engineering in the Fourth Industrial Revolution, pp. 243–271 (2019)
30. Dori, Dov: Object-Process Methodology: A Holistic System Paradigm. Springer, Heidelberg (2002). https://doi.org/10.1007/978-3-642-56209-9

Knowledge-Based Engineering Design Supported by a Digital Twin Platform

Sthefan Berwanger[1,2](✉), Henrique Diogo Silva[1,2](✉), António Lucas Soares[1,2](✉), and Cristiano Coutinho[3](✉)

[1] INESC TEC, Porto, Portugal
{sthefan.berwanger,henrique.d.silva,antonio.l.soares}@inesctec.pt
[2] University of Porto, Porto, Portugal
[3] EFACEC, Porto, Portugal
cristiano.coutinho@efacec.com

Abstract. Data generated throughout the product development lifecycle is often unused to its full potential, particularly for improving the engineering design process. Although Knowledge-Based Engineering (KBE) approaches are not new, the Digital Twin (DT) concept is giving new momentum to it, fostering the availability of lifecycle data with the potential to be transformed into new design knowledge. This approach creates an opportunity to research how digital infrastructures and new knowledge-based processes can be articulated to implement more effective KBE approaches. This paper describes how combining a DT-based Digital Platform (DP) with new engineering design processes can improve Knowledge Management (KM) in product design. A case study of a company in the energy sector highlights the challenges and benefits of this approach.

Keywords: Knowledge-Based Engineering · Digital Platform · Digital Twin · Industry 4.0 · Digital Transformation

1 Introduction

The Industry 4.0 paradigm has promoted the intelligent networking of products and processes along the value chain, enabling more efficient organizational processes that result in innovative customer goods and services. [2,4]. From the technology point of view, Industry 4.0 encompasses developing and integrating emergent and consolidated digital technologies such as the Internet of Things, Cyber-Physical Systems, Artificial Intelligence (AI), Virtual/Augmented Reality, and Cloud Computing.

Within this technological landscape, DT is a concept that builds upon the Industry 4.0 core technologies to be a comprehensive digital representation of a physical system, continuously updated by the data exchange between the counterparts. The DT concept and its technological implementation are closely related to product design and engineering [25]. In a broad approach, recent conceptualizations of the DT of a product encompass the entire product lifecycle

C. Danjou et al. (Eds.): PLM 2023, IFIP AICT 701, pp. 243–252, 2024.
https://doi.org/10.1007/978-3-031-62578-7_23

data and its integration into a single architecture, fostering the optimization of Product Lifecycle Management (PLM) from design to continuous diagnosis and performance analysis, back to design again.

With data analytical methods becoming widespread for organizations to optimize and integrate manufacturing and business processes, the early and continuous use of KBE methods and solutions is required [6]. The effectiveness of KBE in building Computer-Aided Design (CAD) models [16,17] aligned with the developments of semantic Web technologies have allowed KBE to evolve as complementary to CAD systems [5,14], with authors pointing to data organization and sharing standards such as Resource Description Framework and Web Ontology Language to play a vital role in the future of KBE and KBS. Moreover, the joint and articulated addition of DT technologies with KBE and Knowledge-Based System (KBS) can positively impact product development and the organization's recognition as a leader, enhancing its reputation in the market [13].

This exploratory research aims to describe an organization's challenges in a digital transformation process and the general requirements for developing KBE processes supported by an innovative DT-based DP. The research question guiding the case study is "What are the benefits and problems of implementing a KBE strategy supported by a DT-based DP?". The results emphasize the need for early and continuous use of KBE methods and solutions to enable complete optimization and integration of manufacturing and business processes, with the articulated addition of DT technologies with KBE and KBS positively impacting product development.

The remainder of this paper is structured as follows. Section 2 presents a state-of-the-art overview of related works on KBE and DT. Section 3 describes our view of how a DP provides the core services and the needed interfaces to enable knowledge-based services. In Sect. 4, the research methodology and instruments used to collect data are described, and the research findings concerning engineering processes and optimization proposals are presented. Finally, Sect. 5 concludes the paper and highlights future research.

2 Related Work

2.1 Knowledge-Based Engineering

The KBE concept can be synthesized by using KM strategies, tools, and systems to support product design [15]. The goals are to identify relevant knowledge in an organization and to define how to capture, formalize, represent, organize, and store it for better access and sharing. In addition, the ultimate goal is to reduce product development time and costs by automating repetitive and non-creative design tasks and supporting multidisciplinary design optimization in all phases of the design process. The current and most recent notions of KBE relate to the objective of KM in capturing and transmitting knowledge to increase organizational performance. Individual and organizational performance can be improved by preserving and utilizing the assets' present and future knowledge value, including human and automated activities [8]. In sum, KBE is the application of KM to manufacturing design and production. Product design and engineering are inherently knowledge-intensive, requiring structures and processes enabling data and Information Management (IM) to facilitate knowledge extraction and sharing among stakeholders.

2.2 Digital Twin

The most recent concept of DT emerged from [24], defining the DT as a virtual, dynamic model in the virtual world that is entirely consistent with its corresponding physical entity in the real world and can simulate the characteristics of the physical counterpart, behavior, life and performance in a timely fashion. Moreover, [19] defined DT as virtual models of physical objects created digitally to simulate their behaviors in real-world environments. In conclusion, the DT is the digital counterpart of the physical product that aggregates real-time and historical data and information about the product. Furthermore, historical data in this context means that the whole lifecycle of the product is covered. As a step forward in the evolution of the DT concept, [10] builds on previous work that combines DT technologies with semantic models and Data Analytics (DA) tools to suggest a formal description of Cognitive Digital Twin (CDT) as a DT with increased semantic capabilities for detecting the dynamics of virtual model evolution, facilitating the understanding of virtual model inter-relationships, and improving decision-making. These insights can fuel more intelligent product and service behaviors and further develop smart digital product-services systems that can transform customer-supplier relationships and introduce new value propositions [11]. Transformation and integration with business processes become critical to leverage these technologies fully, and KBE provides an approach to this.

2.3 The Role of Digital Twin in KBE

The DT technology is a powerful tool for product development engineers in KBE. The basic DT allows the creation of a digital model of a complex product, simulating and testing several aspects of their designs before they are built, enabling efficient and effective experimental learning. Besides, this reduces the time and cost involved in physical prototyping and allows engineers to identify and address potential problems early in the design process. Advanced DT-centered technologies can further support collaboration and knowledge sharing among engineers and other stakeholders by providing a platform for sharing product data and information, e.g., structural or behavioral design models; data sets from the product operation, or insights from operational data analysis. Such a DT facilitates communication and collaboration across different teams and disciplines, leading to more effective problem-solving and innovation. According to [13], the DT adoption coupled with KBE could lead to partners and the market recognizing organizations as leaders, increasing their reputation. Model-centric design, comprehensive PLM, rapid production of diverse design solutions, and a faster response through a faster knowledge flow were the required skills in this dynamic. In addition, KBE and DT approaches also can help increase knowledge reuse during the engineering phase, which can positively impact the business profits [3].

3 KBE Process Transformation Through DP as Infrastructure

The proliferation of digitalization of business processes has led to the omnipresence of DP in providing digital products and services. The Information Systems (IS) field has long studied DP as digital software systems that act as both innovation and transaction platforms [12,20] by serving as a technological foundation and as a market intermediary [7,9]. In general, a DT can support engineering design processes by: (i) supporting collaboration and coordination of product and process design; (ii) organizing, classifying, and delivering several product design models; (iii) managing the access to data and documents from previous design processes integrated with historical data; and (iv) automating the workflow of some design processes.

With these goals in mind, the DP provides the core services and the needed interfaces to enable knowledge-based services. The DP core provides three main interfaces: (i) user interfaces that allow for the interaction of end-users with all the platform-provided services; (ii) data interfaces for the managing of data; (iii) DT interfaces to establish the bidirectional connections between the platform and the multiple instances of DTs.

By its software nature and modular design, the DP can provide custom sets of foundational tools that, arranged into more complex services, fit the needs of companies and directly interface and be part of business processes. [14] depict a simple KBS as a system capable of receiving inputs, such as customer specifications or product data and processes, and transforming them into outputs, such as CAD models, drawings, processes, or reports, by leveraging product model data from previous configurations, geometries, and engineering knowledge, and external data coming from catalogs of materials descriptions.

Drawing a parallel, a DT can be interpreted as a KBE system: (i) where platform core receives inputs through user interfaces; (ii) which, through the DT interfaces, is capable of accessing historical data from the multiple instances of the DT; (iii) that can leverage data interfaces powered by semantic technologies to interface with multiple external data IS; and (iv) that can interface directly with the organization's business process to deliver the required outputs.

Figure 1 builds on the three KM challenges of knowledge representation, acquisition, and knowledge update [1,23]. These three pillars allow the DP to serve, e.g., as an infrastructure for enterprise customized Machine Learning (ML), DA, and other AI-based services that leverage the existing knowledge to support product design processes better.

Regarding the data architecture, the DT uses ontologies to describe and annotate the domain of knowledge and the rules that constrain it while also leveraging semantic tools, such as knowledge graphs, to semantically link the digital models and allow for the integration of DT models. ML techniques are also crucial for uncovering hidden patterns in the stored knowledge [23]. The several activities and legacy systems that comprise the enterprise's current business processes form the data and information sources that the DP must be able to ingest and manage throughout the entire product's lifecycle. The continuous knowledge update is also a central component of the DP as a KBS. So, this aspect must be part of business processes and strategies.

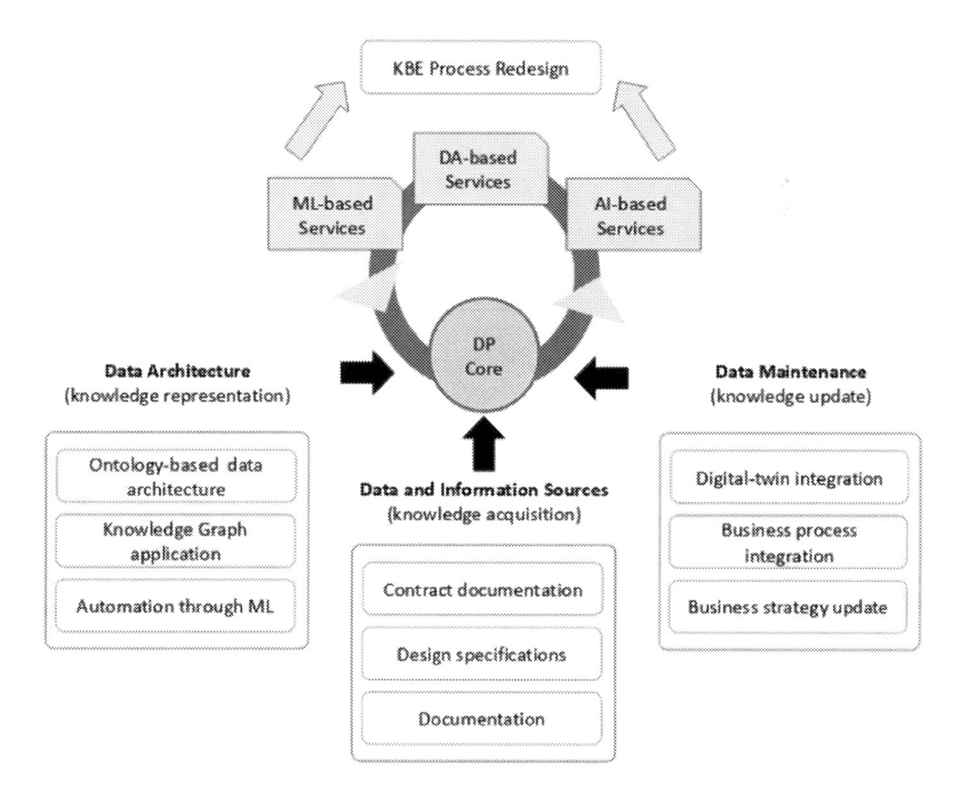

Fig. 1. Interactions between organizational elements and the DP architecture elements.

4 Case Study

4.1 Methodology

The studied company operates in the energy, mobility, and engineering sectors in over 65 countries, whose growth strategy in the international market has been based on the presentation of innovative and quality products. In this case, the product is a Power Transformer (PT), a complex electrical device used in power grids. The relationships between the departments developing the PT are configured as a grid in a matrix structure. It is a sort of organizational administration in which individuals with equivalent skills are pooled for work duties, resulting in more than one manager.

Due to the characteristics of the research, the case study research method was employed to study the context of the engineering process at the company. [21] defined case study research as an empirical inquiry that analyses contemporary phenomena in its real-life setting when the boundaries between phenomenon and context are not readily visible and when numerous data sources are utilized.

The following approach was employed for this work, considering the nature and objectives of the research: (i) establish a research method based on the case study approach; (ii) collect and analyze existing internal documentation; (iii) collect data using focus groups and semi-structured interviews; (iv) analyze the results from the focus group, interviews, and internal documentation; (v) develop improvement proposals for engineering processes and PT design.

4.2 Results and Discussion

This study addresses the research question, "What are the benefits and problems of implementing a KBE strategy supported by a DT-based DP?" As an outcome, we identify and describe the areas where the work contributes.

Current KBE Processes. In the PT design phase, it is an acknowledged requirement to optimize the engineering process by implementing the integration of the IS so that data can be retrieved, updated, and stored at specific flow points and milestones to speed up the completion and validation of the requirements, among other documents. In the same way that elements are stored and recovered, the reuse process is hindered due to the dispersion of elements across legacy systems and the lack of any KM strategy. Furthermore, knowledge is not shared correctly among all teams, especially with new members.

The following instruments were considered the most important for promoting and managing tacit knowledge: mentoring programs to encourage senior workers to train juniors through classes, lectures, hands-on activities, and recording from meetings. Furthermore, most customers the organization consulted avoid sharing the data because they consider it sensitive, reserved, and confidential.

The organization's current, complete engineering processes are complex due to the number of activities, dependencies, and stakeholders involved in different stages, from the start of the design of the PT until its delivery and operation at the customer site. For this reason, specific knowledge-intensive tasks were selected according to their business priority to serve as intervention points.

Mapping KBE Methodologies. KBE requires a KM approach considering various organizational and technical elements, including the industry context, human resources, technological resources, and organizational culture. According to [22], this approach balances an organization's knowledge, resources, and skills with the expertise needed to sell superior products and services to competitors.

The CommonKADS was chosen for implementing KBE supported by the DT platform. Its organization model entails integrating relevant elements and experiences from various sources, including organization theory, business process analysis, and IM, into a coherent and comprehensive package targeted at knowledge orientation in the organization [18]. From this unified view, a detailed organization model was developed to identify challenges and opportunities for a knowledge system implementation and their viability.

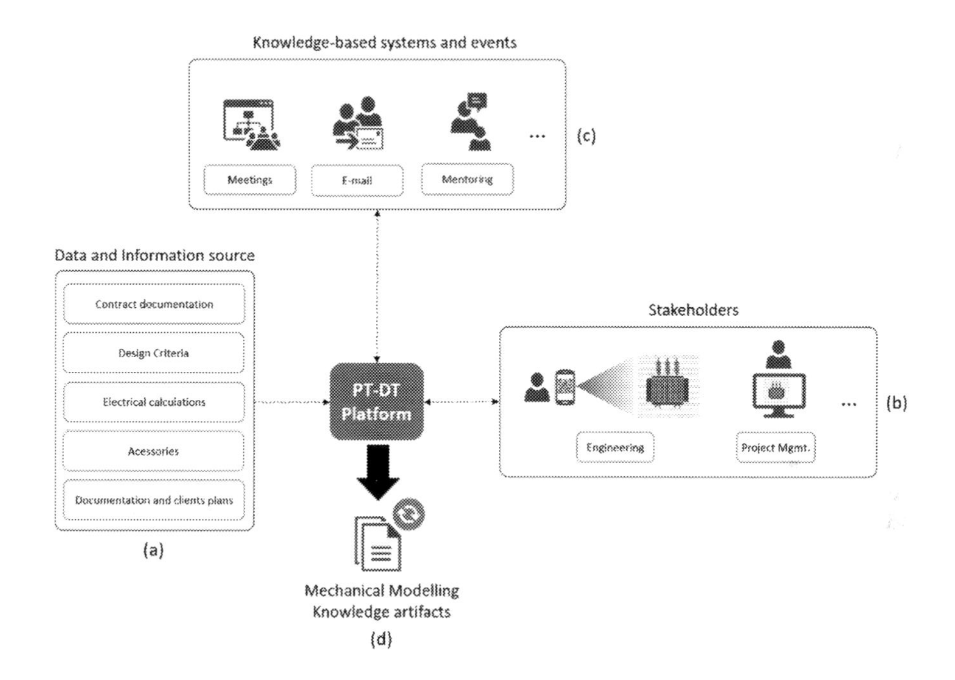

Fig. 2. Redesigned process for mechanical modeling.

The mechanical modeling process is an example of the encountered challenges. It is constantly changing, and a more centralized scenario paired with managing a large amount of information and communication networks is thought to be the goal. New collaborators have a long learning curve regarding the mechanical process. During the training phase, they read the tools' manuals and usually learn by observation and communication with more experienced team members. Much information must be considered, so less experienced designers will have less complex projects allocated to them. Considering these challenges, this task was chosen to be optimized. As presented in Fig. 2, the redesigned mechanical modeling provides input and storage of data and design information from (a) into the DT. This data is already normalized, cleaned, and organized.

Initial data can be queried in (b) by stakeholders via regular computers or augmented reality techniques. The information may not only be retrieved but edited, deleted, and enriched by team collaboration, with versioning control capability offered by the platform. This data and information can also be read and analyzed in meetings and subsequently enriched with the tacit knowledge generated. In the workplace, this tacit knowledge refers to processes and techniques learned through practical experience and training that are later aggregated into the project knowledge artifacts. Finally, the initial and later enriched data and information are stored in knowledge artifacts as in (d), e.g., as media files by the DT, and periodically updated as new information is added, deleted, or edited.

5 Conclusion

The analyzed company faces many challenges during the different phases of the PT lifecycle. In this exploratory research, we argue that a DP managing the DTs of the PTs can assume the role of KBS for centralizing knowledge-sharing practices and infrastructure to provide customized tools and services that can be integrated into current business processes, enabling innovation in products and services.

The study addresses the research question: "What are the benefits and problems of implementing a KBE strategy supported by a DT-based DP?".

It is shown that DP as a KBS, through a KBE approach, can support the organization in this process by (i) during the beginning-of-life stage, integrating teams more closely in the development of complex projects to support information flows and collaboration; (ii) for the middle-of-life, provide ML, DA, and AI-powered services for faster and more precise forecasting, such as failure prediction, to foster mode data-driven decision-making processes; and (iii) in the end-of-life, contributes to the long-term reinforcement of its positioning in favor of sustainability while continuing the knowledge updating process that can greatly benefit future design iterations.

Regarding methods of KBE, interactions with the company stakeholders show that the respondents viewed implementing mentoring programs and recording knowledge from meetings through retrievable means. These tools can help teams acquire new knowledge built on pre-existing knowledge, benefiting the product at any stage of its lifecycle. Furthermore, by improving KBE processes, there is an opportunity to improve the culture of actively sharing information with colleagues.

Finally, data sovereignty concerns were also raised, with most customers refusing to share data of the in-use PT, considering it sensitive and confidential. For customers to be willing to share this data, necessary for complete lifecycle optimization, it would be necessary to establish a data disclosure agreement, security requirements, and rigorous data management policies.

While this paper highlights the importance of KM strategies in supporting the success of the KBE approach, this study has several limitations. Future research should focus on exploring methodologies for implementing improved

KBE, and the selected criteria for implementing the chosen methodology. Additionally, the technical developments of the vision detailed here can serve as an empirical basis for the CDT literature, as few successful implementation cases of CDT can still be found.

Acknowledgments. The project TRF4p0-Transformer4.0 leading to this work is co-financed by the ERDF, through COMPETE-POCI and by the Foundation for Science and Technology under the MIT Portugal Program under POCI-01-0247-FEDER-045926. The second author was additionally funded by the Ph.D. Grant UI/BD/152565/2022 from the Portuguese funding agency, FCT-Fundação para a Ciência e a Tecnologia.

References

1. Abburu, S., Berre, A.J., Jacoby, M., Roman, D., Stojanovic, L., Stojanovic, N.: Cognitive digital twins for the process industry. In: Proceedings of the Twelfth International Conference on Advanced Cognitive Technologies and Applications (COGNITIVE 2020), Nice, France, pp. 25–29 (2020)
2. Anderl, R.: Industrie 4.0-advanced engineering of smart products and smart production. In: Proceedings of International Seminar on High Technology, vol. 19 (2014)
3. Azevedo, M.: Knowledge-based engineering supported by the digital twin: the case of the power transformer at Efacec. Master's thesis, Faculty of Engineering of the University of Porto (2020)
4. Bartevyan, L.: Industry 4.0 - Summary report (2015)
5. Cooper, D., LaRocca, G.: Knowledge-based techniques for developing engineering applications in the 21st century. In: 7th AIAA ATIO Conference 2nd CEIAT International Conference on Innovation and Integration in Aero Sciences, 17th LTA Systems Technology Conference; Followed by 2nd TEOS Forum. Aviation Technology, Integration, and Operations (ATIO) Conferences, American Institute of Aeronautics and Astronautics, September 2007. https://doi.org/10.2514/6.2007-7711
6. Curran, R., Verhagen, W.J., Van Tooren, M.J., Van Der Laan, T.H.: A multidisciplinary implementation methodology for knowledge based engineering: KNOMAD. Expert Syst. Appl. **37**, 7336–7350 (2010). https://doi.org/10.1016/j.eswa.2010.04.027
7. Gawer, A., Cusumano, M.A.: Platform Leadership: How Intel, Microsoft, and Cisco Drive Industry Innovation, vol. 5. Harvard Business School Press, Boston (2002)
8. Girard, J., Girard, J.: Defining knowledge management: toward an applied compendium. Online J. Appl. Knowl. Manag. **3**(1), 1–20 (2015)
9. Hein, A., et al.: Digital platform ecosystems. Electron. Mark. **30**(1), 87–98 (2020). https://doi.org/10.1007/s12525-019-00377-4
10. Lu, J., Zheng, X., Gharaei, A., Kalaboukas, K., Kiritsis, D.: Cognitive twins for supporting decision-makings of internet of things systems. In: Wang, L., Majstorovic, V.D., Mourtzis, D., Carpanzano, E., Moroni, G., Galantucci, L.M. (eds.) Proceedings of 5th International Conference on the Industry 4.0 Model for Advanced Manufacturing. LNME, pp. 105–115. Springer, Cham (2020). https://doi.org/10.1007/978-3-030-46212-3_7

11. Pagoropoulos, A., Maier, A., McAloone, T.C.: Assessing transformational change from institutionalising digital capabilities on implementation and development of Product-Service Systems: Learnings from the maritime industry. J. Clean. Prod. **166**, 369–380 (2017). https://doi.org/10.1016/j.jclepro.2017.08.019
12. Pauli, T., Fielt, E., Matzner, M.: Digital Industrial Platforms. Bus. Inf. Syst. Eng. **63**(2), 181–190 (2021). https://doi.org/10.1007/s12599-020-00681-w
13. Rebentisch, E., Rhodes, D.H., Soares, A.L., Zimmerman, R., Tavares, S.: The digital twin as an enabler of digital transformation: a sociotechnical perspective. In: 2021 IEEE 19th International Conference on Industrial Informatics (INDIN), pp. 1–6. Institute of Electrical and Electronics Engineers (IEEE), October 2021. https://doi.org/10.1109/indin45523.2021.9557455
14. Reddy, E.J., Sridhar, C.N.V., Rangadu, V.P.: Knowledge based engineering: notion, approaches and future trends. Am. J. Intell. Syst. **5**(1), 1–17 (2015). https://doi.org/10.5923/j.ajis.20150501.01
15. Rocca, G.L.: Knowledge based engineering: between AI and CAD. Review of a language based technology to support engineering design. Adv. Eng. Inform. **26**(2), 159–179 (2012). https://doi.org/10.1016/j.aei.2012.02.002
16. Rosenfeld, L.W.: Solid modeling and knowledge-based engineering. In: Handbook of Solid Modeling, pp. 91–911. McGraw-Hill, Inc., USA, June 1995
17. Sandberg, M., Boart, P., Larsson, T.: Functional product life-cycle simulation model for cost estimation in conceptual design of jet engine components. Concurr. Eng. **13**(4), 331–342 (2005). https://doi.org/10.1177/1063293X05060136
18. Schreiber, G., et al.: Knowledge Engineering and Management. The MIT Press (1999). https://doi.org/10.7551/mitpress/4073.001.0001
19. Tao, F., et al.: Digital twin and its potential application exploration. Jisuanji Jicheng Zhizao Xitong/Comput. Integr. Manuf. Syst. CIMS **24**(1), 1–18 (2018). https://doi.org/10.13196/j.cims.2018.01.001
20. Tiwana, A.: Evolutionary competition in platform ecosystems. Inf. Syst. Res., 266–281 (2015). https://doi.org/10.1287/isre.2015.0573
21. Yin, R.K.: Case Study Research: Design and Methods. SAGE Publications Inc, Thousand Oaks (2008)
22. Zack, M.H.: Developing a knowledge strategy. Calif. Manage. Rev. **41**(3), 125–145 (1999). https://doi.org/10.2307/41166000
23. Zheng, X., Lu, J., Kiritsis, D.: The emergence of cognitive digital twin: vision, challenges and opportunities. Int. J. Prod. Res. (2021). https://doi.org/10.1080/00207543.2021.2014591
24. Zhuang, C., Liu, J., Xiong, H.: Digital twin-based smart production management and control framework for the complex product assembly shop-floor. Int. J. Adv. Manuf. Technol. **96**(1–4), 1149–1163 (2018). https://doi.org/10.1007/s00170-018-1617-6
25. Zhuang, C., Liu, J., Xiong, H., Ding, X., Liu, S., Wen, G.: Connotation, architecture and trends of product digital twin. Comput. Integr. Manuf. Syst **23**(4), 53–768 (2017). https://doi.org/10.13196/j.cims.2017.04.010

Industrialization of Site Operations Planning and Management: A BIM-Based Decision Support System

Félix Blampain[1,2](✉), Matthieu Bricogne[1], Benoît Eynard[1], Céline Bricogne[2], and Sébastien Pinon[2]

[1] Université de technologie de Compiègne, 60203 Compiègne Cedex, France
{felix.blampain,matthieu.bricogne,benoit.eynard}@utc.fr
[2] Spie Batignolles, 92000 Nanterre, France
{felix.blampain,celine.bricogne,
sebastien.pinon}@spiebatignolles.fr

Abstract. Construction industry faces many challenges as socio-economical needs evolve and Building Information Modeling (BIM) disrupts practices. BIM processes and solutions are mainly suited for the design stage of a project and experts are currently working to bridge the gap with the construction stage. Improving on BIM practices could be done by creating a more industrialized way of handling construction operations by learning from other industrial fields. This could help to manage quality, delay, and cost more precisely, while considering new indicators (e.g.: GHG emissions). BIM could be improved by developing a construction stage specific data structure inspired by Product Lifecycle Management (PLM) systems. This requires to dynamically manage data for different stakeholders before and during the operations. This paper presents a conceptual data framework to develop a BIM-based decision support system to plan and manage construction operations. A 4D digital mock-up is gradually enriched to support construction processes studies resulting in a *As Planned* view of the building. Process engineering work is carried through iterative loops and supported by knowledge-based indicators. A cross-disciplinary workflow allows the use of new production methods (e.g.: off-site modular construction). An *As Built* view of the mock-up is concurrently created as operations advance and changes occur, feeding back the knowledge-based indicators.

Keywords: Building Information Modelling · Product Lifecycle Management · Knowledge-Based Engineering · Enterprise Information System

1 Introduction

The Architecture, Engineering, and Construction (AEC) industry is transforming its ways to adapt to contemporary needs. Professionals must balance the management of costs, time, and quality with news demands, e.g.: evolving standards, sustainable development [1, 2]. Changes are needed in designing and constructing buildings. This could

Published by Springer Nature Switzerland AG 2024
C. Danjou et al. (Eds.): PLM 2023, IFIP AICT 701, pp. 253–262, 2024.
https://doi.org/10.1007/978-3-031-62578-7_24

be done by industrializing the production of buildings. Industrialized construction can be defined by new building methods, e.g.: off-site modular construction [3], and by precisely managing a project information during a building lifecycle [4, 5]. Industrializing the construction stage of a building project can be done by tackling site management activities. The core activities of site management are operations planning and monitoring, both rely on an interdisciplinary workflow involving process engineering practices, budget management, resources management i.e.: workers, machines and tools, materials, and sub-systems. Combining those, we can describe a site or a building under construction through operations workflows or the flows of resources in and around a site [6]. This gives a product-process-resources (PPR) point of view of the work [7].

Building Information Modeling (BIM) is at the center of this transformation as it helps professionals managing information and improves productivity [8]. Manufacturing industry practices is a source of inspiration to develop knowledge management practices for AEC projects. Manufacturing information management practices are centered around the holistic concept of Product Lifecycle Management (PLM) [9]. Learning from PLM could improve BIM practices [10]. A first step to industrialize building production could be made by combining a PLM-based data management approach with site management BIM applications [11].

This paper presents a first approach to define a knowledge-based decision support system for site operations management. We focus on site planning and monitoring activities with a BIM-based and product-process-resource-oriented workflow. We investigate what are the main components of an industrialized site management information system from a functional, roles, and data point of views.

In Sect. 2 we present the scientific background relevant for this work. In Sect. 3 we present the *As-Is* system and detail the information system *To Be* by defining its key components. Conclusions and future work are discussed in Sect. 4.

2 Scientific Background

2.1 Challenges in Structuring Information for Site Operations Management

Construction sites are complex production systems [12]. Operation scheduling and monitoring are important tasks to make sure everything is done right. Site management often rely on basic tools: Gantt charts, spreadsheets, and 2D drawings. These tools and their uses have not changed significantly in the last 30 years [13, 14]. Site management methods and digital tools are being developed to improve productivity. Digital solution focus primarily on linking a digital mock-up (DMU) with a project schedule, creating a 4D BIM process [15]. The main challenge is to account for the recurring changes that occur on the construction site, and the data visualization depending on stakeholders' point of view [16, 17]. Commercial solutions are emerging but their adoption rate remains low [15]. BIM is seen as a new way to deal with construction operations while integrating a PPR point-of-view [18]. Adapting BIM practices to the site management means to develop a digital thread between the design and construction stages, and new workflows to account for the new information management capabilities [5, 12]. This information management approach is known as Building Lifecycle Management [4]. It draws from the Product Lifecycle Management approach developed in the manufacturing industry

[19]. In [20] we detail our views on why and how to leverage a PLM-inspired approach to improve on BIM practices for site management, and how to account for 4D BIM capabilities. This would improve managing projects that are becoming more complex from a technical, digital, and organizational point-of-view [21].

3 Methodological Framework for an Information System Supporting Site Operations Management

BIM and site management practices could be improved by developing a greater industrial working approach. This can be achieved by developing a knowledge-based information system for site management activities. In [20] we presented a first draft of a BIM-based decision support system that could help professionals to plan and monitor site operations. In this section we define and detail how this system could be frame. It is built around a process engineering BIM mock-up that is connected to other enterprise information systems.

3.1 As-Is Organization

Planning and managing construction operations are complex activities involving many stakeholders. Before operations starts, planning work is mainly done by process engineers. Their work focuses on analyzing the feasibility of a project and devising processes that account for technical and financial constraints. This work generally depends on the engineers' experiences. Process engineers cooperate with other professionals to plan construction operations. The general foreman and site foreman play a key role in defining processes. They review deliverables and provide feedback on technical solutions and on-site context, e.g., site logistic constraints, workers expertise, sub-contractors skills. Equipment manager and purchaser also help optimizing construction work by allocating and buying the needed materials, machines, tools, and subcontracted services, and human resources manager dispatch workers. The general foreman activities currently represent a choke point to manage engineering and construction activities [20]. A new work organization is needed.

3.2 Developing a New Information System

Framing the System Functions
Enterprise information systems (EIS) are complex organizations defined by their social, technical, and economical dimensions [22]. Creating a new EIS implies to structure a new work dynamic with different shared responsibilities to use it properly. Doing so means to define the constitutive elements of the system and their relations from different points-of-view [23]. Several methodologies and tools exist to define and build an EIS, e.g., The Open Group Architecture Framework (TOGAF) [24], the GRAI methodology [25], or the Zachman framework (ZF) [26]. We choose the ZF because it is recognized as a *de facto* standard [27], and is practical enough to be used for a research study.

The ZF is based on a 6×7 matrix: columns represent the basic questions of engineering work, and lines represent the concretization of the work according to a specific point-of-view. Each box must consider the content of the ones on its sides and above and below to create a systemic picture of the enterprise and its EIS. Each box then represents a primary component of the EIS [27].

In Table 1 we detail our comprehension of the system to be developed according to the ZF. We choose to focus on few boxes that represent the core description of the system. Boxes are defined through iteratives interviews of experts. Three boxes are refined using diagrams to improve comprehension in Figs. 1, 2, and Table 2.

Lines are originally titled with names associated with AEC actors [28]. We changed them to avoid misinterpretation because this research work is conducted within a construction company. The new names seek to reflect the point-of-view of the EIS they represent.

Table 1. Zachman framework of the system to be

	Why	When	Who	How	What	Where
Strategic				Profoundly change working practices		
Operational	Improve productivity. Efficiently use resources. Reduce carbon footprint.	Project lifecycle		Capitalize on engineering knowledge. Strengthen expertises. Transform design and construction activities.		
Fucntional			Strengthen engineering practices. Site management team focuses on production. Minimize design activities during construction stage.			
Technological			See figure 1	See figure 2		
Technical					See table 2	

Strategic – How

The experts we interviewed concur that AEC work practices must change to tackle the many challenges facing the AEC industry. This can only be done by profoundly reshaping working practices within the company and changing how the industrial ecosystem operates.

Operational - Why

Changes are motivated by a will to improve the industry productivity while developing sustainable ways to build and use resources. Companies must adapt to climate change and economic change, while remaining cost-effective.

Operational - When

Changes of practices must occur on the entire lifecycle of a construction project. Companies must reorganize themselves to build differently.

Operational - How

Managing engineering knowledge can leverage new building practices and help to better manage complex engineering projects.

Functional - Who

Knowledge management practices should help to improve existing working practices and developing new ones. Site management activities must be reshaped to focus solely on production management. Design activities are to be managed independently and construction can start when the building design is considered mature enough and less prone to changes.

Defining the System's Work Organization
Technological - Who

In Fig. 1 we suggest a new work division across the lifecycle of an AEC project so to avoid choking point. Boxes represent major AEC activities. We numbered the activities to better understand the work sequence from start (#1) to finish (#6). A project lifecycle starts with the design of the building. We suggest creating a new role with the engineering manager who should coordinate and manage technical analysis for the project in and out of the construction company. Once the design is set, process engineer can plan site activities. The engineering manager and process engineer team up to optimize the building from a design and construction point-of-view. Materials, machines, tools, and sub-systems can be bought or allocated from the company resources stocks, during processes development while working forces can be dispatched or sub-contracted. We suggest creating the role of logistic coordinator to manage physical resources just as a HR manager dispatch workers between sites. The logistic coordinator interacts with the general foreman to adapt the allocation of resources to the site as operations develop. With this work division, the general foreman and site foreman can focus entirely on managing construction operations. When construction operations are finished, the building is handed over to the client.

Defining the System's Basic Data Architecture.
Technological - How

The work division suggested in Fig. 1 helps us to define the major functions of our EIS. Figure 2 represents a functional architecture view of the system to be. This representation is not based on any standard for information architecture. Blue boxes represent the main functions. They are numbered in the same order as the activities in Fig. 1. White boxes represent sub-functions of each main function.

Work is centered around the DMU of the project. Specialized views of the DMU are gradually developed for specific activities. Information managed in the DMU are exchanged with other information systems when needed to carry-out specific functions. Engineer manager develops an *As Designed* view at the beginning of the project. It serves as a basis to develop a preliminary planning of the project. When design activities are finished, process engineer develop an *As Planned* view and a precise master planning of construction activities. Engineers optimize the project as they develop their solution through iteration cycles. Thanks to the quantity take-off (QTO) obtained, resources can be acquired and dispatched to prepare operations. Data from the *As Planned* DMU are exchanged with the Enterprise Resource Planning (ERP) and the central purchasing service. When operations start the general foreman and site foreman develop the *As Built* view of the DMU and compare it to the *As Planned* view. Depending on the operations development, resources and workers dispatching can be adapted and the *As Planned* view can be reconfigured to take into account new and up to date parameters, e.g.: delivery period, lifecycle assessment indicators, change orders. When operations are over, the building is handed over to the client, and the project information are archived for legal and contractual purposes, and for knowledge management purpose. We add a change management function to the system to control data evolution through the entire project lifecycle.

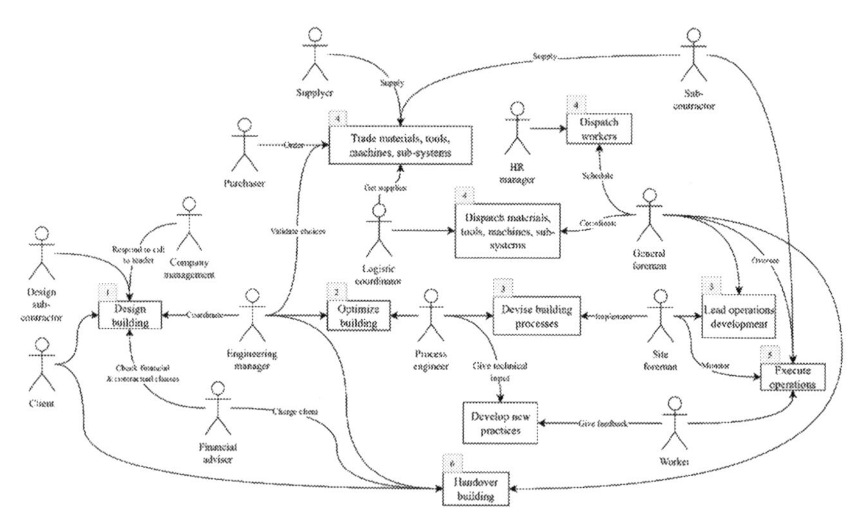

Fig. 1. Operation planning and monitoring tasks organization of the system *to be*

Technical – What

To further detail our understanding of this system. We present in Table 2 an example of the type of data to be used as a work basis. We also mention their possible interactions with other information system or data bases. We detail the data needed by the function *Devise building processes* as it is a central point of the system. Each sub-function can be associated with a technological solution. We then list the type of data needed by each function. For example, QTO are obtained by managing the geometric data of the *As Planned* DMU. We define operating condition based on the QTO, the associated time ratio for each type of element, and the tools and machines needed to execute building operations. The detailed planning can be linked with the HR information system or the central purchasing system to assess the resources available for the project.

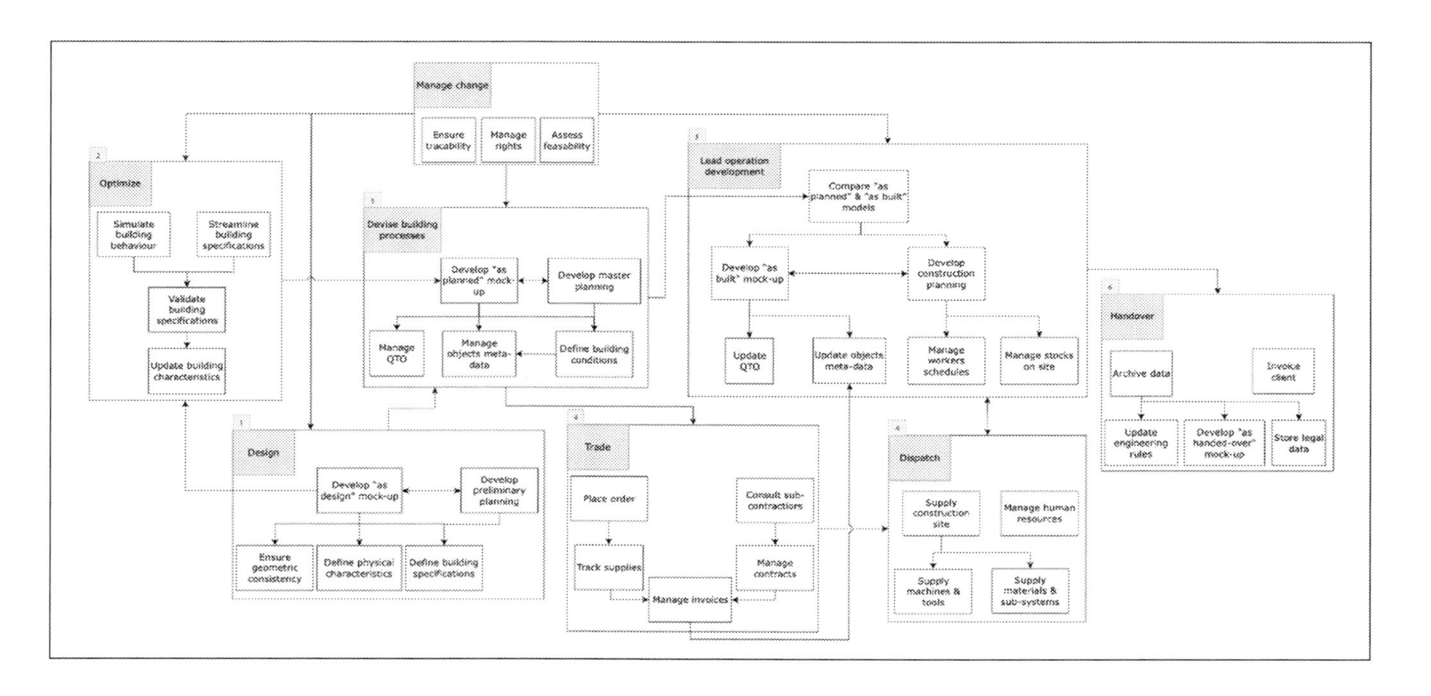

Fig. 2. Operation planning and monitoring functional architecture of the system to be

Table 2. Example of data typology for the system to be

<table>
<tr><td>Function</td><td colspan="4">Devise building processes</td></tr>
<tr><td>Sub-function</td><td colspan="2">Develop as planned mock-up</td><td colspan="2">Develop master planning</td></tr>
<tr><td>Technological solution</td><td colspan="2">4D BIM model</td><td colspan="2">Planning solution (e.g., Gantt chart)</td></tr>
<tr><td rowspan="2">Data</td><td>Manage QTO</td><td>–Objects geometry</td><td rowspan="2">Define operating conditions</td><td rowspan="2">–Time ratios
–Activities sequences
–Version
–Link with QTO
–Link with HR system
–Link with logistic data base
–Link with central purchase service</td></tr>
<tr><td>Manage objects meta-data</td><td>–Physical properties
–Version
–Lifecycle Assessment
–Link with planning
–Link with central purchase service</td></tr>
</table>

4 Conclusion and Future Work

The architecture, engineering, and construction industry is in need of changes. Projects are becoming more complex, and new demands must be met. To adapt, construction companies are industrializing their work practices, notably site management activities. Industrialization can be achieved by combining a project digital mock-up to other information systems. This paper presented a conceptual framework of an information system for planning and monitoring of site operations. We identified the core functions of the system. We presented a new enterprise architecture to leverage this system. We also presented a functional architecture of the system and the main types of data to be managed by the system.

This work is part of a greater research project. We intend to investigate further the composition of the system we introduced by precisely identifying the data needed by the system and how they interact to process a given function, or defining performance indicators to support decision making. We ambition to test it on a life-size use-case with our industrial partner.

Acknowledgement. The authors would like to express their gratitude to the input and feedback given by Spie batignolles experts.

References

1. Won, J., Cheng, J.C.P.: Identifying potential opportunities of building information modeling for construction and demolition waste management and minimization. Autom. Constr. **79**, 3–18 (2017). https://doi.org/10.1016/j.autcon.2017.02.002

2. Wuni, I.Y., Shen, G.Q.: Critical success factors for management of the early stages of prefabricated prefinished volumetric construction project life cycle. ECAM **27**, 2315–2333 (2020). https://doi.org/10.1108/ECAM-10-2019-0534
3. O'Connor, J.T., O'Brien, W.J., Choi, J.O.: Critical success factors and enablers for optimum and maximum industrial modularization. J. Constr. Eng. Manage. **140**, 04014012 (2014). https://doi.org/10.1061/(ASCE)CO.1943-7862.0000842
4. Bricogne, M., Eynard, B., Troussier, N., Antaluca, E., Ducellier, G.: Building lifecycle management: overview of technology challenges and stakeholders. In: International Conference on Smart and Sustainable City. p. 47. IET, Shanghai (2011). https://doi.org/10.1049/cp.2011.0284
5. Harkonen, J., Mustonen, E., Haapasalo, H.: Construction-related data management: classification and description of data from different perspectives. Int. J. Manag. Knowl. Learn. **8**, 195–220 (2019)
6. Garcia-Lopez, N.P., Fischer, M.: A construction workflow model for analyzing the impact of in-project variability. In: Construction Research Congress 2016, pp. 1998–2007. American Society of Civil Engineers, San Juan (2016). https://doi.org/10.1061/9780784479827.199
7. Terzi, S., Bouras, A., Dutta, D., Garetti, M., Kiritsis, D.: Product lifecycle management – from its history to its new role. IJPLM **4**, 360 (2010). https://doi.org/10.1504/IJPLM.2010.036489
8. Sacks, R., Eastman, C., Lee, G., Teicholz, P.: BIM Handbook - A Guide to Building Information Modeling for Owners, Designers, Engineers, Contractors, and Facility Managers. Wiley, Hoboken (2018)
9. Saaksvuori, A., Immonen, A.: Product Lifecycle Management. Springer, Heidelberg (2008). https://doi.org/10.1007/978-3-540-78172-1
10. Jupp, J.R.: Incomplete BIM implementation: exploring challenges and the role of product lifecycle management functions. In: Bernard, A., Rivest, L., Dutta, D. (eds.) PLM 2013. IAICT, vol. 409, pp. 630–640. Springer, Heidelberg (2013). https://doi.org/10.1007/978-3-642-41501-2_62
11. Blampain, F., Bricogne, M., Eynard, B., Bricogne, C., Pinon, S.: Digital thread and building lifecycle management for industrialisation of construction operations: a state-of-the-art review. In: Gerbino, S., Lanzotti, A., Martorelli, M., Mirálbes Buil, R., Rizzi, C., Roucoules, L. (eds.) Advances on Mechanics, Design Engineering and Manufacturing IV, JCM 2022. Lecture Notes in Mechanical Engineering. Springer, Cham (2023). https://doi.org/10.1007/978-3-031-15928-2_77
12. Mäki, T., Kerosuo, H.: Site managers' daily work and the uses of building information modelling in construction site management. Constr. Manag. Econ. **33**, 163–175 (2015). https://doi.org/10.1080/01446193.2015.1028953
13. Sriprasert, E., Dawood, N.: Next generation of construction planning and control system: the LEWIS approach. In: Proceedings IGLC-10 (2002)
14. Bolshakova, V., Guerriero, A., Halin, G.: Identification of relevant project documents to 4D BIM uses for a synchronous collaborative decision support. In: Creative Construction Conference, pp. 1036–1043. Budapest University of Technology and Economics (2018). https://doi.org/10.3311/CCC2018-134
15. Boton, C., Kubicki, S., Halin, G.: The challenge of level of development in 4D/BIM simulation across AEC project lifecyle. A case study. Procedia Eng. **123**, 59–67 (2015). https://doi.org/10.1016/j.proeng.2015.10.058
16. Boton, C., Kubicki, S., Halin, G.: Designing adapted visualization for collaborative 4D applications. Autom. Constr. **36**, 152–167 (2013). https://doi.org/10.1016/j.autcon.2013.09.003
17. Naticchia, B., Carbonari, A., Vaccarini, M., Giorgi, R.: Holonic execution system for real-time construction management. Autom. Constr. **104**, 179–196 (2019). https://doi.org/10.1016/j.autcon.2019.04.018

18. Dashti, M.S., Reza Zadeh, M., Khanzadi, M., Taghaddos, H.: Integrated BIM-based simulation for automated time-space conflict management in construction projects. Autom. Constr. **132**, 103957 (2021). https://doi.org/10.1016/j.autcon.2021.103957
19. Aram, S., Eastman, C.: Integration of PLM solutions and BIM systems for the AEC industry. In: International Symposium on Automation and Robotics in Construction and Mining, Montreal, Canada (2013). https://doi.org/10.22260/ISARC2013/0115
20. Blampain, F., Bricogne, M., Eynard, B., Bricogne, C., Pinon, S.: Digital solution for planning and management of construction site operations: a proposal of BIM-based software architecture and methodology. In: EduBIM, p. 176. Eyrolles (2023). https://hal.utc.fr/hal-03792058/document
21. Whyte, J., Tryggestad, K., Comi, A.: Visualizing practices in project-based design: tracing connections through cascades of visual representations. Eng. Proj. Organ. J. **6**, 115–128 (2016). https://doi.org/10.1080/21573727.2016.1269005
22. Romero, D., Vernadat, F.: Enterprise information systems state of the art: past, present and future trends. Comput. Ind. **79**, 3–13 (2016). https://doi.org/10.1016/j.compind.2016.03.001
23. ISO: ISO/IEC/IEEE 42010:2022 Software, systems and enterprise — Architecture description (2022)
24. Dedic, N.: FEAMI: a methodology to include and to integrate enterprise architecture processes into existing organizational processes. IEEE Eng. Manag. Rev. **48**, 160–166 (2020). https://doi.org/10.1109/EMR.2020.3031968
25. Chen, D., Vallespir, B., Doumeingts, G.: GRAI integrated methodology and its mapping onto generic enterprise reference architecture and methodology. Comput. Ind. **33**, 387–394 (1997). https://doi.org/10.1016/S0166-3615(97)00043-2
26. Zachman, J.A.: A framework for information systems architecture. IBM Syst. J. **26**, 276–292 (1987). https://doi.org/10.1147/sj.263.0276
27. Lapalme, J., Gerber, A., Van der Merwe, A., Zachman, J., Vries, M.D., Hinkelmann, K.: Exploring the future of enterprise architecture: a Zachman perspective. Comput. Ind. **79**, 103–113 (2016). https://doi.org/10.1016/j.compind.2015.06.010
28. Sowa, J.F., Zachman, J.A.: Extending and formalizing the framework for information systems architecture. IBM Syst. J. **31**, 590–616 (1992). https://doi.org/10.1147/sj.313.0590

On Considering a PLM Platform for Design Change Management in Construction

Hamidreza Pourzarei[1,2](✉), Conrad Boton[1], and Louis Rivest[2]

[1] Construction Engineering Department, Ecole de Technologie Superieure, Montreal, Canada
hamidreza.pourzarei.1@ens.etsmtl.ca, conrad.boton@etsmtl.ca
[2] Systems Engineering Department, Ecole de Technologie Superieure, Montreal, Canada
louis.rivest@etsmtl.ca

Abstract. Design change (DC) which refers to any type of design or construction modification made once a contract is awarded is an important issue in today's construction. DCs usually would lead to an increase in the cost and time overrun of the project. It is thus important to manage them by using an effective management process. Various design change management processes (DCM) have been proposed by different research studies as well as different DCM processes are used in practice (real world). In addition, IT tools have an important role in the DCM process that could provide various functionalities to facilitate the DCM process. However, according to the scientific literature, there is still a need for improvement/offer of a collaborative platform in construction. This article aims to evaluate whether a platform, typically categorized as a PLM (Product Lifecycle Management) tool, is capable of meeting the collaboration requirements of DCM. 3DExperience is a cloud-based collaborative platform that is used in PLM-supported industry. This collaborative platform is a connected online environment where all the design, collaboration, and data management capabilities are stored in a single user interface. More precisely, this article investigates how PLM platform functionalities could address the needs of the DCM process in construction. By assessing the research findings of this article, it is demonstrated that PLM platforms have the potential to be utilized for DCM. Such an evaluation could lead to improved productivity in the DCM process within construction.

Keywords: Building Information Modeling · Product Lifecycle Management · 3DExperience · Design Change Management · Engineering Change Management

1 Introduction

The goal of design change management (DCM) is to establish an organized and effective approach to recording and managing design change (DC) [1]. DCM plays an important role in controlling and overseeing modifications to a building design throughout its construction phases. It ensures that changes are executed in an orderly and methodical way and that the resulting changes are precise, consistent, and abide by applicable standards and regulations [1, 2].

Published by Springer Nature Switzerland AG 2024
C. Danjou et al. (Eds.): PLM 2023, IFIP AICT 701, pp. 263–276, 2024.
https://doi.org/10.1007/978-3-031-62578-7_25

It is important to mention that building information modeling (BIM) is a collaborative approach that encourages effective communication and information management between all parties involved in a construction project [3, 4]. It appears to have the potential to solve some persistent construction difficulties (interoperability, information flow optimization, etc.), which would lead to improved productivity [5].

Using a collaborative platform for DCM allows various stakeholders to communicate as well as work together effectively on design changes. This can help to ensure that all necessary parties are informed of DCs and that any issues or concerns are addressed in a timely manner [6]. Additionally, a collaborative platform can help to streamline the DCM process, making it more efficient and effective [1, 2]. The involved stakeholders (e.g., designers, engineers, etc.) could access and share the data and allow for real-time updates in a collaborative platform. In other terms, a collaborative platform could facilitate the communication and collaboration between the involved departments and teams for both internal and external exchanges [2, 7].

Although various tools and methods have been proposed by different research studies [1, 6], it seems there is still a need for improvement/offer of a collaborative platform to address the aforementioned needs [1]. BIM platforms are commonly used in construction projects to support various activities, including DCM. However, some scientific literature [1, 2] advocates that PLM platforms are more advanced and therefore it is important to consider whether they could also be used to support construction activities.

This article intends to address the following research question "*Can a PLM platform be utilized to support design change management in construction?*" Therefore, the objective of this paper is to evaluate the potential of a specific PLM platform to support the DCM process in construction projects. By examining the features and capabilities of such a platform, we aim to determine its suitability for managing design changes in construction projects. 3DExperience is typically used in the PLM-supported industry and we selected it as the specific PLM platform for this article. This PLM platform was chosen as the primary focus of this research due to its significance and prominence within the field. This platform is a software platform that developed by Dassault Systems[1], which is used for product design, simulation, analysis, and manufacturing. It comprises various applications for different phases of the product development process (e.g., CATIA for design, SIMULIA for simulation, and ENOVIA for data management).

It is important to mention that such a platform provides tools for design change management, which can be used to control and manage changes to the product design throughout its development and manufacturing. It helps to ensure that changes are made in a controlled and systematic manner, and that the resulting changes are accurate, consistent, and compliant with relevant standards and regulations [8].

The aim of this article is to assess the potential of a PLM platform for design change management within construction. It examines the various functionalities of the platform and how they can be utilized to enhance communication, collaboration, and data management in the design change process, ultimately improving the overall efficiency and effectiveness of the construction process. This article, therefore, provides some understanding of how a PLM platform can be utilized to support the design change management process in construction and its potential benefits.

[1] https://www.3ds.com/

This article is structured in six sections. The second section provides a brief overview of existing literature on design change management. The third section details the research methodology used in this study. The fourth section presents the findings of the study, focusing on the results of using 3DExperience as a collaborative platform for a typical DCM process. The fifth section discusses these findings, providing insights and analysis on the implications of the results. The final section concludes the article.

2 Background Research

The definition of design change (DC) states that it refers to any modification made to the design or construction of a project after the contract has been awarded and signed. An interesting aspect of this definition is that it highlights the timing of the DC, which occurs after the contract has been signed. This similarity between the definition of DC and engineering change is noteworthy [1], as both involve changes made to the design or construction of a project after a certain stage in the process has been completed. The primary objective of DCM is to ensure that all design changes are recorded accurately and managed efficiently, so that they do not cause delays, budget overruns or other issues throughout the project lifecycle [1].

Design change management helps to identify, evaluate and implement design changes while ensuring that they are consistent with the product's original design intent, technical specifications and regulatory requirements. It helps to minimize the potential for errors and inconsistencies, as well as reduce the risk of delays in the design process [1, 2, 6, 9–12].

Design change management also helps to ensure that all necessary parties are informed of changes, and that any issues or concerns are addressed in a timely manner. This can help to minimize the potential for confusion and misunderstandings, and can ensure that the final product meets the needs of all stakeholders [1, 10].

Design change management also provides a historical record of all design changes, which can be useful for future reference and for compliance with regulations [1]. The effectiveness of using the DCM approach can vary significantly depending on the specific characteristics of the project. Factors such as the nature, type, complexity, and size of the project, as well as the types of contracts involved, can all play a role in determining how successful the use of DCM will be [11].

To automate such a process, one must first comprehend the information that needs to be changed. In addition, to enhance its efficiency, one should also have a clear understanding of how the information should be classified, organized, interconnected, and managed [13]. However, the overall improvement of the change process cannot be achieved unless all types of business information are integrated, structured, and made easily accessible to all involved parties [13].

Hence, a collaborative platform can significantly enhance the effectiveness of the DCM process by enabling a variety of functionalities such as facilitating information sharing, communication, and ensuring that relevant information is accessible at the appropriate time [7]. Despite the numerous techniques and tools that have been proposed for this purpose [1], research studies have shown that there is still a gap in this area, and there is a pressing need for a collaborative platform that can effectively address the requirements of the DCM process [1]. This platform should be designed to bridge the gap between the different parties involved in the process and provide a centralized location for all the necessary information and tools.

3 Methodology

This research article presents a methodology with four key stages. The first stage entails a thorough examination of the existing scientific literature to understand and analyze the DCM process in construction that utilizes BIM technology. The purpose of this literature review and analysis stage is twofold: to describe the main components of the DCM process and compare them with the ECM process in PLM-supported industry, and to determine the necessary functionalities needed to adequately support the DCM process. The main actions in this stage are literature review and analysis, and requirement gathering.

The second stage of the methodology focuses on extracting and analyzing the tools and functionalities provided by the selected PLM platform to support the ECM process. The end goal is to determine the requirements necessary to effectively execute the ECM process within the environment of the selected PLM platform. The main actions in this stage are platform analysis, functionality identification, and requirement determination.

The third stage of the methodology involves a comparison of the extracted DCM process in construction and the ECM process in PLM side. The aim of this stage is to identify and evaluate similarities and differences between the two processes. These comparisons contribute to mapping the offered functionalities to meet the requirements of the DCM process. The main actions in this stage are process comparison and functionality mapping.

The final stage of the methodology highlights how the functionalities provided by the selected PLM platform can address the requirements of the DCM process. This stage provides a visual representation of the functionalities offered by evaluated platform and illustrates how they can be utilized to satisfy the needs of the DCM process. The end goal of this stage is to present a clear understanding of the capabilities of the PLM platform in supporting the DCM process. The main actions in this stage are functionality demonstration and capability presentation.

It's worth mentioning that this article presents the outcomes of the aforementioned methodology, but it does not present all the stages of the methodology.

4 A PLM Collaborative Platform

The objective of this section is to evaluate the different functionalities of a specific PLM platform that can be utilized in managing design changes. To accomplish this, the DCM and ECM are compared in the construction and PLM-supported industry to identify similarities and differences. In the second step, the functionalities of the selected PLM platform used in ECM are briefly presented, along with their potential use in the DCM process. The highlighted functionalities will be aligned with the specific requirements of DCM, providing a comprehensive understanding of how this platform can be effectively employed to manage design changes and meet the needs of construction.

4.1 Comparison of the DCM in Construction and ECM in PLM-Supported Industry

In this step, it is necessary to compare the DCM process from construction and the ECM process from the PLM-supported industry. It is worth noting that in our previous research [1], we conducted a comparative analysis of DCM in construction and ECM in the PLM-supported industry. The comparison and data presented in this article aimed at identifying the main activities and requirements of these two processes. In addition, this study concentrates on the initial four phases of DCM and ECM since they comprise the majority of activities in these processes. Additionally, the table below compares the features of the DCM process utilized in construction with those of the ECM process in the PLM-supported industry [14]. The table presented below outlines the DCM process extracted from the research of [1] and the ECM process extracted from the research of [14] (Table 1).

Table 1. Comparison of the DCM in Construction and ECM in PLM-supported industry

Phase	DCM process in Construction	ECM process in PLM-supported industry
Phase 1	**Initiate**	**Request**
	✔ Initiate the request for information (RFI) ✔ A design change request (DCR) is collected or initialized ✔ The DCR should address the reason for the change, the priority, the type of change, and which components are likely to be affected	✔ The engineering change request (ECR) is submitted
Phase 2	**Evaluate**	**Instruction**
	✔ The DCR impact analysis and feasibility studies are conducted ✔A set of potential solutions for the design change is defined ✔One solution will be selected and analyzed by the professionals (change management team) ✔Update mark-up drawings and documents	✔The request is analyzed to determine whether it is worthwhile to make the change ✔The change management team analyzes and proposes solutions ✔One solution is chosen ✔Update mark-up drawings and documents
Phase 3	**Approve**	**Execution**
	✔Based on the evaluation of the proposed change, the change management team (including professionals and client) decides whether to approve or reject the change	✔The change management team prepare the documents (drawings, specifications, etc.) for the chosen solution ✔Based on the evaluation of the proposed change, the change management team decides whether to approve or reject the change
Phase 4	**Implement**	**Notification/Application**
	✔Released the latest version of the documents (e.g., design change notice) for implementation ✔Notify the involved teams ✔Implement the proposed change ✔Monitor and track implementation	✔The solution for the change is carried out at the company level over the specified time frame

In the initial phase of both processes, a problem is identified. In the case of ECM, these problems can arise from design reviews, manufacturing, or even in the field that would lead to ECR. This definition is consistent with the concept of a DCR described in the scientific literature. Similarly to a DCR in literature, any member of the organization can raise an issue. Once identified, the problem can be submitted, tracked, prioritized, and resolved [8]. It's appropriate to note that a request for information (RFI) is a type of technical request used to clarify an issue. If the clarification provided to respond the RFI is sufficient to resolve the issue, the case will be closed. However, if further clarification is needed, it will result in a request for a DCR.

The instruction phase in the PLM side is similar in nature to the evaluate phase in the DCM process. Both phases involve assessing and analyzing proposed changes in order to determine their impact. Specifically, the primary task of this phase for both processes is to perform a thorough examination of the requested change, including an analysis of its potential effects on the over-all project or product. In essence, both the instruction phase in the PLM-supported industry and the evaluate phase in the DCM process are focused on ensuring that any proposed changes are carefully considered and evaluated before being implemented.

The third phase of both the DCM and ECM processes requires obtaining approval from the change management team. A key difference between the two processes is that during this phase of the DCM process, there are usually multiple meetings, negotiations, and discussions between teams and departments to determine the cost of the change. Once the change management team approves the chosen solution, the change order moves to the implementation phase. This phase is critical in ensuring that any changes are approved efficiently and in a timely manner.

Not surprisingly, phase four of both the DCM and ECM processes concludes by emphasizing the implementation of documented solutions across the company within a specific time frame. The updated versions of the documents will be shared with the relevant departments to facilitate the implementation.

It is worth noting that there are several notable similarities between the different phases of both the DCM and ECM processes. These similarities could suggest that the use of PLM solutions can be considered for implementing effective DCM practices within construction.

In the subsequent section, we will delve into the various functionalities offered by 3DExperience and describe how they can be applied to meet the specific requirements of the DCM process.

4.2 A PLM Platform for DCM

This section provides an understanding of the functionalities of a PLM platform that can be utilized to support the DCM process. It is divided into four parts, corresponding to the four phases of the D/ECM process in 3DExperience: issue management, change request, change order, and change action [8]. The characteristics of this process as well as the functionalities of the platform were gathered from two sources: the 3DExperience user manual [8] and the results of our previous research [1]. It is important to note that while the ECM process presented has four phases, in real-world scenarios, the ECM process may vary, with as few as one phase or as many as four phases depending on user needs. However, for the purposes of this research, the full ECM process, comprising all four phases, is considered. This section also examines how these functionalities can be tailored to the DCM process.

4.2.1 Phase 1: Issue Management—Initiate

An issue is a problem that can be raised by anyone in the field [8]. This phase encompasses a range of different functionalities, which are presented in Fig. 1. This figure helps visualize the various components, which are classified using three distinct colors. Firstly, the central yellow node indicates the name of the phase, which in this case is "issue management." Surrounding this node are five green nodes, each representing a different functionality within this phase. Finally, the eleven blue nodes located around the green nodes represent sub-functionalities of the functionalities. These groups are also applicable to the subsequent phases.

Legend of the Figure (which is applicable for all phases):

- Yellow node: indicate the phase.
- Green node: indicate the functionalities.
- Blue node: indicate the sub-functionalities.

During the Initiate phase of the DCM process, these five key functionalities can be utilized. These include opening an issue, submitting an issue, assigning an issue, analyzing the issue, and disposing of the issue. Each of these functionalities has its own set of sub-functionalities that can be customized to fit the specific needs of the project.

The issue encountered in this platform is reminiscent of the RFI/DCR encountered in construction. RFIs are initiated by any user by simply opening a new issue and attaching

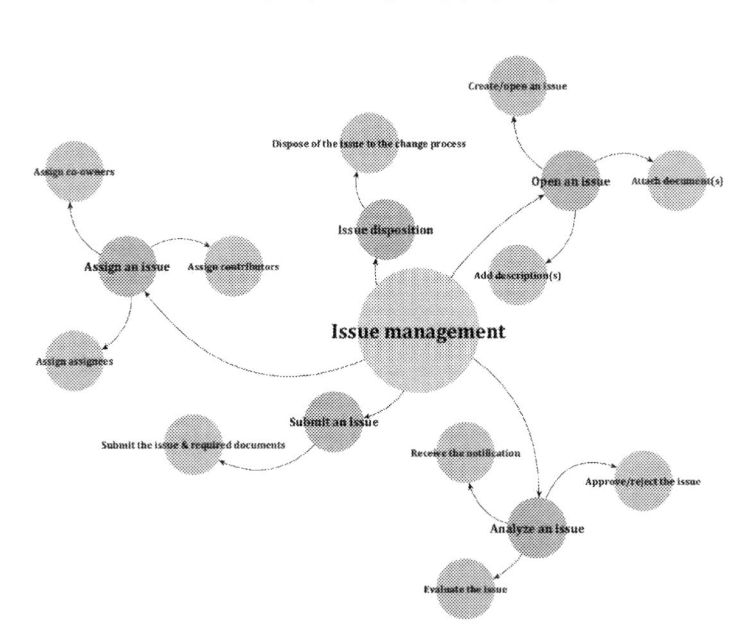

Fig. 1. Functionalities and sub-functionalities of the issue management phase

any relevant documents or explanations for the project team to access. Professionals can then be designated to respond to the RFI and determine whether a DCR is necessary. They perform a pre-feasibility study and provide feedback to the change management team. In this way, the RFI process ensures that all necessary information is gathered and assessed before making any design changes.

4.2.2 Phase 2: Change Request—Evaluate

It is at this phase where a thorough analysis of the requested change is conducted to evaluate its potential impact on the project. So it involves the use of various functionalities to perform the impact analysis effectively, which are depicted in the figure presented below. The figure comprises 11 green nodes representing different functionalities, and 34 blue nodes representing sub-functionalities. These elements work to support the impact analysis process and help to identify potential impacts that proposed changes may have on different aspects of a project (Fig. 2) .

This phase of the DCM process is a crucial phase as it allows for a thorough analysis of the requested change and the proposed solutions. It is during this phase that the change management team can utilize the functionalities offered by this platform to gain a deeper understanding of the impact of the requested change. These functionalities include change assessment and impact analysis, which can help in identifying the potential risks and benefits associated with the change, as well as assist in determining the best course of action.

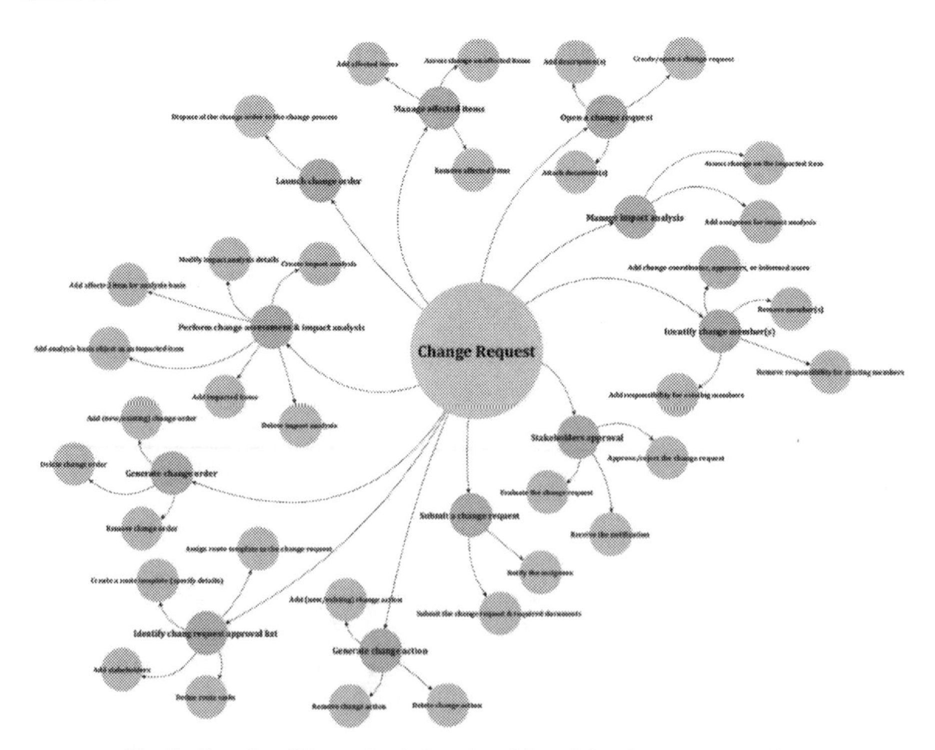

Fig. 2. Functionalities and sub-functionalities of the change request phase

Furthermore, collaboration functionalities (e.g. notify the assignees) are also available for identifying and assigning responsibilities to specific team members. This allows for more effective communication and coordination within the team, ensuring that all members are aware of their roles and responsibilities. Other important activities that take place during this phase include managing affected items, generating a change action plan, managing the impact analysis, identifying and gaining approval from stakeholders, generating a change order, and implementing the approved change order. By managing affected items, the team can ensure that all items that will be impacted by the change are identified and addressed.

4.2.3 Phase 3: Change Order—Approve

The third phase of the DCM process may benefit from functionalities depicted in the accompanying figure, which comprises 12 green nodes representing different functionalities and 38 blue nodes representing sub-functionalities (Fig. 3).

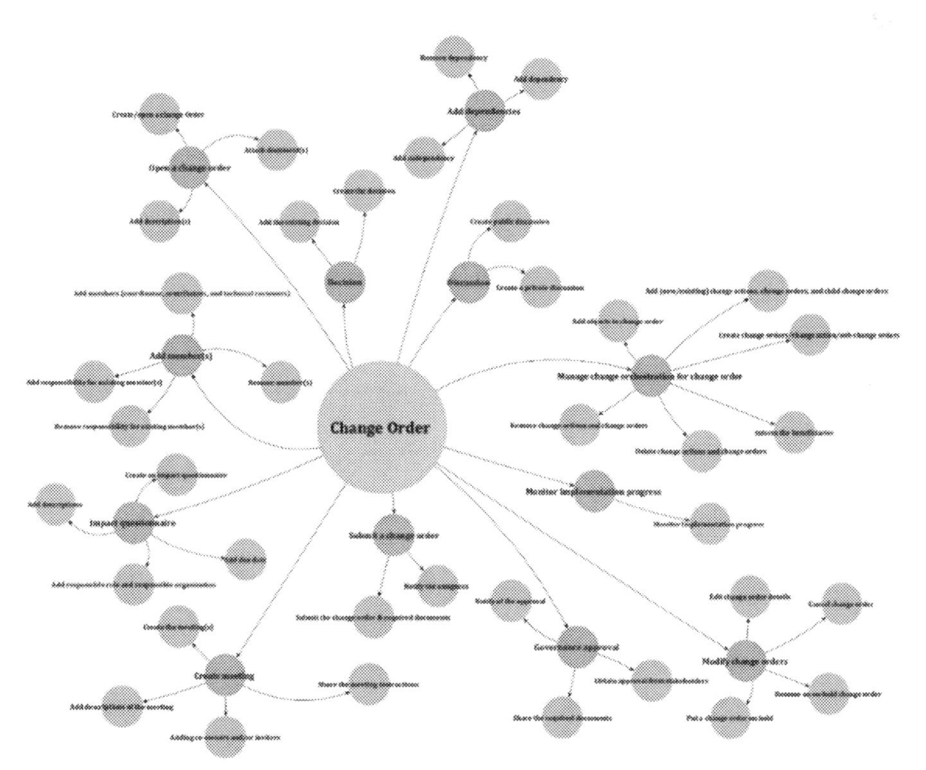

Fig. 3. Functionalities and sub-functionalities of the change order phase

During this phase, the change management team must make important decisions regarding the proposed change, and negotiate with stakeholders to gain their approval. Additionally, collaboration functionalities are available for effective communication and negotiation with stakeholders, ensuring that all parties are aware of the requested changes and have an opportunity to provide feedback and approval.

Some of the functionalities that can be applied during the process include creating meetings, discussion sessions, and decisions, as well as gaining governance approval. Creating meetings and discussion sessions allows for effective communication and coordination within the team. These functionalities allow the team to schedule and hold meetings and discussion sessions, and to invite relevant stakeholders to participate.

Creating decisions functionality allows the team to document important decisions made during the process, and governance approval functionality allows the team to gain approval from relevant stakeholders and governing bodies. This is important to ensure that the requested changes are in compliance with established guidelines and regulations. Overall, the offered functionalities are allowing for effective communication and coordination, as well as making the process more efficient and effective.

4.2.4 Phase 4: Change Action—Implement

The activities of this phase include opening a change action, submitting a change action, proposing a change, adding attachments and dependencies, making a decision, and gaining governance approval. The offered functionalities are specifically designed to assist with these activities and are illustrated in the accompanying figure, which includes 7 green nodes representing different functionalities and 21 blue nodes representing sub-functionalities (Fig. 4).

Fig. 4. Functionalities and sub-functionalities of the change action phase

The Implementation phase of the DCM process requires a number of activities to be completed in order to implement the requested change. These activities include releasing the latest version of the documents, notifying the involved teams, and monitoring and tracking the implementation [1].

Releasing the latest version of the documents is an essential step in ensuring that all parties have access to the most up-to-date information. Notifying the involved teams is another important activity in the Implementation phase. The platform allows the team to quickly and easily notify the relevant stakeholders of the requested change, making sure that everyone involved is informed about the requested changes and given the chance to offer their opinions and give their consent.

Monitoring and tracking the implementation is a critical activity in ensuring that the requested changes are implemented as intended. The platform allows the team to track the progress of the implementation, identify any issues that arise, and take appropriate action to resolve them. This ensures that the requested changes are implemented efficiently and effectively.

5 Discussion

While construction has adopted BIM as a collaborative approach for enhancing communication and information management, the literature suggests that there is still a need for a better collaborative platform [15]. One potential solution is to utilize the current PLM platform within construction. The implementation of a PLM platform can significantly impact the management of design changes in construction. This is due to the platform's ability to provide a collaborative and integrated environment that enables various stakeholders, such as architects, engineers, and contractors, to effectively communicate and work together on design changes. The figure below illustrates the integration of the DCM process discussed in Sect. 4.1 with the PLM functionalities presented in Sect. 4.2. The purpose of this figure is to depict the potential adaptation of PLM functionalities to the DCM process, thereby mapping their compatibility.

It is worth noting that the compilation of functionalities and sub-functionalities draws inspiration from the research of [16], and is subsequently generalized. This list effectively categorizes the functionalities and sub-functionalities according to the requirements of the DCM process.

The figure above displays the DCM process stages in the first column, while the second column presents the main activities associated with the DCM process. The third column showcases the categorized functionalities of the PLM platform, with their respective sub-functionalities listed in the fourth column. It is important to note that the numerical values assigned to each activity/functionality indicate the number of relationships it has with others.

The Fig. 5 highlights two notable observations. Firstly, activities such as 'impact analysis,' 'solution definition,' 'solution analysis and selection,' and 'final evaluation' have the potential to utilize the most functionalities of the PLM platform. Secondly, 'sharing,' 'access management,' 'file management,' and 'workflow management' emerge as the primary functionalities that can be leveraged in the context of the DCM process. It is important to note that this mapping represents a sample DCM process, and the

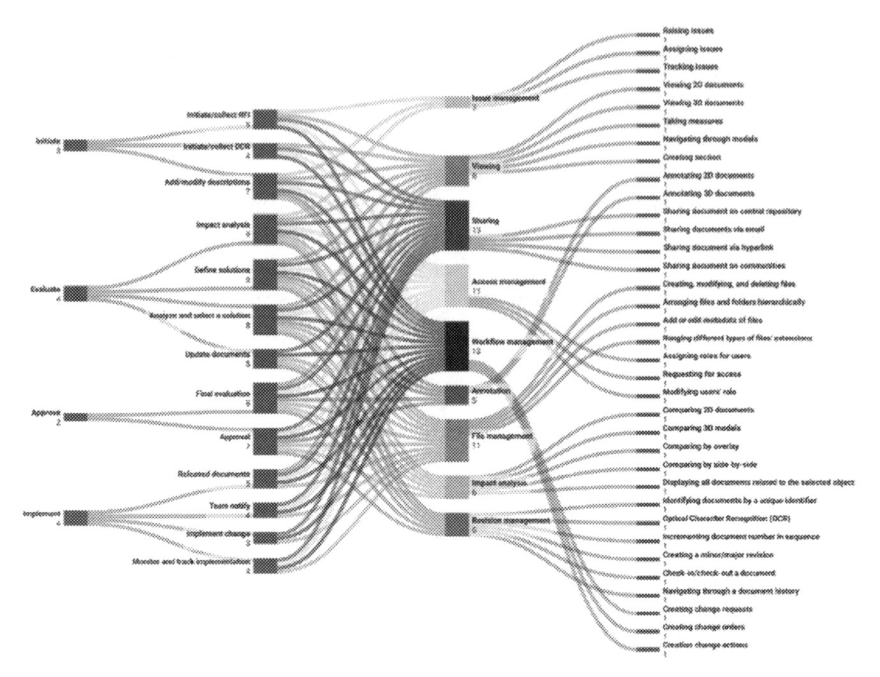

Fig. 5. Exploring the Relationship between PLM Functionalities and the DCM Process

relationship between DCM activities and functionalities may vary depending on specific company requirements.

The functionalities above could be used in the DCM process to increase productivity in the DCM process. One of the key benefits of using PLM platforms in construction is its ability to provide real-time access to building design data. This allows all stakeholders to access and share the same data, which can help to reduce the risk of errors and inconsistencies, and minimize the potential for delays in the design process. Additionally, the platform's data management capabilities can help to ensure that all necessary parties are informed of changes, and that any issues or concerns are addressed in a timely manner.

The evaluated PLM platform provides tools for design change management, which can be used to control and manage changes to the building design. This can help to ensure that changes are made in a controlled and systematic manner, and that the resulting changes are accurate, consistent, and compliant with relevant standards and regulations.

On the other hand, it is important to highlight that incorporating such a PLM platform into construction may present certain limitations that can be broadly categorized into three areas: complexity, integration difficulties, and cost. The platform may have a complex interface requiring a learning curve for users to be able to use its features and functions. Integrating such a platform with other existing design systems and tools may be challenging and require time and effort. Lastly, implementing a PLM platform can be a costly solution for companies, which might require budget allocation for implementation and maintenance.

6 Conclusion

The Design Change Management process plays a crucial role in construction. Its primary goal is to monitor and control design changes throughout the project's lifecycle. Utilizing a collaborative platform for DCM is vital as it provides a communication and collaboration hub for all project stakeholders to effectively manage design changes. This leads to a well-informed team, faster resolution of concerns and issues, and ultimately, a smoother project delivery.

The purpose of this article is to evaluate the potential of a PLM platform for DCM within construction. The findings of this article can be divided into two categories. Firstly, we compared the DCM process from construction with the ECM process from PLM-supported industry to identify similarities and differences. Since there are more similarities than differences, the article then outlines how a PLM platform can support the DCM process in construction. The analysis indicates that the PLM platform (e.g. 3DExperience) can be a potential platform as a collaborative solution for construction as examined from the DCM process standpoint.

References

1. Pourzarei, H., Ghnaya, O., Boton, C., Rivest, L.: Comparing engineering/design change management and related concepts in BIM- and PLM-supported industry from the literature. Res. Eng. Des. (2022). No. Submitted in Journal of Research in Engineering Design
2. Hao, Q., Shen, W., Neelamkavil, J., Thomas, R.: Change management in construction projects. NRC Inst. Res. Constr. NRCC-50325 (2008)
3. Jupp, J.R.: Incomplete BIM implementation: exploring challenges and the role of product lifecycle management functions. In: Bernard, A., Rivest, L., Dutta, D. (eds.) PLM 2013. IAICT, vol. 409, pp. 630–640. Springer, Heidelberg (2013). https://doi.org/10.1007/978-3-642-41501-2_62
4. Sacks, R., Eastman, C., Lee, G., Teicholz, P., Handbook, B.I.M.: A Guide to Building Information Modeling for Owners, Designers, Engineers, Contractors, and Facility Managers. Wiley, Hoboken (2018)
5. Boton, C., Rivest, L., Forgues, D., Jupp, J.R.: Comparison of shipbuilding and construction industries from the product structure standpoint. Int. J. Prod. Lifecycle Manag. **11**(3), 191 (2018). https://doi.org/10.1504/IJPLM.2018.094714
6. Motawa, I.A., Anumba, C.J., Lee, S., Peña-Mora, F.: An integrated system for change management in construction. Autom. Constr. **16**(3), 368–377 (2007)
7. Isaac, S., Navon, R.: Feasibility study of an automated tool for identifying the implications of changes in construction projects. J. Constr. Eng. Manag. **134**(2), 139–145 (2008)
8. Biovia, D.S.: 3DEXPERIENCE User Assistance. Dassault Systèmes (2022). https://www.3ds.com/support/documentation/users-guides/
9. Sun, M., Fleming, A., Senaratne, S., Motawa, I., Yeoh, M.L.: A change management toolkit for construction projects. Archit. Eng. Des. Manag. **2**(4), 261–271 (2006)
10. Mejlænder-Larsen, Ø.: Using a change control system and building information modelling to manage change in design. Archit. Eng. Des. Manag. **13**(1), 39–51 (2017)
11. Hwang, B.-G., Low, L.K.: Construction project change management in Singapore: status, importance and impact. Int. J. Proj. Manag. **30**(7), 817–826 (2012). https://doi.org/10.1016/j.ijproman.2011.11.001

12. Ibbs, W.: 'Project Change Management.' Construction Industry Institute Special Publication, Austin, Texas (1994)
13. Guess, V.C.: CMII for Business Process Infrastructure. Holly Publishing Company, Lancaster (2002)
14. Maurino, M.: La gestion des données techniques: technologie du concurrent engineering. Masson, Paris (1993)
15. Pourzarei, H., Rivest, L., Boton, C.: Cross-Pollination as a comparative analysis approach to comparing BIM and PLM: a literature review. In: Nyffenegger, F., Ríos, J., Rivest, L., Bouras, A. (eds.) PLM 2020. IFIP, vol. 594, pp. 724–737. Springer, Cham (2020). https://doi.org/10.1007/978-3-030-62807-9_57
16. Ghnaya, O.: Comparaison des outils informatiques pour la gestion des modifications de conception/d'ingénierie entre les industries soutenues par le BIM et le PLM, Master thesis, École de technologie supérieure (2023)

BIM Technology Application to Propagate the Knowledge and Information for Design Changes and Modifications Throughout the Product Development Cycle

José Roberto Alcântara Lobo, Anderson Luis Szejka(✉), and Osiris Canciglieri Junior

Industrial and Systems Engineering Graduate Program (PPGEPS), Pontifical Catholic University of Parana (PUCPR), Curitiba, Brazil
anderson.szejka@pucpr.br

Abstract. A design change may be considered a modification in a particular aspect of this design which has already been released and delivered (production phase) or even during its conception and development phase. Depending on the complexity and stage in the life cycle of the product where the change process is triggered, these changes may take from a few hours to several months to re-adapt the referred design. By applying concepts such as Parallelism, Concurrent Engineering, or most recently, Digital Manufacturing, it is imperative to guarantee that the resulting information about design changes is accurately propagated, reaches all involved stakeholders. Regarding Information and Knowledge Management, the emerging of new technologies such as Cloud Computing, Web Services or more specifically BIM (Building Information Modeling) have boosted the integration of all information projects in collaborative online environments. This paper further explores possibilities to use BIM (Building Information Modeling) technology for organizing and collaboration of the information and knowledge that comes with design changes through the entire products' lifecycle.

Keywords: Design Change · Product Life Cycle Management · Building Information Modeling

1 Introduction

A new product design is a process that implies stages, from the conception (focus on creation) to its manufacture (focus on manufacturing) and subsequent maintenance and disposal (focus on operation). Considering the uncertainties, decision points, and improvement opportunities, which occur throughout this cycle, concept changes, functionalities, and constructive details are inherent to the process. In short, changes to how planned, how designed, how built, how sold, how operated, and how maintained [1].

According to [2], the need of change in a product development is caused by i) Correcting a design error which becomes evident until either testing and modeling or

Published by Springer Nature Switzerland AG 2024
C. Danjou et al. (Eds.): PLM 2023, IFIP AICT 701, pp. 277–286, 2024.
https://doi.org/10.1007/978-3-031-62578-7_26

customer use discloses it; ii) A change in customers' requirements leading to part of the product redesign or iii) a material or manufacturing method change. This may be caused by material unavailability, a supplier change, or compensating a design error. There are always consequences associated with these causes. Among them: Difficulty in measuring the change's impact [3], the need for input data review, reworking on already manufactured components/parts/subsystems [4], or the financial impacts caused by non-estimated costs. A change occurs throughout the entire product life cycle, from the moment when a concept is first selected up to when it finally goes out-of-service, yet activity differs significantly based on the stage of the product's lifecycle.

No matter which stage, many stakeholders (or actors) are involved, either within organization like the engineers, designers and planners, or external stakeholders, mainly customers and suppliers. Furthermore, as pointed out by [5], records need to be kept on when the changes have been done, and what were the changes, and when these changes are effective. Also important is the statement by [6]: "interested stakeholders engaged at the design, manufacturing, operation, and maintenance steps of a product are entirely independent from each other". Summarizing, in the change context, there are formalization, availability, and collaboration demands within the product cycle.

Significant collaboration has been given by [7] that "is critical to understand product (built object) data, master data, transfer data and, especially, data governance: who produces, modifies, utilizes or owns the data and how these roles possibly change during a built object lifecycle".

On this basis, numerous methods have been developed to manage a built object data modification where one of the most relevant areas being worked on through interoperable file formats across multiple platforms [8]. Along these lines, at the end of the 1990s, a structure emerged aimed at avoiding interoperability problems in the execution of projects: the IFC (Industry Foundation Classes).

An IFC file is an open and non-proprietary file format. It can be used to exchange and share data, during the design, construction, management, and maintenance phases, between all figures and the various applications developed by different software houses without the need to support native (proprietary) files [9]. Often associated with the IFC terminology, technology is increasingly used in construction projects called BIM (Building Information Modeling). According to [10], a Building Information Model is a digital representation of a construction project's physical and functional characteristics. In this model, there is an equivalence between a real object and an equivalent digital object. In a simplified way, the relationship between IFC and BIM is that, in technological terms, it is possible to export a BIM model to an IFC format. By sharing all information in an open format, such as IFC, all actors in the development project can access relevant information when needed so that everyone can work together efficiently and collaboratively.

Thus, considering that the management of design changes is a socio-technical system [11], the efficient use of information and communication technology (ITC) tools is a success factor in reducing, in a way collaboratively, the impacts of changes within the product development process. Therefore, this article aims to evaluate procedural, organizational, and technological aspects that organizations must observe to use BIM models effectively in their product development processes.

2 Design Change in the Product Lifecycle

Conceptually, there have been several studies on aspects of design change, which have defined this subject. According to [5], a design change is a product component modification that occurs after it has gone into production.

In a more comprehensive way [11] defines design changes as "changes to parts, drawings or software that have already been released during the product design process, regardless of scale change". One last, by [12] defines "change as any kind of project document modification by either the owner, owner's agent, or project engineer". These three assertions above provide an insight about the scope, the impacts, and the stakeholders involved in a project change task. In terms of scope, it is important to introduce one more concept, where project changes are part of, which is PLM (Product Lifecycle Management).

PLM has emerged since the late 1990s aiming at covering the entire product's useful life cycle, i.e., from the initial idea to its disposal [13]. The "life cycle" approach has now made engineering information increasingly accessible to all stakeholders, whether internal or external. Not only engineering information, but also applications, processes, people, work methods, and equipment [8]. There is a business development trend to focus on the organization's key competencies, leading to an enhanced collaboration between partners, suppliers and contractors A relevant aspect to highlight is related to stakeholders, whether internal or external. There is a business development trend to focus more on the organization's key competencies, leading to an enhanced collaboration between partners, suppliers, and contractors [1].

However, to enable an organization's-controlled integration with external agents, collaboration tools and methods had to be further developed. Along this path, there is another extremely strong concept: PDM (Product Data Management), which has become a cornerstone for organizations' digital operations systems [14].

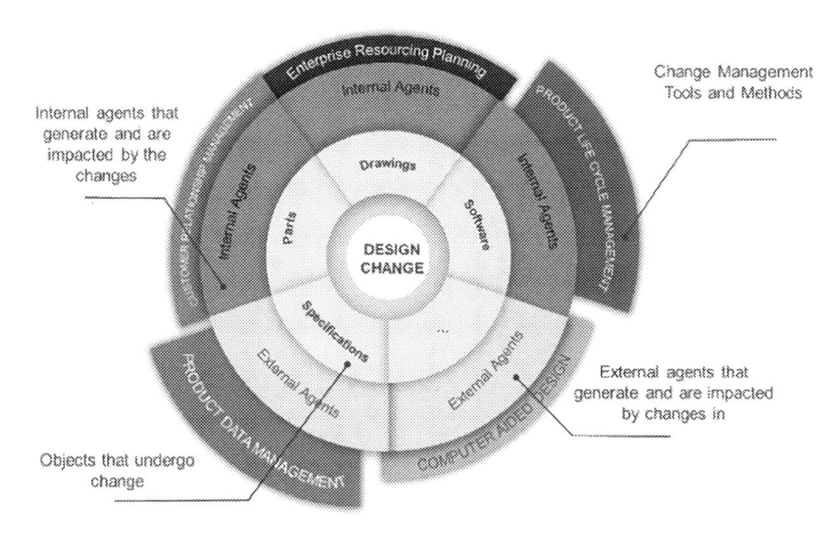

Fig. 1. Relations within the Design Change Management process

The PLM and PDM framework also include other technological tools, such as ERP (Enterprise Resource Planning), CRM (Customer Relationship Management), and CAD (Computer Aided Design). Figure 1 is a graphic representation to show how all these elements are related to each other within the Design Change Management process.

3 Building Information Modeling (BIM)

All developed technologies and methods, summarized on the previous section, described the non-integrated participation of stakeholders involved in design change management. By "nonintegrated" is meant the information exchange between the components, from the IT point of view, is not performed within a single data platform. Information about geometry and components' position, for example, will be stored within CAD tools. The basic data such as units of measurement or quality requirements, on the other hand, belong to the domain of ERP software. The design project planning is also contained within the ERP linked to PM tools. Such "discontinuities" in the database interfaces make it harder to determine two critically elements of a design change: their impacts and consequences on the design project.

Professor Charles M. Eastman created the concept of BDS (Building Description System) in 1974, which would be a system to improve the strengths of a building project and reduce its weaknesses [15]. The first use of the term Modeling Building Information, in 1992 by [16], was a paper that discussed multiple views on building modeling, which has been transformed into BIM (Building Information Modeling). Since then there has been a shift of paradigms for handling project aspects in an integrated way [17]. Despite not being a new initiative, it has spread only through the availability of higher-capacity and affordable data processors on the market [18], BIM models have brought a new approach of project delivering, by fostering people, systems, business structures, and practices integration in a collaborative process [19]. BIM is a combination of both geometrical and functional, qualitative, and quantitative contents. BIM model may be summarized according to Fig. 2: through people using technological procedures, organization accomplishes more collaborative work.

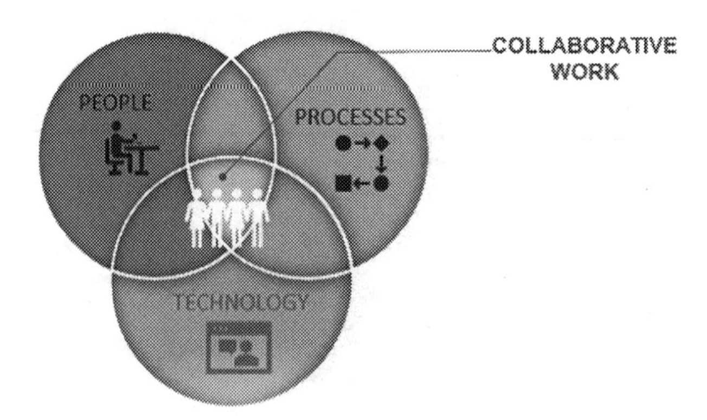

Fig. 2. Simplified BIM model

Technology encompasses the required infrastructure for operation, as well as the software and hardware, data traffic and security, besides the users' training. Any technology-related choice must be properly evaluated, including the organizational business model, the business strategies, and financial investment capacity. On the other hand, focusing on people is also a key part of the implementation strategy [18]. Given that BIM has collaborative work as one major potential, the staff team members should be skilled in both internal and external team relationships. Furthermore, their ability to be flexible for changes and have cognitive capability to take up new concepts and ITC practices.

Other skill to be observed and developed by the teams who are involved with BIM models is the communication ability: to identify errors or potential enhancement and to report them at the right time to the right person, and by a very effective way as a whole, with the appropriate information and detail level, in order to support decision-making and required actions to fix or improve the process [18]. A virtual process itself is as good as the people who operate it. If the staff is not well prepared for the resources, if BIM tool operators do not have expertise and transdisciplinary knowledge about the projects, if designers have no experience of the construction execution, and if the other staff involved in the project work in an isolated and non-collaborative way, new technology will not reach its optimum level. The BIM process focus is related to the information structure linked to activities, components, elements, and processes which may occur throughout the cycle. BIM process adoption, thus, implies the company's operational processes to make the implementation process structuring viable.

If a BIM model is approached from the interoperability standpoint, [19] then, BIM approach provides a comprehensive definition as a "methodology of managing essential construction design and the design data in digital form throughout its entire lifecycle". This means that beyond a 3D viewer and detailing, BIM models offer benefits including cost savings, delivery time acceleration, and opportunities for engineering, planning and supply, and waste reduction throughout all design and construction phases. Figure 3 illustrates on interoperability between several elements of a BIM model's processes.

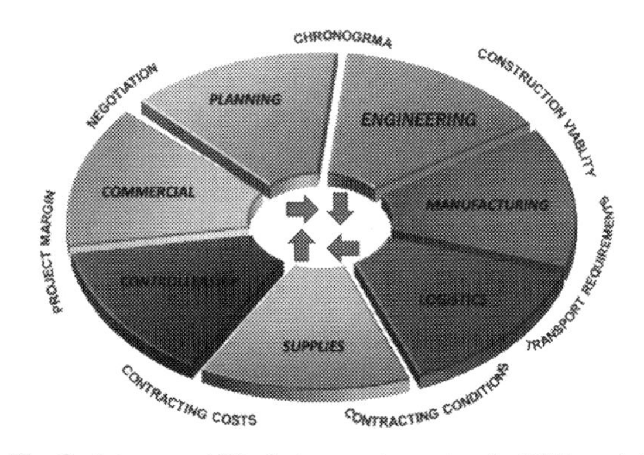

Fig. 3. Interoperability between elements of a BIM model

Technologically a BIM model consists of the combination of information content, geometry, site surveys, and functional data on performance, materials, and quantities.

For many experts, BIM involves dynamic management of data and information that are generated by and used during the design lifecycle [20]. Figure 4 illustrates integration between different technologies involved in information management.

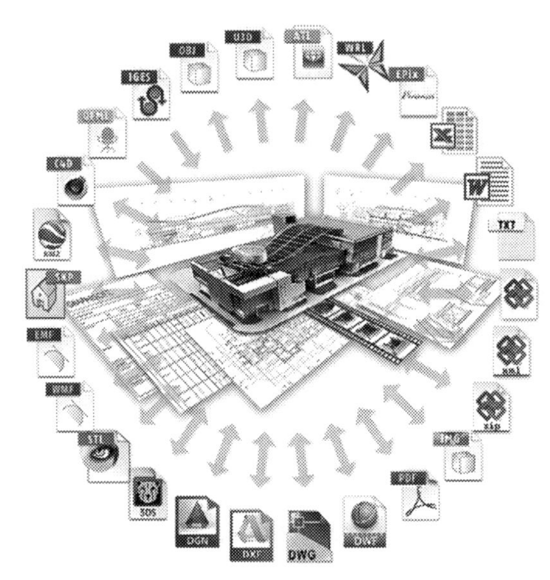

Fig. 4. Integration between different technologies involved in information management.

4 BIM Modeling Within Design Changes

Concerning changes identification and comparison between different versions for the same design, current 2D and 3D CAD software can very efficiently identify and indicate geometry or component position changes. Some of these include Autodesk® Revit®, Vico Doc Set Manager™ or Solibri Model Checker™ [21]. Once design project changes are taken into the framework of business processes, ERP and DMS (Document Management System) softwares are also equipped with the ability to detect and report on these changes. For example, component changes with purchase orders or production orders at the manufacturing stage. Also, identification of the installed base of components/products or systems that a project change will impact.

There have been many studies done, on this research stream, identifying and propagating the design changes. Yet they are focused on building construction projects. For example, [22] has done a comprehensive study on Pharmaceutical Sciences Building design, which is built in Vancouver, Canada, at UBC (University of British Columbia). Following a similar approach, [23] has developed a model where changing and impacting were the position on a building's walls and doors. Also [21] developed important research, for what the author called "green features", focused on energy efficiency and conservation aspects.

But, once arguing an adaptive and integrative BIM ought to be able of maintaining the data consistency throughout the whole construction model [22] BIM model dealing with product lifecycle phases becomes advisable.

When considering a BIM model creation approach to design changes throughout the product lifecycle, there are important elements to work on: i) change points, ii) change consequences, and iii) change propagation.

By change points, is meant what has to be observed when identified a need for design change. Changes consequences are the impacts that the change points will undergo. Last, by change propagation, is meant what kind of adjustments must be made to products' lifecycle agents/objects after design changes have been approved. Graphically Fig. 5.

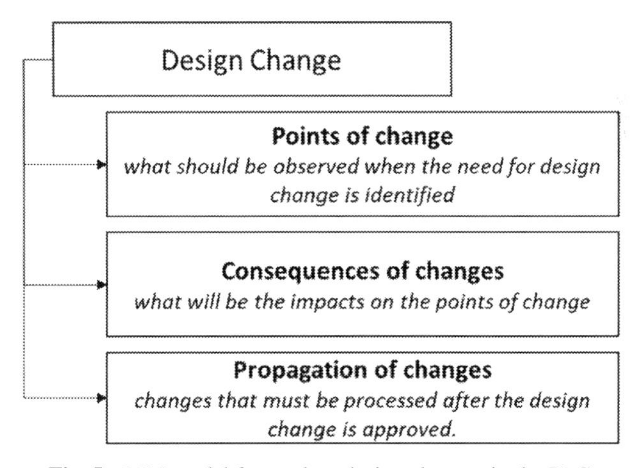

Fig. 5. BIM model focused on design changes in the PLC.

These three elements take place in two distinct phases within the design change process: 1) decision-making - whether a change will or will have to be implemented and 2) updating - to make the updates on the design objects. During the decision-making step, it's important be clear: i) why this change has to be made, ii) what change agents - whose areas/parts have to evaluate the change impacts throughout the whole product cycle, iii) which objects must be evaluated by actors, iv) check list - what has to be checked, v) analysis - what has to be checked on the identified objects by the actors, v) analysis - what has to be checked on the identified objects by the actors, vi) stakeholders - what areas/parties have to be reported after all analysis has been performed, vii) decision - which area/departments have the approval authority to make the change. Sequentially at the update stage, it is identified: i) which areas/stakeholders are supposed to make the updates that have been identified and approved, ii) which objects must be maintained (be preserved) and, closing the cycle, the areas/stakeholders who must be communicated about the changes made. Figure 6 presents the fundamental steps of the design change using BIM Modelling.

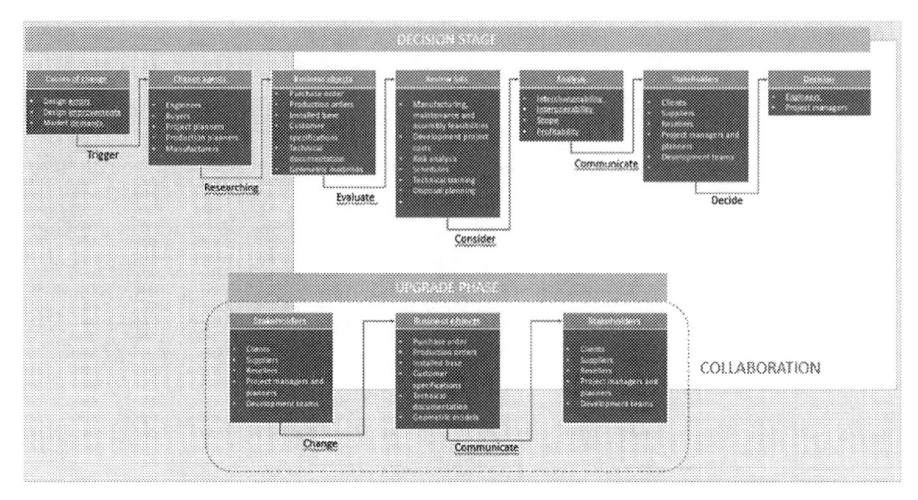

Fig. 6. Proposed BIM Modeling within Design Changes.

5 Conclusion

Brazilian Federal Government has officialized the National Strategy for Building Information Modeling (BIM) Dissemination, or BR BIM Strategy, which aims at promoting a proper environment for the methodology investment and its dissemination in Brazil. One of the stipulated goals is increasing BIM use 10-fold, so 50% of the civil construction GDP may have adopted the methodology until 2024 [24]. Such information provides some perspective about the significance of this methodology use in construction project management. Productivity gains, sustainability, control, transparency, and financial return are quite considerable. These gains, obviously, will demand not only financial outlays (capital investments), but also investments in training, reorganization of work teams, among others. With respect to models' application and an efficient change management that occurs throughout product development projects, working with BIM technology has been very efficient as well. One major reason for this is the opportunity, as several studies and practical applications have demonstrated, to organize the information coming from the different parts that integrate the complete lifecycle and products development cycle and therefore to increase mitigating change effects on projects.

Acknowledgements. The authors would like to thank the Pontifical Catholic University of Parana (PUCPR), the National Council for Scientific and Technological Development (CNPq) and Coordination for the Improvement of Higher Education Personnel (CAPES) for the financial support of this research.

References

1. Halttula, H., Haapasalo, H., Silvola, R.: Managing data flows in infrastructure projects - The lifecycle process model. J. Inf. Technol. Constr. **25**, 193–211 (2020)
2. Ullah, I., Tang, D., Wang, Q., Yin, L., Hussain, I.: Managing engineering change requirements during the product development. Concurr. Eng. **26**, 171–186 (2018). https://doi.org/10.1177/1063293X17735359
3. Eltaief, A., Remy, S., Louhichi, B., Ducellier, G., Eynard, B.: Comparison between CAD models using modification ratio calculation. Int. J. Comput. Integr. Manuf. **32**(10), 996–1008 (2019)
4. Jokinen, L., Leino, S.P.: Hidden product knowledge: problems and potential solutions. Procedia Manuf. **38**, 735–744 (2019)
5. Wright, I.C.: A review of research change management: product design into engineering implications for (1997)
6. Zhang, Y., Shi, L., Ren, S., Zhang, D.: A model-driven dynamic synchronization mechanism of lifecycle business activity for complicated and customized products. Procedia CIRP **83**, 748–752 (2019). https://doi.org/10.1016/j.procir.2019.04.234
7. Halttula, H., Haapasalo, H., Silvola, R.: Managing data flows in infrastructure projects - the lifecycle process model. J. Inf. Technol. Constr. **25**, 193–211 (2020). Web
8. Habib, H., Menhas, R., McDermott, O.: Managing engineering change within the paradigm of product lifecycle management. Processes **10**(9), 1770 (2022)
9. ISO 16739:2013 - Industry Foundation Classes (IFC) for data sharing in the construction and facility management industries (2013)
10. Eastman, C.M.: BIM Handbook: A Guide to Building Information Modeling for Owners, Managers, Designers, Engineers, and Contractors. Wiley, Hoboken (2008)
11. Schuh, G., Prote, J.P., Luckert, M., Basse, F., Thomson, V., Mazurek, W.: Adaptive design of engineering change management in highly iterative product development. Procedia CIRP **70**, 72–77 (2018)
12. Moayeri, V., Moselhi, O., Zhu, Z.: BIM-based model for quantifying the design change time ripple effect. Can. J. Civ. Eng. **44**(8), 626–642 (2017)
13. Lee, S.G., Ma, Y.-S., Thimm, G.L., Verstraeten, J.: Product lifecycle management in aviation maintenance, repair and overhaul. Comput. Ind. **59**, 296–303 (2008). https://doi.org/10.1016/j.compind.2007.06.022
14. Bilello, P.A.: Product lifecycle management: 21st century paradigm for product realization. Comput. Aided Des. **39**(2), 173–174 (2007)
15. Eastman, C.: General purpose building description systems. Comput. Aided Des. **8**, 17–26 (1976). https://doi.org/10.1016/0010-4485(76)90005-1
16. Van Nederveen, G.A., Tolman, F.P.: Modelling multiple views on buildings. Autom. Constr. **1**, 215–224 (1992). https://doi.org/10.1016/0926-5805(92)90014-B
17. Eastman, C., Teicholz, P., Sacks, R.: BIM Handbook: A Guide to Building Information Modeling for Owners, Managers, Designers, Engineers, and Contractors. Wiley-Blackwell, Hoboken (2018)
18. BIM Design Process: Brazilian Agency for Industrial Development, vol. 1, 82 p. (2017)
19. Succar, B.: Building information modelling framework: a research and delivery foundation for industry stakeholders. Autom. Constr. **18**(3), 357–375 (2009). Web
20. Georgiadou, M.C.: An overview of benefits and challenges of building information modeling (BIM) adoption in UK residential projects. Constr. Innov. **19**, 298–320 (2019). https://doi.org/10.1108/CI-04-2017-0030
21. El-Diraby, T., Krijnen, T., Papagelis, M.: BIM-based collaborative design and socio-technical analytics of green buildings. Autom. Constr. **82**, 59–74 (2017). https://doi.org/10.1016/j.autcon.2017.06.004

22. Pilehchian, B., Staub-French, S., Nepal, M.P.: A conceptual approach to track design changes within a multi-disciplinary building information modeling environment. Can. J. Civ. Eng. **42**, 139–152 (2015). https://doi.org/10.1139/cjce-2014-0078
23. Moayeri, V., Moselhi, O., Zhu, Z.: BIM-based model for quantifying the design change time ripple effect. Can. J. Civ. Eng. **44**, 626–642 (2017). https://doi.org/10.1139/cjce-2016-0413
24. National Strategy for the Dissemination of Building Information Modeling – BIM. Brazilian Government. Ministry of Industry, Foreign Trade and services (2018)

Towards a Multi-view and Multi-representation CAD Models System for Computational Design of Multi-material 4D Printed Structures

Hadrien Belkebir[1(✉)], Romaric Prod'hon[1], Sebti Foufou[2,3], Samuel Gomes[1], and Frédéric Demoly[1]

[1] ICB UMR 6303 CNRS, Université de Technologie de Belfort-Montbéliard, UTBM, Belfort, France
hadrien.belkebir@utbm.fr
[2] ICB UMR 6303, Université de Bourgogne, Dijon, France
[3] Computer Science, University of Sharjah, Sharjah, UAE

Abstract. The emerging technology of 4D printing combines additive manufacturing and active materials under energy stimulation to create objects with shape and/or property-changing capacities. Designing such structures requires careful consideration of the transformation's specifications, shape and structure, stimulation strategy, and materials selection, thereby integrating multiple perspectives constraints and knowledge. The long-term objective aims to develop a computational design synthesis for 4D printing framework that comprises generative, evaluation, and recommendation procedures. To do so, these procedures require a suitable information backbone aligned with the involved stakeholders and the design process. Therefore, we propose a multi-view and multi-representation system in a computer-aided design (CAD) environment to support generation and synthesis while streamlining design intents. An implementation is made through an CAD add-on and a case study is introduced to demonstrate its applicability.

Keywords: CAD representation · 4D Printing · Computational design

1 Introduction

4D printing is an emerging technology that combines additive manufacturing and active materials under energy stimulation, such as heat, light, electric/magnetic fields, moisture, solvent, pH, and mechanical energy, leading to adaptive, transformable, deployable, or self-assembling objects and structures. Consequently, 4D printing opens the door to innovative applications in architecture, automotive, space, and biomedical domains, among others [1]. Beyond rapid progress in single-active material 4D printing, multi-material 4D printing has gained growing attention over the last five years. This strategy involves combining active and passive materials to achieve a desired shape change, providing additional freedom in the design stage to build objects with enhanced mechanical and actuation performance [2]. However, working with active and passive materials induces

C. Danjou et al. (Eds.): PLM 2023, IFIP AICT 701, pp. 287–297, 2024.
https://doi.org/10.1007/978-3-031-62578-7_27

new challenges for spatial arrangement of materials in accordance with the stimulation strategy [3]. For example, Roudbarian et al. [4] integrate a shape memory polymer (SMP) inside an elastomeric matrix to change the deflection profile of the structure when exposed to specific temperatures. Similarly, hydrogels are used to control the shape change due to their swelling/deswelling effect when exposed to multiple stimuli such as water, solvent, light, etc. [5]. By controlling the internal structure and combining multiple active and passive materials together, possibly being hard and soft, it is possible to program a wide range of shape-change behaviors called transformations, such as folding, bending, twisting, or contraction/expansion. In addition, the meta-structure of a 4D printed part, namely the spatial arrangement of void, active and passive material in the structure that behave like a metamaterial, also has a significant impact on the object's behavior [6].

This new way of thinking and designing objects leads to a tremendous number of design solutions specifically during the embodiment design phase. For instance, the geometry definition, transformations specifications, materials selection and space arrangement, meta-structure selection and fabrication techniques involves multiple actors with different expertise, concerns, knowledge, and viewpoint [7]. These stakeholders involve diverse levels of perspective and abstract representations of the 4D-printed object. Among the stakeholders involved, the product/system architect specifies transformation functions and needs a suitable computer-aided design (CAD) representation. One current intuitive representation manipulated by both architects and designers introduces skeleton modeling. This representation allows to quickly define kinematics of an object and support mechanical design. Other CAD representations can also be found in the literature such as tree-based representations, graph-based representations, boundary-representation (B-rep), 3D meshes, or CSG representation. Feature-based representation [8] are notably suited at capturing design intents but can be complex to implement and use. Since skeletons are efficient for data manipulation and automatic generation and intuitive for mechanical experts, skeleton has been used as a backbone in this article for representing both kinematics of the object and mechanical design space.

The geometric definition of the object is rather a matter of concern of designer. The mechanical engineer, with the support of the material expert, brings more attention to the spatial arrangement of mechanical properties and materials, and requires a dedicated representation supporting advanced simulation. Finally, the process planner or maker in the context of 3D printing must select the suitable or more viable fabrication technique to be used in function of the shape, requirements in terms of quality and materials used in the overall structure.

While interconnected, these multiple perspectives or views, may sometimes conflict with each other as a change within one representation may impact others. For instance, a change of the product architect's model can exert influence on the geometric definition, consequently leading to the need for adjustments in other interconnected aspects of the design. Providing such an interdisciplinary multi-view and multi-representation tool for the actors involved in the computational design synthesis for 4D printing is challenging and strategic in the context of finding the right material and distribution of materials to realize a targeted transformation, usually named inverse design problem. This representation backbone must encompass concurrent knowledge integration as well as multi-scale representation and relies on computational design synthesis (CDS) [9], which is the fact of creating, generating and optimizing design by algorithm mean. The latter is a design methodology that leverages computational methods and algorithms to explore, generate, and optimize design solutions [10]. It involves the use of digital tools and techniques to generate and analyze large number of potential design solutions in an efficient and systematic manner, with the goal of finding the optimal solution for a given design problem. CDS encompasses a wide range of techniques, including optimization algorithms, generative design, simulation, and parametric modeling. It allows designers to quickly and easily test and evaluate numerous potential design solutions, and to make informed decisions based on data-driven insights and analysis.

Considering the wide design space involved and the high complexity arising from 4D printing, CDS seeks to speed up the design process. In the context of 4D printing, CDS has been successfully applied to the locomotion of soft robots as enhanced by van Diepen and Shea [11]. More specifically, generative mechanisms for 4D printing may include automatic generation of materials distribution at the voxel level based on artificial intelligence (AI) [12], automatic generation of meta-structures, and automatic skeleton extraction. Non-exhaustive evaluation mechanisms may include AI-based driven simulations, process suitability evaluation, skeleton or geometry analysis, and toolpath optimization. Recommendation mechanisms may consist of automatic material or process recommendation based on formal ontology [13], materials distribution, or meta-structure recommendation.

2 Multi-view and Multi-representation Model

The proposed multi-view and multi-representation model aim to capture the diverse abstraction levels of 4D printed objects and structures in the embodiment design phase is composed of four distinct views, namely as specified, as designed, as structured, and as manufactured (or printed). These views are related to the concerns of the actors involved and introduce specific representations, such as skeleton-based, geometry-based, voxel-based, and layer-based respectively. The subsequent sections provide description of the representations used and their usefulness in the design of 4D-printed structures.

2.1 Skeleton-Based Representation

The skeleton-based-representation supports the specifications of the transformation functions with a rough design structure. The object is represented by bones and joints. A joint is a connection between multiple bones and can be represented by a point, whereas a bone is a structural element linking two joints. The model of the skeleton is shown in Fig. 1.

To simulate 4D printing-induced motion, specified transformations can be applied at the skeleton level. Different types of transformations are available for the product architect and can be allocated either to the bone (i.e., bending, twisting, stretching, expansion, and contraction primitives) or the joint (i.e., folding primitive). The transformation primitives and their related high-level parameters are illustrated in Fig. 1. For instance, the twisting transformation is defined by one angle parameter which is the twisting angle.

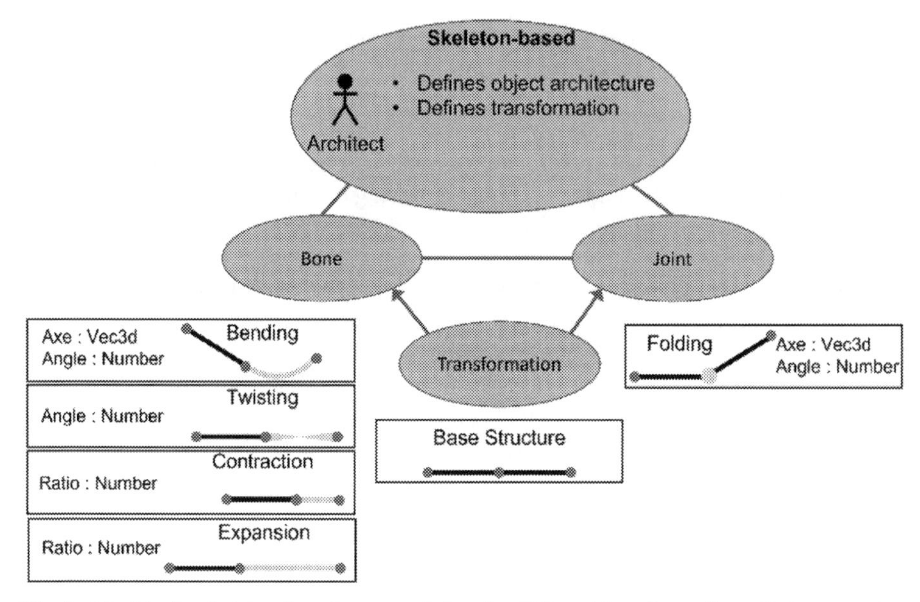

Fig. 1. Model of a skeleton-based representation for different transformations, their parameters and where they are applied.

The appropriate rotation is applied along the bone axis and will impact the overall skeleton, thus the geometry by successive kinematics. The bending transformation can be specified by a bending angle and a bending axis which is defined as a 3D vector starting from the joint. At this stage, this representation does not consider the object's geometry but rather focuses on capturing the main behavior or intended transformation to achieve. Using skeleton-based representation, the architect defines transformation requirements for each bone or joint creating a target scenario for the overall behavior of the 4D printed object. This simple and abstract representation serves as a basic core structure for other subsequent representations.

2.2 Geometry-Based Representation

Then the geometric representation is introduced to support the designer activity, which consists of the definition of a design space built upon/with the design skeletons. The design space is progressively built by introducing the boundary and the rough shape of the object. It can be defined using common CAD geometry volume like B-rep or mesh. These elements are defined by vertices (3D points) and connecting faces (triangles or square) in the framework as illustrated in Fig. 2, and allow the designer to create functional surfaces and envelopes-volumes. Each bone and joint of the skeleton-based representation is tied with a part or volume of this representation. With this representation, the designer can also specify locally or globally abstract design intent or properties. For instance, the designer may specify a region of the object/structure to be transparent, rough, colored specifically or prehensible, another region to be a functional surface that need to keep a certain shape and with specifics feature to permit future assembly. Various abstract properties are available and may also come from existing knowledge captured in ontologies [13].

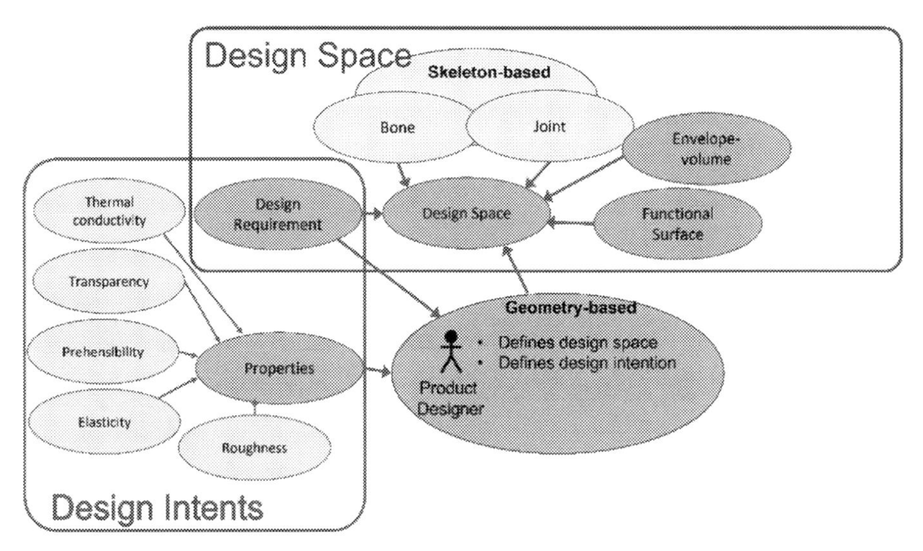

Fig. 2. Geometry-based representation with connection to the skeleton-based representation, design space elements and design intention elements

The geometry-based representation also serves as a foundation for animating the overall geometry of the structure with specified transformations represented by the skeletons [14]. Realistic mechanical rendering and body collision are not handled at this stage as it only provides insight to the designer into how the rough geometry will behave, and which design requirements must be met and where. This close connection between the skeleton- and geometry-based representations induces streamline communication between the product architect and the designer, as the skeleton model control the transformation of the part shape.

2.3 Voxel-Based Representation

The voxel-based representation is intended to be used by both material experts and mechanical engineers. Together, they must choose appropriate distribution of active and passive materials and where to apply them to the 4D printed structure. Another concern about topological optimization is also addressed here by choosing region where no matter will be placed.

The meta-structure defines the regular or irregular arrangement of matter or voids inside a 3D design space. Incorporating voids in the structure may result in weakening the object's mechanical structure but is also widely discussed in the literature as the basic idea behind topology optimization. If performed correctly, topology optimization delivers a lightweight part with a similar mechanical resistance than a massive part. More specifically in the context of 4D printing, such void meta-structures may render the actuation easier by providing multiple entry points for the stimulus to penetrate deep inside the structure. Also, thinner structures may result in increased deformations which could be suitable for 4D printing.

To ease dynamic material tuning on the voxel-based representation, the object is broken down into volume elements called voxels. Each voxel can be either void, active, passive, or even a meta-structure as illustrated in Fig. 3. Moreover, this kind of representation allow the possibilities to switch between an "abstract" representation where a single voxel is in fact a more complex distribution, and a more "concrete" view where each voxel is rendered as existing inside the structure.

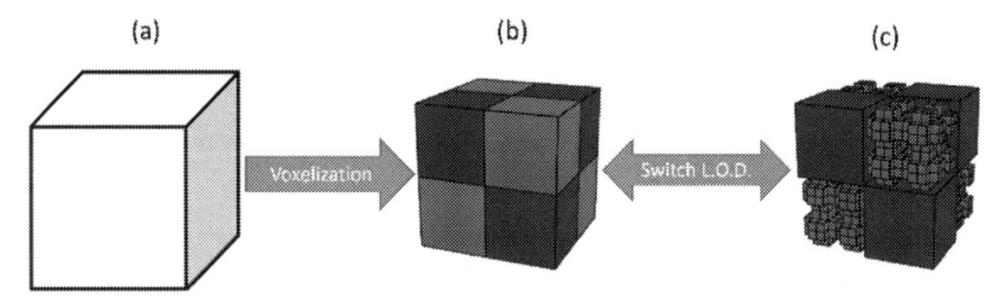

Fig. 3. Example of voxelization and matter distribution with several levels of detail (L.O.D). (a) Initial design space to voxelate. (b) Abstract distribution of active (red) and passive (blue) matter, where red voxels are defined as meta-structure. (c) Resulting distribution of the previous representation showing all the details of the meta-structure.

The global allocation of voxels and void regions defines the material distribution. Each material has a specific color allowing the material expert and the mechanical engineer to easily tune and display the material distribution. The object decomposition into voxels is based on a recursive hierarchical data structure called octree. The octree data structure provides the flexibility to choose between a simple, high-level representation or a more detailed, finer representation, depending on the needs of the current design phase. Decomposing a 3D design space into voxels of a given size is called the voxelization. Moreover, with this representation, partial or adaptive voxelization can be applied only to regions of interest where specified transformations must be fulfilled. Such regions are clearly defined inside both skeleton and geometry-based representations using the transformation location and the properties applied onto the design space.

With a such a representation, it becomes easier to simulate and analyze different scenarios, such as the deformation of the object to different loading conditions or to environmental stimuli. Realistic mechanical rendering of the structure in 4D printing has been widely studied and voxel-based representation is particularly suitable for mechanical computation and simulation of 4D printed object's behavior using direct stiffness method [6, 15]. Fast mechanical simulation of the 4D printed structure is available for the mechanical engineer and material expert. An adaptation of VoxSmart [16], a fast behavior simulator of smart material has been adapted and used to allow its connection and usage directly on the voxel-based representation of this framework.

Moreover, the simulation helps to analyze, compare, and confront the simulation with the required prescribed transformation at the skeleton level to check if the requirements

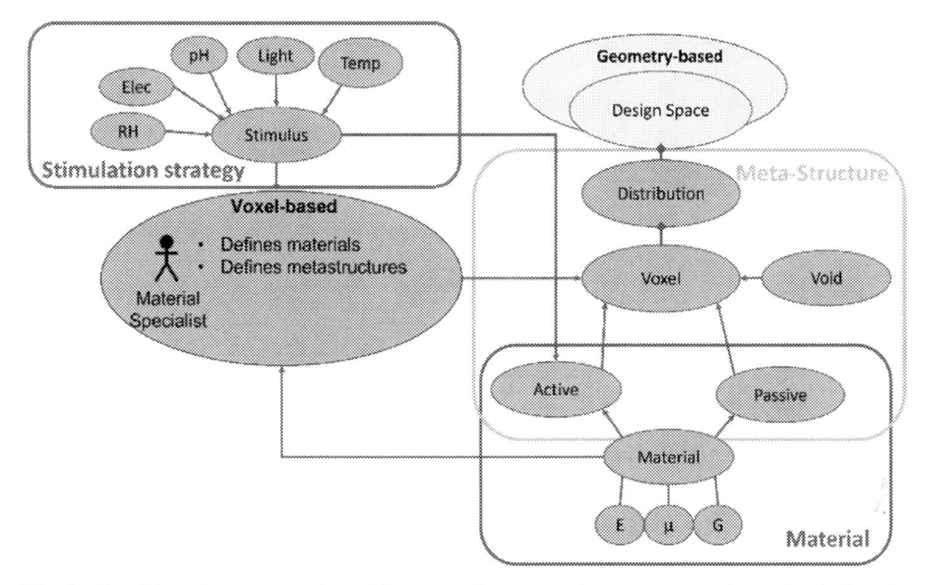

Fig. 4. Voxel-based representation with connection to previous representation, stimulation elements, meta-structure elements and material elements (E: Young's modulus, υ: Poisson's ratio, G: Shear modulus)

are fulfilled showing an intricated relationship between multiple stakeholders as shown in Fig. 4.

2.4 Layer-Based Representation

So far, the multiple representations presented above do not consider the additive manufacturing technique, which can be seen as the final step in the design for 4D printing process. This is where the process planner or maker must provide its expertise and select the appropriate printing technology. Indeed, as pointed out above, the materials selection inherently impacts the technical processes usable, generating potential design conflicts between designer, the process planner, and the material expert.

Moreover, not all materials can be printed with all existing additive manufacturing technologies. Among the available AM processes, on can use direct ink writing, fused filament fabrication, digital light processing, or material jetting to print 4D objects/structures. Some technologies require a printing path as they are based on material extrusion, some others work directly on 3D files like STL file format. The first benefit of the layer-based representation consists of proposing different outputs depending on the technique used. As 4D printing is inherently a layer-by-layer process, it is logical and intuitive for the process planner to provide a layer-by-layer representation for the 4D-printed structure. A layer-by-layer representation is much more convenient to analyze and customize toolpath with extrusion-based processes or analyze support creation or requirement. The layer-by-layer approach can also be efficiently combined with voxel-based modeling as shown in Bader et al. [17], therefore connecting again the process planner with material expert and mechanical engineer.

3 Case Study

The digital chain of the proposed representations in the context of computational design synthesis for 4D printing is illustrated through a case study: a little man which makes a step by raising and lowering opposite arm/shoulder and bending opposite legs/hips backward and forward.

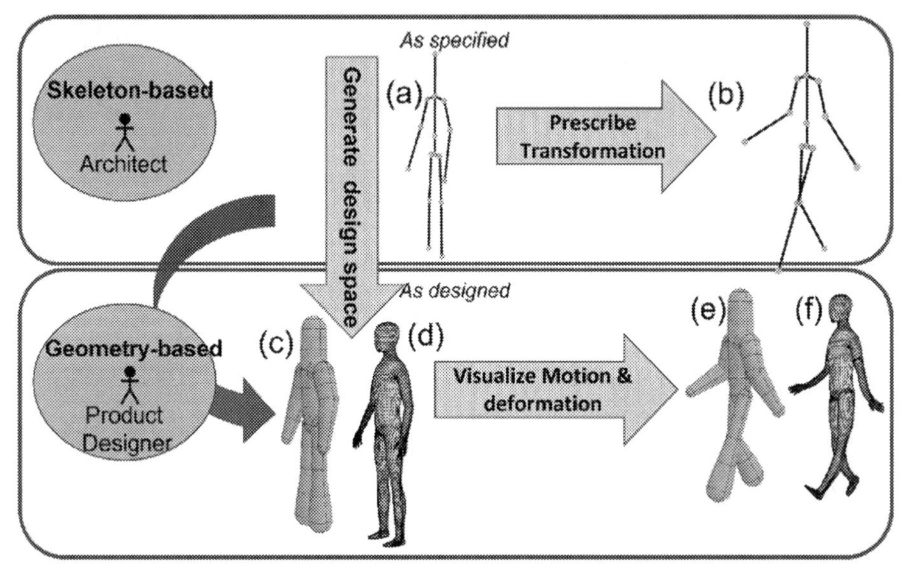

Fig. 5. Skeleton- and geometry-based representations of the little man. (a) The initial skeleton and its joints. (b) Skeleton with motion prescribed. (c) Generated design space from skeleton. (d) Complex mesh associated with the design space. (e) Generated activated design space driven by skeleton motion. (f) Complex mesh deformed by activated design space.

First, the skeleton-based representation of the model is created, then, the different transformations are created and applied as shown in Fig. 5 (a) and (b). For instance, a first transformation of bending of $-20°$ is applied on the left hips toward its V-axis, which lead to the raise of the leg. Then a bending of $30°$ is also applied to the right knee on its V-axis to make the foot of the little man going back. Finally, two bending of -45 and $45°$ are applied on right elbow and left shoulder respectively, forming the as specified view. Then, the product designer can generate the design space by assigning spheres of different radius on each joint to create an initial volume for working illustrated in Fig. 5 (c) and (e). The designer can also refine the shape of the model by adding or creating a more detailed body inside the initial design space as shown in Fig. 5 (d) and (f), forming the as designed view.

Starting from the previous view, the material specialist can easily select region to voxelize and applying a materials distribution and associated meta-structure to the model to achieve the desired transformation as illustrated in Fig. 6 (a). To realize bending transformation, different distribution of material using a hydrogel as active material and

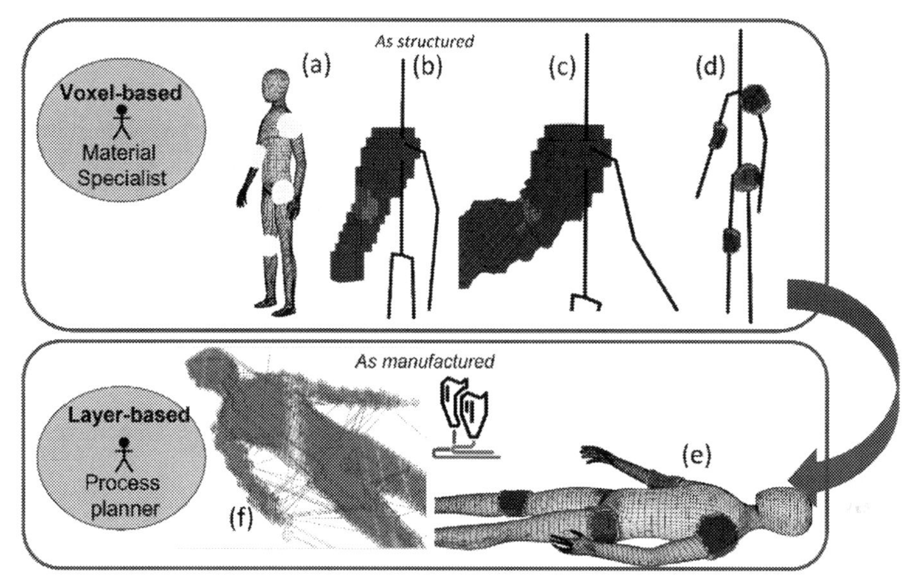

Fig. 6. Voxel and layer-based representations of the little man. (a) Regions of interest are in yellow. (b) Distribution of materials in the subpart (elbow) of the little man. (c) Resulting motion when simulated in VoxSmart. (d) Visualization of the four distributions. (e) Material jetting configuration with separated part for each material. (f) FFF Toolpath for printing

polylactic acid (PLA) as passive material are defined as depicted in Fig. 6 (b). With the help of the first two views, the material specialist can simulate the deformation behavior of a subpart of the object inside VoxSmart [16] as visualized in Fig. 6 (c) to validate the transformation thus forming the as structured view. Finally, the as structured view is exploited to generate the as manufactured view to provide the printing files which can be STL files for each material for a material jetting printer or G-Code path for a FFF printer as described in Fig. 6 (e) and (f) respectively.

4 Conclusion

The proposed approach of a multi-view and multi-representation system for 4D printing design has shown the potential to synthetize and centralize different stakeholder viewpoints with their own concerns and outlook into the same digital workflow. It also shows the possibility to extend and connect existing work into the workflow, allowing further development and extension of this work. Moreover, CDS serves as the foundation for future developments of a more complete framework, where additional sources of intelligence, such as knowledge database or machine learning can be integrated to provide generative, evaluation and recommendation mechanisms adapted to 4D printing. At different levels, these mechanisms can assist the stakeholders in the rapid determination of a complex design solution. Currently, the stakeholder needs to know how active materials behave and where to assign them, but in the future, a more complete CDS framework will automatically match stakeholder needs resolving conflicts automatically between prescribed transformations, design intents, material selection and arrangement, and additive manufacturing technique selection. As a perspective, machine learning can be used to generating counter-intuitive material distributions or meta-structures, improving mechanical computation for 4D printing simulation or prescribing a material distribution to achieve a target transformation. This work opens exciting possibilities and demonstrates the great potential of future works joining 4D printing design with artificial intelligence to unlock full potential of technology.

References

1. Kuang, X., et al.: Advances in 4D printing: materials and applications. Adv. Funct. Mater. **29**, 1805290 (2019). https://doi.org/10.1002/adfm.201805290
2. Shahrubudin, N., Lee, T.C., Ramlan, R.: An overview on 3D printing technology: technological, materials, and applications. In: Procedia Manufacturing, pp. 1286–1296. Elsevier B.V. (2019)
3. Bickel, B., et al.: Design and fabrication of materials with desired deformation behavior. ACM Trans. Graph. **29**, 1 (2010). https://doi.org/10.1145/1778765.1778800
4. Roudbarian, N., Baniasadi, M., Nayyeri, P., Ansari, M., Hedayati, R., Baghani, M.: Enhancing shape memory properties of multi-layered and multi-material polymer composites in 4D printing. Smart Mater. Struct. **30**, 105006 (2021). https://doi.org/10.1088/1361-665X/ac1b3b
5. White, E.M., Yatvin, J., Grubbs, J.B., Bilbrey, J.A., Locklin, J.: Advances in Smart Materials: Stimuli-Responsive Hydrogel Thin Films. Wiley (2013)
6. Athinarayanarao, D., et al.: Computational design for 4D printing of topology optimized multi-material active composites. NPJ Comput Mater. **9**, 1 (2023). https://doi.org/10.1038/s41524-022-00962-w
7. Demoly, F., Dunn, M.L., Wood, K.L., Qi, H.J., André, J.C.: The status, barriers, challenges, and future in design for 4D printing. Mater. Des. **212**, 110193 (2021). https://doi.org/10.1016/j.matdes.2021.110193
8. Case, K., Gao, J.: Feature technology: an overview. Int. J. Comput. Integr. Manuf. **6**, 2–12 (1993). https://doi.org/10.1080/09511929308944549
9. Roucoules, L., Demoly, F.: Multi-scale and multi-representation CAD models reconciliation for knowledge synthesis. CIRP Ann. **69**, 137–140 (2020). https://doi.org/10.1016/j.cirp.2020.04.089
10. Cagan, J., Campbell, M.I., Finger, S., Tomiyama, T.: A framework for computational design synthesis: model and applications. J. Comput. Inf. Sci. Eng. 5(3), 171–181 (2005). American Society of Mechanical Engineers Digital Collection
11. van Diepen, M., Shea, K.: A spatial grammar method for the computational design synthesis of virtual soft locomotion robots. J. Mech. Design Trans. ASME. **141** (2019). https://doi.org/10.1115/1.4043314
12. Sossou, G., Demoly, F., Belkebir, H., Qi, H.J., Gomes, S., Montavon, G.: Design for 4D printing: modeling and computation of smart materials distributions. Mater Des. **181**, 108074 (2019). https://doi.org/10.1016/j.matdes.2019.108074
13. Dimassi, S., et al.: A knowledge recommendation approach in design for multi-material 4D printing based on semantic similarity vector space model and case-based reasoning. Comput. Ind. **145**, 103824 (2023). https://doi.org/10.1016/j.compind.2022.103824
14. Sumner, R.W., Zwicker, M., Gotsman, C., Popović, J.: Mesh-based inverse kinematics. ACM Trans. Graph. **24**, 488–495 (2005). https://doi.org/10.1145/1073204.1073218
15. Hiller, J., Lipson, H.: Dynamic simulation of soft multimaterial 3D-printed objects. Soft Robot. **1**, 88–101 (2014). https://doi.org/10.1089/soro.2013.0010

16. Sossou, G., Demoly, F., Belkebir, H., Qi, H.J., Gomes, S., Montavon, G.: Design for 4D printing: a voxel-based modeling and simulation of smart materials. Mater. Des. **175**, 107798 (2019). https://doi.org/10.1016/j.matdes.2019.107798
17. Bader, C., et al.: Making data matter: voxel printing for the digital fabrication of data across scales and domains. Sci. Adv. **4**, eaas8652 (2018). https://doi.org/10.1126/sciadv.aas8652

A State of the Art of Collaborative CAD Solutions

Hugo Locquet[1,2(✉)], Louis Rivest[1], and Matthieu Bricogne[2]

[1] École de Technologie Supérieure, Montréal, Canada
hugo.locquet.1@ens.etsmtl.ca

[2] Université de Technologie de Compiègne, 60200 Compiègne, France

Abstract. Despite the ever-growing need for speed in conception, computer-aided design (CAD) has stayed confined to the PLM paradigm, which lets only a single user at a time access and modify a CAD document. A new type of CAD, called collaborative CAD, has emerged to overcome this conundrum. Tasks can be parallelized with much more ease when multiple users can access a document simultaneously. Two main currents exist in collaborative CAD. The first one is synchronous work, which involves multiple users working in the same document simultaneously. The second one is asynchronous work, which involves each user working in a copy of the document and then all users' work being merged at the end. In this paper, the workflows of PLM-based CAD and collaborative CAD are compared with a concrete example, the capabilities and limitations of current collaborative CAD solutions are identified and analyzed, and future perspectives regarding this work are presented.

Keywords: Computer Aided Design · Collaborative Computer Aided Design · Synchronous Computer Aided Design · Asynchronous Computer Aided Design

1 Introduction

A new approach to product design, called collaborative computer-aided design (CAD) is emerging. It has the advantage of letting multiple designers work on a CAD document simultaneously. This speeds up the design process and makes it possible for multiple users with different competencies to contribute to the design at the same time. To make an analogy with word processing and spreadsheet software, collaborative CAD is akin to Microsoft 365 Word[1], Google Docs[2], Collabora Office[3] or Nextcloud Office[4], which allow multiple users to modify a document together in real time. To continue with the

[1] https://support.microsoft.com/en-us/office/collaborate-on-word-documents-with-real-time-co-authoring-7dd3040c-3f30-4fdd-bab0-8586492a1f1d, last consulted 20/02/2023.

[2] https://support.google.com/docs/answer/2494822?hl=en&co=GENIE.Platform%3DAndroid, "Share & collaborate on a file with many people", last consulted 20/02/2023.

[3] https://www.collaboraoffice.com/, last consulted 20/02/2023.

[4] https://nextcloud.com/office/, last consulted 20/02/2023.

Published by Springer Nature Switzerland AG 2024
C. Danjou et al. (Eds.): PLM 2023, IFIP AICT 701, pp. 298–308, 2024.
https://doi.org/10.1007/978-3-031-62578-7_28

same comparison, the classic version of Microsoft Word, or Open Office, both of which allow only one user at a time to edit a document shared through SharePoint or a cloud service, would correspond to the classic way of designing a product, using a product lifecycle management (PLM) system. PLM aims to streamline the flow of information about a product and its related processes throughout the product's lifecycle [1].

One of PLM's numerous tools and functionalities is product data management (PDM), whose purpose is to act as an electronic vault for all product-related documentation. In a PDM platform, a user typically needs to *check out* a document to edit it. When they are finished editing the document, they *check in* the document to return it to the vault for the next person to use. Said user is the sole editor of the document while they have it checked out. If anyone tries to modify a document while it is checked out by another user, they will see an error message. This is called *pessimistic* conflict management [2]. The way in which documentation is managed is different from one software program to the next, as it is demonstrated later. This tedious way of managing documents can induce backlogs when complex documents like CAD models are involved. Each designer must wait their turn to work on a document. The serial way of designing seems dissonant with the current era of simultaneous engineering.

Collaborative CAD, which is also known as multi-user CAD (MUCAD) or *optimistic* conflict management, has been developed to be more compatible with simultaneous engineering. Its goal is to reduce, if not render obsolete, the need to check a document into and out of a PDM platform. Not only does collaborative CAD reduce design time, but users are more satisfied performing CAD collaboratively than alone according to Zhou, Phadnis and Olechowski [3].

There are two ways of working in collaborative CAD: *synchronously* and *asynchronously*. Synchronous work means that users are working on the same CAD document together in real time, with different parameters defining each user's area of work, like users being restricted to working on different paragraphs in a collaborative word processing solution. Asynchronous work involves users each working in a copy of the CAD document and their work being incorporated in the main document at different moments using various methods.

The objective of this article is to give a precise state of the art of collaborative CAD in terms of both research and some of the current commercial software. That is done through a literature review, the reporting of the results of tests conducted on three CAD software programs – 3DEXPERIENCE, Fusion 360 and Onshape – to explore their synchronous work capabilities, and a further study and analysis of Onshape's collaboration capabilities, in that order.

2 Literature Review: Advancements in Collaborative CAD

By diverging from the pessimistic approach and its document unicity, collaborative CAD opens itself to conflicts between copies. Hepworth et al. established two kinds of conflicts for feature-based systems [2]: *semantic* and *syntactic*. Semantic conflicts happen when two users don't have the same understanding of the original design intent. Since each interpretation is valid, the conflict must be resolved by getting the two parties to come to an understanding. Syntactic conflicts revolve around the fact that two copies are no longer geometrically sound because a change has been made. Three types of syntactic conflicts have been identified. The first one is called *feature/self* and is the result of a feature being modified in both copies. When the copies are merged, both versions of the feature cannot coexist as they are supposed to represent the same thing. The second is called *parent/child* and is the result of a parent feature having been changed and a child of that feature not being able to be computed anymore. The third type of syntactic conflict is called *child/child*. It is similar to the feature/self conflict but different in that the feature didn't exist prior to the work of both users. The features created are children of the same parent and cannot coexist as they respond to the same design intent, but with a different interpretation.

The main framework proposed for the simultaneous use of a CAD document by multiple users is to create an editable copy of the document for each user. In most cases, the CAD document is stored on a server, and multiple users can access it. The conflict management method and communication between the server and a user can differ from one document to the next, but the layout of the framework is often similar to what is shown in Fig. 1.

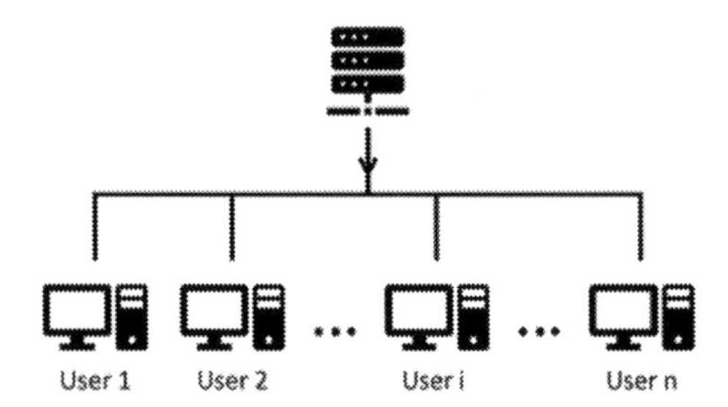

Fig. 1. Framework for multiple users to access the same document

The *first level* of conflict management in collaborative CAD proposes an approach that is neither pessimistic nor optimistic, but rather *hybrid*, and borrows from both of those approaches. It adopts the optimistic approach's ability for multiple users to access a document and the pessimistic approach's lock feature. In the hybrid approach, however, said lock exists *within* a CAD document. This method moves the collaboration limit's atomic value from the document to a semantic part of it.

Hepworth et al. proposed to use the CAD document feature as the atomic value. This means that when a user locks one or more feature(s) in the framework, this information is relayed to the server, which transmits it to the other users by preventing them from editing the locked features, as is illustrated in Fig. 2 [2]. To make this solution more user-friendly, Moncur et al. proposed some visual improvements such as indicating on the CAD representation the areas that are locked, and a user interface that decomposes a document into its features and reports in real time their state (locked or not) and the user with edit permission [4].

Red et al. took a similar approach, but with more flexibility in terms of the atomic value. In their approach, there are three levels of decomposition, classified in order of complexity they bring [5]. The first one, at the lowest level, is feature-based. In this

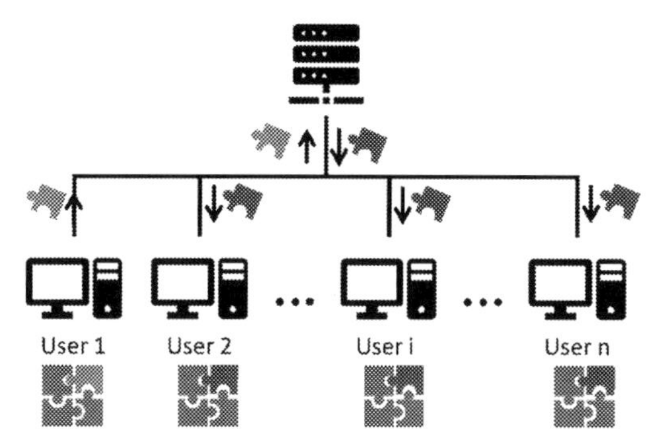

Fig. 2. Feature-level lock

case, the elements of the model serve no purpose because the user determines which features to lock. The second one, at the middle level, is based on the experience and role of the user. In this case, some semantics are needed, as the location that is locked must correspond to the user. For example, if a bearing engineer opens the file, the bearing components in it are automatically locked. The third and most complex one is based on the specifications. Each part of the CAD model is decomposed and locked based on the engineering specifications. Take a simple part like a screw, for example. (It does not need multiple users to be designed, this is for illustration purposes.) Its specifications would be the type and size of head, its diameter, and the length and type of thread. The specifications could be separated into two topics: the head and the body. Hence, one user could lock all the features composing the head, while another could do the same with the body.

The *second level* of conflict management eliminates all notions of checking in and out. Instead, the communication between the server and the users increases in frequency. All copies of a CAD document are connected to the server at all times. Every modification made in one copy is carried over to the others to constantly keep all users working in the same version, as is stated by Hepworth et al. If two operations were to happen simultaneously, the first one to arrive to the server is the only one sent to update the other copies. The second operation is sent back to the user who performed it and needs to be fitted to the updated version of the document [6]. A solution to respond to this hurdle is proposed by Jing et al. They propose to store the document containing the second operation as a local copy. One of the problems encountered is the naming of the topological entities. If the new version of the document modified topological entities used in the local copy, the parent-child link would be broken. Two solutions are proposed to remedy this. The first one is to maintain a link between two topological entities with identical geometry but different names. The second is to create a common name for topological entities that were originally one and the same, as shown in Fig. 3, where face F1 and edge E1 have been split into F1.1 and F1.2, and E1.1 and E1.2, respectively [7]. Cheng et al. improved upon this by identifying every operation so the metadata of the CAD document contains a history of all the operations performed. In the case of a conflict, it is transferred to a history buffer, and all the conflicting operations are tried together by an algorithm to propose a solution for the conflict [8].

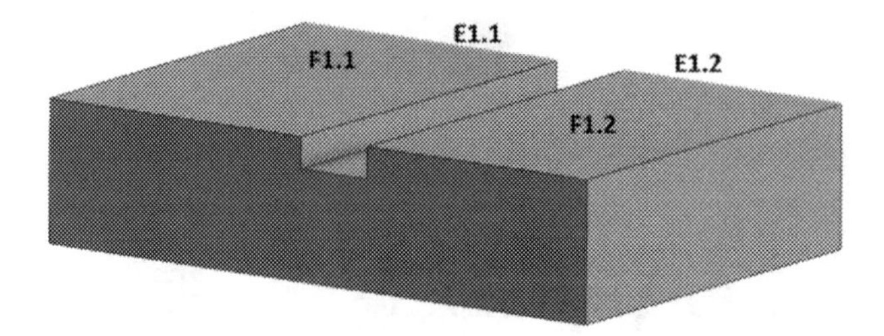

Fig. 3. Naming new topological entities based on their common ancestors

In that case, the first operation to arise would prevail. For Yu et al., however, prioritization is incomplete and should depend on the context. They propose to assign each operation a weight based on four criteria: designer authority, time order, operation type and model compatibility. The importance of these criteria can be tuned accordingly to the needs of the project [9].

All these solutions have been presented to resolve conflicts. But Hepworth et al. went one step further and proposed a solution to reduce the occurrence of conflicts. For context, they used the first level of conflict management, in which the feature is the atomic value. They incorporated two tools into a CAD software program. The first is a chat tool to facilitate communication while users modify a CAD document. The second is a task management tool whose purpose is to organize tasks beforehand. Doing so enables each user to know their own and others' areas of work and thus limits conflicts [10]. In addition, Stone et al. proposed a method to determine the optimal number of users needed to work on the same CAD model. Knowing this information would prevent a redundancy of tasks for the tool presented above. This method uses the taxonomy of the CAD document being worked on and the feature as the atomic value. The output is a tree describing the document. An optimal number of users can be determined from the shape (height and width) of the tree and the number of branching levels [11].

All these solutions have been presented in a synchronous CAD environment, but the collaborative CAD can also be performed asynchronously. While an analogy could be made between the previous methods and text editing software, the following method stems from the IT domain, like Git https://git-scm.com/book/en/v2/Git-Branching-Basic-Branching-and-Merging[5]. A lot of programming involves collaborative work on software files, and a lot of tools exist to manage this. Bricogne et al. explain this from a slightly different point of view than the IT one [12]. The tool highlighted in this case is branch and merge (B&M). Branch, as its name implies, creates a new branch in the CAD document's versioning tree. Merging can be done using two different methods: *2-way merge* and *3-way merge* [13]. In two-way merge, the two branches of the document are compared, and the differences highlighted. Depending on the algorithm used by the merge tool, the entirety of one document may overthrow the other, the differences may be able to be reviewed one by one, with the user choosing which operations to keep, or a rule (timeline, designer rank, etc.) may determine which operations prevail. In three-way

[5] https://gitscm.com/book/en/v2/GitBranchingBasicBranchingandMerging, last consulted 05/05/2023.

merge, branches are compared between themselves and to the node they originate from. The merging options are similar to those available for 2-way merging.

As is explained above, collaborative CAD is split in two main categories: synchronous and asynchronous. The former is the most similar to PLM-based CAD and vastly more well-known than the latter. Tests were conducted on some CAD software programs to explore their collaboration capabilities and similarities with the literature. The testing results are reported next and bring to light another way of designing a product using currently available software with concrete examples.

3 Experimenting with Collaborative CAD in Synchronous and Asynchronous Scenarios

In this chapter, the testing methodology used is explained, and then the results of tests highlighting synchronous and asynchronous collaborative CAD are explored.

3.1 Methodology

To unravel the collaborative capabilities of the three software programs chosen, a test case was created that is as similar as possible to an industrial scenario. The software programs considered were chosen for their collaboration capabilities. Siemens's Teamcenter[6] and NX, PTC's Windchill[7] and Creo, 3DEXPERIENCE[8], and Fusion 360[9] all use a PLM-based architecture and therefore take a pessimistic approach. 3DEXPERIENCE and Fusion 360 were chosen from these programs for their integrated PDM-CAD solutions as well as their availability for testing. They are compared to Onshape, which takes an optimistic approach, as it is demonstrated below.

In the test case, independent factors have compelled two departments within the same aerospace company to modify a CAD document and two small changes made to different aspects of a part have resulted in a conflict. The first user, an engineer from the Engineering department (EE), has to locate and modify the parts that are affected by a certain issue. In this case, the wall thickness of part P and all of its series has to be modified. EE then modifies all items in the P series to thicken their walls. In parallel, an engineer from the Manufacturing department (EM) needs to redesign some parts to fit them to new tools. Part P is one of the parts in question, as its corner and fillet radii were designed for the former tools. EM is tasked with applying the change to the whole P series. EE and EM are therefore modifying the same part simultaneously.

The test involved two users working simultaneously, and both of their screens were recorded. Since the purpose of the test was to verify the programs' collaboration capabilities, not their performance, there are no issues with it being conducted by a single pair of participants.

[6] https://plmcoach.com/teamcenter-plm-architecture/, last consulted 20/02/2023.

[7] https://www.ptc.com/en/products/windchill/architecture-and-deployment, last consulted 20/02/2023.

[8] https://www.3ds.com/cloud/plm-innovation-platform, last consulted 20/02/2023.

[9] https://www.autodesk.com/autodesk-university/class/Bring-It-All-Together-Fusion-Between-PDM-and-PLM-2019, last consulted 20/02/2023.

The test was run twice for the pessimistic programs. The first time it was run, the robust steps used in the industry were followed, meaning the part was reserved, the modifications were made, and then the part was released. The purpose of doing this was to represent this way of working and have a baseline against which to compare the optimistic approach. The second time the test was run, a collaborative CAD-type workflow was used to observe first-hand the pessimistic software's limitations. This means that the two users tried to modify the same document simultaneously, and the software's response was observed. The test was run only once for the optimistic software because the industry-standard workflow does not exist for it.

3.2 Exploration of Synchronous Work

Software with a Pessimistic Approach

The tests conducted confirm that although the workflow from one software program to another is different, the pessimistic way of working is the core of the paradigm used by 3DEXPERIENCE and Fusion 360. In the first run of the test, 3DEXPERIENCE had the following workflow: a document has to be checked out from the PDM before it can be opened in the CAD software. Once the document has been opened, modifications can be made and saved. Afterwards, the user needs to go back into the PDM to check the document back in before the next user can check it out and follow the same procedure, as Fig. 4 shows. An explicit checkout process can be used. As for Fusion 360's workflow, the user only has to open the document and modify it. The action of checking the document out and in is implicit. It happens the moment a user creates a new feature or edits an existing one, as shown in Fig. 4.

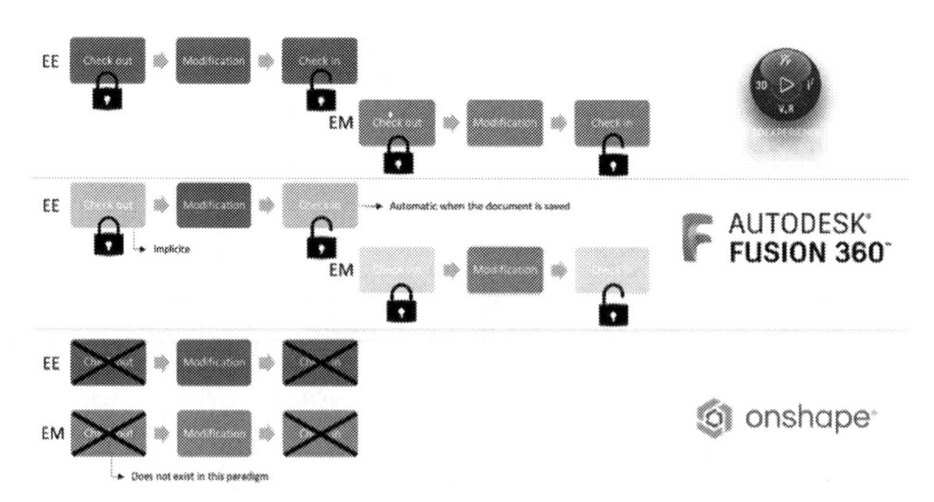

Fig. 4. Diagram of the synchronous workflow

In both cases, the document can be opened by other users simultaneously, but cannot be modified. The second run of the test, in which two users open the document and modify the same part simultaneously, highlights this. In 3DEXPERIENCE, both users could open and modify the document; however, only the first to open the document could save their changes made to the document. The other user received an error message saying that the document is already opened by another user. The only way they can save their modifications is by either creating a new version or saving the file as a new document. In both cases, the work of the two users cannot be easily combined. This happens because the check out and in steps were not properly done in accordance with the recommended guidelines. Had the users gone through the PDM, the first user would have checked out the document and the second user would have seen that the document was already checked out. By bypassing the checkout process, the software never warned the second user that their work could not be saved before they made changes and tried to save them. This result shows that the software relies heavily on its protocols and the publisher's vision of the workflow. If a single step is missed, a lot of work can be lost. This means that all users need to learn and understand the basics of PLM to efficiently use 3DEXPERIENCE. However, the program may be able to support implicit checkout with a customized configuration. The standard version was used for testing.

As for Fusion 360, the checkout process is implicit and happens directly in the CAD software instead of in a PLM platform. When the first user modifies a document, all other users are automatically locked out of the document. If a second user tries to modify the document, an error message immediately warns them that the document is already being modified by someone else. In addition, an icon appears in the CAD document indicating it is being modified. This approach lets users browse a little more freely by incorporating various discrete fail-safes. Less-thorough knowledge of PLM and version control are required to use Fusion 360.

Software with an Optimistic Approach

When testing Onshape, the two users could access and modify the document at the same time. Each user could see changes happen in real time as they worked. Additionally, both users could see which feature the other was working on, with an icon displayed next to the feature in the tree. The work took less time to complete as both tasks were done at the same time, as illustrated previously in Fig. 4. For testing purposes, both users tried to open the same feature simultaneously and modify the same value. When the first user opened the value window and modified it, the second could see it change in real time. The same thing occurred for a sketch, which has a smaller atomic value than the smallest atomic value proposed in the literature. While it was not used in this specific test, a tool does exist that lets a user adopt the point of view of any other user.

3.3 Exploration of Asynchronous Work

One of the software programs chosen also features asynchronous working tools and a branch-and-merge solution. As stated on its website[10], Onshape was inspired by Git-flow version management and its strong agile philosophy. It creates a lifecycle timeline for every document. Its representation is based on three core elements: a dot, which corresponds to the version; a thick line, which corresponds to the main document; and

[10] https://learn.onshape.com/learn/article/gitflow-version-management, last consulted 20/02/2023.

lines branching off of the thick line, which correspond to the branches, as Fig. 5 shows. A branch stems from a version of the main document and contains all the history up to that version. All future modifications made to the main document will not impact the branch, and vice versa. A user then applies all modifications needed to the branch. Any branch can be compared with another or with the main document at any moment. Comparison is both graphic, with a 3D model representing the differences, and textual, with all the features modified or created in a branch or deleted from a branch mentioned, as illustrated in Fig. 5.

When the modifications done on a branch are sufficiently mature, it can be merged with another branch or the main document. The merging order is important, as the software's behavior is dependent on it. Two entities are recognized: the source, which is the branch containing the modifications, and the target, which is the line it is applied to. When a merge happens, there are three possibilities. The first is to overwrite the target with the source. The second is to combine the target and the source by either searching for a common ancestor and applying the modifications, or adding/removing features. The third is a mirror of the first possibility, as the target overwrites the source.

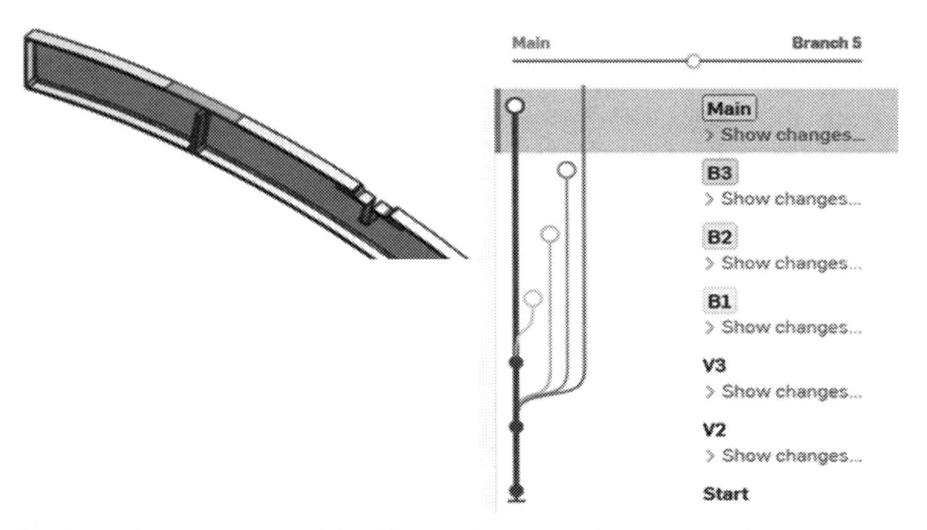

Fig. 5. Graphic representation of the differences between two branches (left); Branches of a part in Onshape (right)

The Shortcomings of Asynchronous Work

The two main shortcomings of asynchronous work are the level of detail the compare tool is limited to an imprecise atomic value, and the absence of a link between the elements of the model and the intent of the designer, both of which are explained below.

A description of Onshape's document structure is necessary to explain these points. A document groups everything about the part/product it represents, including its features and 3D model, part assembly constraints, drawings, and materials. All of this is decomposed into tabs of different attributes. For example, there is a specific tab called

Part Studio in which one or more parts can be created or modified, and another tab called Assembly where all assembly constraints and animations are created and executed. When a merge is requested, and the three choices are proposed. To make an analogy with synchronous work, the tab seems to be the atomic value Onshape uses for its merge tool. The atomic value could be reduced if the element in the tabs could be chosen like the features compared in Fig. 5. Additionally, when comparing a branch with another branch or the main document, all the features are listed as they appear in the history tree. If users didn't name them as they worked, a complex part may have numerous different features with generic names, which renders comparison incredibly difficult. A folder tool exists when editing a part, to be able to classify a group of features. However, the folder is not accounted for in comparison. This means any shred of intent the designer input into the document is lost. To continue with the analogy to IT, this would be equivalent to omitting all the commentary in a software file.

4 Conclusion

A lot of research has been done exploring the possible future of computer-aided design (CAD). CAD has been used for a long time in the PLM paradigm, most often in combination with a PDM platform to ensure version control is as robust and secure as possible. Most attempts to create an alternative that allows multiple users to edit a document at the same time have involved introducing an atomic value, which lessens the rigidity of PDM, but they have essentially been adaptations of the pessimistic way of approaching CAD. With its Onshape software, PTC proposes a novel solution that dissociates PDM and CAD by relying on Gitflow version management to handle the version control of CAD documents. It has created a new paradigm that supports both completely synchronous and asynchronous CAD. The testing done to compare the pessimistic and optimistic CAD software programs showed a clear difference in the workflow and the time needed to edit a document. The asynchronous branch-and-merge solution showed another possible way of collaborating, in which users can each work in a separate copy of a document and pool their efforts by merging their work afterward.

Another aspect of these last two solutions for collaborative CAD is their connectivity capacity. While a synchronous collaborative CAD solution requires a constant connection to the server and the network to interact with other users, an asynchronous CAD solution does not. Once a branch has been created, the asynchronous solution does not require connectivity for modifications, only for merging.

To conclude, three approaches to CAD are presented in this article. The first one, and the most widely used, is the PLM-based pessimistic approach. It revolves around the unicity of information and the robustness of use to the detriment of certain collaboration capabilities. The second one is the optimistic approach, which prioritizes collaboration between users even if it induces some conflicts between their work. Finally, the hybrid approach is a solution that falls in between the pessimistic and optimistic approaches. While it emphasizes collaboration, it limits itself to avoid conflicts by implementing certain sets of rules. Two ways of working – synchronously and asynchronously – are also highlighted for the optimistic and hybrid approaches.

References

1. Mas, F., Arista, R., Oliva, M., Hiebert, B., Gilkerson, I., Rios, J.: A review of PLM impact on US and EU aerospace industry. Procedia Eng. **132**, 1053–1060 (2015). https://doi.org/10.1016/j.proeng.2015.12.595
2. Hepworth, A.I., Tew, K., Nysetvold, T., Bennett, M., Jensen, C.G.: Automated conflict avoidance in multi-user CAD. Comput.-Aided Des. Appl. 11(2), 141–152 (2014). https://doi.org/10.1080/16864360.2014.846070
3. Zhou, J., Phadnis, V., Olechowski, A.: Analysis of designer emotions in collaborative and traditional computer-aided design (2019)
4. Moncur, R.A., Jensen, C.G., Teng, C.C., Red, E.: Data consistency and conflict avoidance in a multi-user CAx environment. Comput.-Aided Des. Appl. 10(5), 727–744 (2013). https://doi.org/10.3722/cadaps.2013.727-744
5. Red, E., Marshall, F., Weerakoon, P., Jensen, C.G.: Considerations for multi-user decomposition of design spaces. Comput.-Aided Des. Appl. **10**(5), 803–815 (2013). https://doi.org/10.3722/cadaps.2013.803-815
6. Hepworth, A., DeFigueiredo, B., Shumway, D., Fronk, N., Jensen, C.G.: Semantic conflict reduction through automated feature reservation in multi-user computer-aided design. In: 2014 International Conference on Collaboration Technologies and Systems (CTS), Minneapolis, pp. 56–63. IEEE, Mai (2014). https://doi.org/10.1109/CTS.2014.6867542
7. Jing, S.x., He, F.z., Han, S.h., Cai, X.t., Liu, H.J.: A method for topological entity correspondence in a replicated collaborative CAD system. Comput. Ind. **60**(7), 467–475 (2009). https://doi.org/10.1016/j.compind.2009.02.005
8. Cheng, Y., He, F., Wu, Y., Zhang, D.: Meta-operation conflict resolution for human–human interaction in collaborative feature-based CAD systems. Clust. Comput. **19**(1), 237–253 (2016). https://doi.org/10.1007/s10586-016-0538-0
9. Yu, M., Cai, H., Ma, X., Jiang, L.: Symmetry-Based Conflict Detection And Resolution Method Towards Web3D-Based Collaborative Design. https://doi.org/10.3390/sym8050035
10. Hepworth, A., Halterman, K., Stone, B., Yarn, J., Jensen, C.G.: An integrated task management system to reduce semantic conflicts in multi-user computer-aided design. Concurr. Eng. Res. Appl. **23**(2), 98–109 (2015), https://doi.org/10.1177/1063293X15573595
11. Stone, B., et al.: Methods for determining the optimal number of simultaneous contributors for multi-user CAD parts. Comput.-Aided Des. Appl. **14**(5), 610–621 (2017). https://doi.org/10.1080/16864360.2016.1273578
12. Bricogne, M., Rivest, L., Troussier, N., Eynard, B.: Concurrent versioning principles for collaboration: towards PLM for hardware and software data management. Int. J. Prod. Lifecycle Manag. **7**(1), 17–37 (2014). https://doi.org/10.1504/IJPLM.2014.065457
13. Bricogne, M., Troussier, N., Rivest, L., Eynard, B.: Angile design methods for mechatronics system integration. In: Bernard, A., Rivest, L., Dutta, D. (eds.) Product Lifecycle Management for Society. IFIP Advances in Information and Communication Technology, vol. 409, pp. 458-470. Berlin, Heidelberg (2013). https://doi.org/10.1007/978-3-642-41501-2_46

KARMEN: A Knowledge Graph Based Proposal to Capture Expert Designer Experience and Foster Expertise Transfer

Jean René Camara[1,3](✉), Philippe Véron[2], Frédéric Segonds[1], Esma Yahia[2], Antoine Mallet[3], and Benjamin Deguilhem[3]

[1] LCPI, Arts et Métiers Institute of Technology, HESAM Université, 75013 Paris, France
jean-Rene.CAMARA@ensam.eu

[2] LISPEN, Arts et Métiers Institute of Technology, HESAM Université, 13617 Aix-en-Provence, France

[3] Capgemini Engineering R&D – AEROPARK 3 Chemin de Laporte, 31300 Toulouse, France

Abstract. [Context] At the cusp of Industry 4.0 and against a backdrop of fierce competition, manufacturing companies must design and manufacture increasingly complex and cost-effective products. Human resources must therefore preserve and maintain their knowledge and the intellectual heritage of their experts.

[Problem] In the next few years, there will be a lack of skilled resources in the manufacturing industry due to retirements. Let's also mention the turnover of consultants working within these companies. It is essential to implement solutions today in order to protect the intellectual heritage of tomorrow. This paper ambition to answer to how can the knowledge of these experts be captured and used, and how knowledge graph could be a suitable tool to achieve this objective.

[Proposal] This article proposes a methodology for implementing **KBE** (**K**nowledge **B**ased **E**ngineering) solutions. This methodology called **KARMEN** (**K**nowledge **A**ccess **R**equest for **M**anufacturing and **E**ngineering by **N**etwork graph) is based on an **FBS** type ontology (**F**unction, **B**ehavior, **S**tructure) as well as on the exploitation of **K**nowledge **G**raphs.

A use case of redesigning a mechanical part for metal additive manufacturing will be presented. Besides, an experimental protocol will be specified to capture the knowledge of business experts within a graph-oriented database built on **Neo4J**.

Finally, it will demonstrate that navigation within a knowledge graph can be a powerful tool for knowledge transfer and support in designing novice profile.

Keywords: Knowledge Management · KBE · Knowledge Graph · Ontology · Neo4J

1 Introduction

[Context] Knowledge management is essential for the competitiveness of companies operating in the international market [1]. It represents an essential advantage for the preservation of the intellectual heritage of the company in a context where the manufacturing sectors needs to produce faster and at lower cost.

Published by Springer Nature Switzerland AG 2024
C. Danjou et al. (Eds.): PLM 2023, IFIP AICT 701, pp. 309–322, 2024.
https://doi.org/10.1007/978-3-031-62578-7_29

According to Hawisa et al. [2], "Manufacturing companies must have a good knowledge of their products and processes to be competitive. This is increasingly important as products become more complex" [3].

In particular, **KBE** (**K**nowledge **B**ased **E**ngineering) is part of this approach to leverage the knowledge and the expertise of technical experts within the company by focusing on the management of engineering knowledge. Several **KBE** systems are able to exploit process and product engineering knowledge with the goal of reducing the time spent on repetitive tasks, increasing time for creativity and reducing the cost of product development [3].

In this competitive environment, manufacturing industries must produce faster and at lower cost. As noted by Kim et al. [4], access to appropriate information at the appropriate time is a crucial issue for companies. It enables the efficient operation of various stakeholder activities, as well as the acquisition of new knowledge that can create value.

[Problem] In the next few years, the manufacturing industry will face a shortage of skilled workers due to retirements. It is important to start implementing solutions to pre-serve the industry's intellectual heritage for the future. The purpose of this paper is to explore how the knowledge of these experts can be captured and utilized, and how a knowledge graph could be a suitable tool to achieve this goal.

[Proposal] In this paper, a new methodology for implementing **KBE** solutions is proposed. This methodology called **KARMEN** (**K**nowledge **A**ccess **R**equest for **M**anufacturing and **E**ngineering by **N**etwork graph) is based on an **FBS** (**F**unction, **B**ehaviour, **S**tructure) type ontology and on the use of **K**nowledge **G**raphs. As noted by Giro et al. [5], the **FBS** ontology provides a uniform framework for classifying processes and includes higher level semantics in their representation.

A use case for redesigning a mechanical part for metal additive manufacturing is presented. Besides, an experimental protocol is specified to capture the knowledge of business experts in a graph-oriented database built on **Neo4J** software [6]. This scientific article, investigates how can a **K**nowledge **G**raph be created and proposed to capture and make available complex knowledge from real engineering design activities [7]. The following research question is investigated:

- How can the tacit knowledge of design experts be captured within a Knowledge Graph?

To answer these research question, the authors of this paper, first define the de-sign activities, product lifecycle phase and trace the origins of **KBE**. They explore the use and application of knowledge graphs in engineering design. This allows introducing and explaining the proposed **KARMEN KBE** methodology through four main phases. Lastly, an experimentation is conducted to validate the methodology through a case study of additive manufacturing.

2 Related Work

In this section, a review of the related work on design activities and product lifecycle phase, **KBE** methodologies, **K**nowledge **G**raph construction is presented.

2.1 Design Activities and Product Lifecycle Phase

Design can be considered as a learning process, in which knowledge is collected, synthesized, and organized to achieve an outcome [7–9]. Another viewpoint of design according to **ISO 9000** organization [8], defines design as: "set of processes that transform requirements into specified characteristics or a specification of a product, process, or system." It is impossible to discuss product design without also discussing product lifecycle as they are closely interconnected.

The meaning of term 'lifecycle' generally indicates the whole set of phases, which could be recognised as independent stages to be passed followed performed by a product, from 'its cradle to its grave'. By adopting a well-known model, the product lifecycle can be defined by three main phases [9]:

- **B**eginning **O**f **L**ife (**BOL**).
- **M**iddle-**O**f-**L**ife (**MOL**)
- **E**nd-**O**f-**L**ife (**EOL**)

According to Terzi et al. [10], the concepts of product design, manufacturing, repair, and recycling have evolved significantly since the dawn of human civilization. They are now complex and require a significant amount of knowledge to be properly executed. Despite advances in the tools and methods to design and support products, the core principle remains unchanged: identify the customer needs and create a product that meets those needs.

According to Robinson's calculations, engineers spend over 55% of their work time acquiring or sharing knowledge, making the organization and structuring of knowledge a vital aspect of engineering practice [11]. These statistics demonstrate the importance of knowledge management within design activities. The following section focuses on tools for **KBE**.

2.2 KBE Methodologies

Various definitions of **KBE** can be found in the literature. For this paper, the selected definition is the one presented by La Rocca [3] as "Knowledge Based Engineering is a technology based on the use of dedicated software tools called **KBE** systems, which are able to capture and systematically reuse product and process engineering knowledge, with the final goal of reducing time and costs of product development by means of the following:

- Automating repetitive and non-creative design tasks.
- Supporting multidisciplinary design optimization at all the stages of the design process".

The core component of the system is the product model, which stores knowledge about the product and the process. Data from external databases are fed into the system. The input to the **KBE** system is usually according to the customer's specifications, and various outputs are produced during processing. The system software is object-oriented, allowing it to perform calculations on demand [12]. According to Verhagen [13] there are a number of **KBE** methods that support the development of **KBE** applications and systems. By far the best known of these is the **MOKA** (**M**ethodology and software tools **O**riented to **K**nowledge-Based Engineering **A**pplications) [14].

This method proposes the implementation of a **KBE** in 6 steps.

- Identify
- Justify
- Capture
- Formalize
- Package
- Activate

The aim of **MOKA** is to provide (i) a 20–25% reduction in the costs and time associated with KBE application development (ii) an efficient methodology to capture and formalize product and process knowledge and (iii) an **IT** tool that supports the capture, representation of knowledge.

The **MOKA** methodology utilizes both an informal and formal model. The informal model utilizes **ICARE** forms, which stands for **I**llustrations, **C**onstraints, **A**ctivities, **R**ules, and **E**ntities. These forms are used to break down and store pieces of knowledge. The formal model uses **MML** (**M**oka **M**odeling **L**anguage, a variation of **UML**) to organize and structure the elements of the informal **ICARE** model. It is an ontological approach that enables the translation of the informal model into a formal model.

However, the **MOKA** methodology encounters the following problems [15]:

- **MOKA** is product-oriented rather than process-oriented.
- **MOKA** focuses solely on supporting the knowledge engineer, not the end-user.
- **MOKA** is unable to account for the maintenance and reusability of knowledge

According to Camara et al. [16] it is essential today to propose new methodology of capitalization thanks to the help of AI tools. Artificial intelligence algorithms such as neural networks allow training models to process heterogeneous data and extract new knowledge. These new AI tool opportunities will facilitate knowledge capture and representation throughout the design lifecycle process.

2.3 Knowledge Graphs in Engineering Design

According to Sowa [17] a semantic network or net is a graph structure for representing knowledge in patterns of interconnected nodes and arcs. Sowa identifies six common types of semantic networks, each of which is described below:

- Definitional networks
- Assertional networks
- Implicative networks.
- Executable networks
- Learning networks.
- Hybrid networks.

In the study, the "Definitional networks" type will be used to represent the knowledge graph. Indeed, Huet et al. [18] claim that knowledge graphs break down silos and focus on interactions. It is proposed to use the semantic network to model, store, and represent all product design information and expert resource knowledge. This semantic network will constitute the knowledge graph.

The benefits of using a knowledge graph are numerous. Here are some examples:

- Visualization: A knowledge graph offers a visual representation of the relationships between different entities and concepts, making it easier to understand and explore knowledge.
- Search and navigation: Queries on the graph can extract relevant information and explore the relationships between different elements.
- Implicit knowledge capture: The relationships between entities in the graph can represent implicit knowledge, making it actionable.
- Collaboration and knowledge sharing: A knowledge graph facilitates collaboration and the sharing of knowledge within an organization.

Thus, a user will be able to use the knowledge graph to navigate and search for information. He will also be able to consult his dashboard to access the relevant **KPI**. Queries and data analysis algorithms will be used to enrich the knowledge graph.

2.4 Related Work Synthesis

The analysis of the related works emphasizes that there is a need and a strong interest in proposing new solutions for capturing, sharing and transfering product lifecycle design knowledge within the enterprise. The knowledge and experience of experienced design engineer is valuable and represents the intellectual heritage of the company. Today, companies lack tools to manage product-related knowledge. Product life cycle information is stored in different software specific to each department in the company. Therefore, it is crucial to keep all the knowledge of the experts, to share it and to pass on to the newly hired designers. In the next chapter of this article, a new method and solution for implementing a **KBE** System called **KARMEN** is proposed.

3 KARMEN KBE Proposal

In this section, the proposed **KARMEN** method for capturing, modelling, and sharing the knowledge of an experienced designer, in a knowledge graph is described. A specific use case on redesigning mechanical part for additive manufacturing serves to validate the proposed approach. The **KARMEN** method consists of 4 main phases: Find, Acquire, Model, Exchange. The functional modelling language **IDEF0**. [19], is used to elucidate the **KARMEN** main functions with a well-structured graphical component among boxes, arrows, rules, and diagrams. A box represents a function activity and describes what happens in the function, as shown in (Fig. 1).

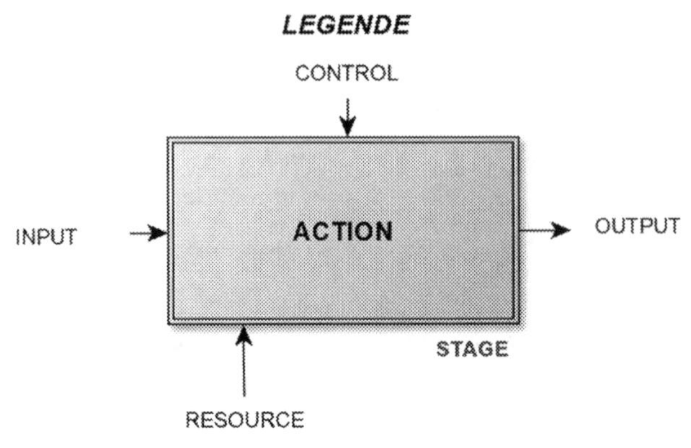

Fig. 1. IDEF0 diagram.

3.1 Additive Manufacturing and Use Case

The use case deals with additive manufacturing (Fig. 2). Indeed, it is about redesigning and optimizing a part for additive manufacturing. **A**dditive **M**anufacturing (**AM**) brings new design potential compared with traditional manufacturing [20]. Additive manufacturing is removing the limitations of traditional manufacturing methods, allowing designers to create almost any shape, enabling the use of fully optimized lightweight designs without compromising on performance. **D**esign **F**or **A**dditive **M**anufacturing (**DFAM**) is analogously defined as the design for manufacturability applied to **AM** [21]. It is a design approach that takes into account the characteristics and constraints of additive manufacturing, such as material limitations, mechanical properties, and manufacturing processes, in order to maximize the benefits of this technology. **DFAM** can help reduce costs, minimize waste, and improve the performance of parts produced through additive manufacturing. However, this manufacturing process requires expertise to fully exploit the potential of the additive manufacturing value chain. This expertise and knowledge is often known by only a few experts within the company.

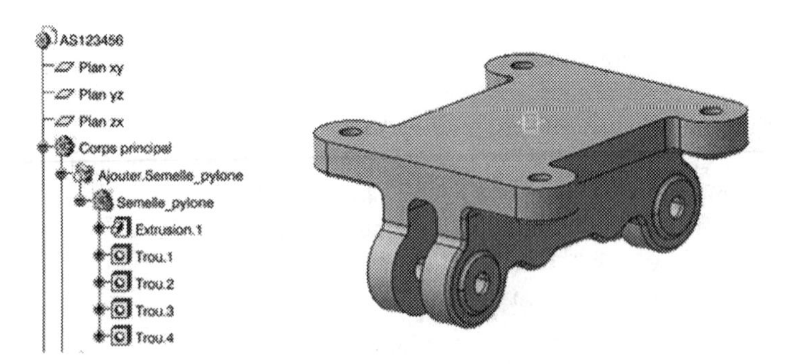

Fig. 2. Part and his CSG tree to redesign for additive manufacturing.

The diagrams below in (Fig. 3) presents the **KARMEN** methodology. All phases enable knowledge to be captured, structured, formalised and shared chronologically. In the following section, the four steps of the **KARMEN** methodology are described in detail.

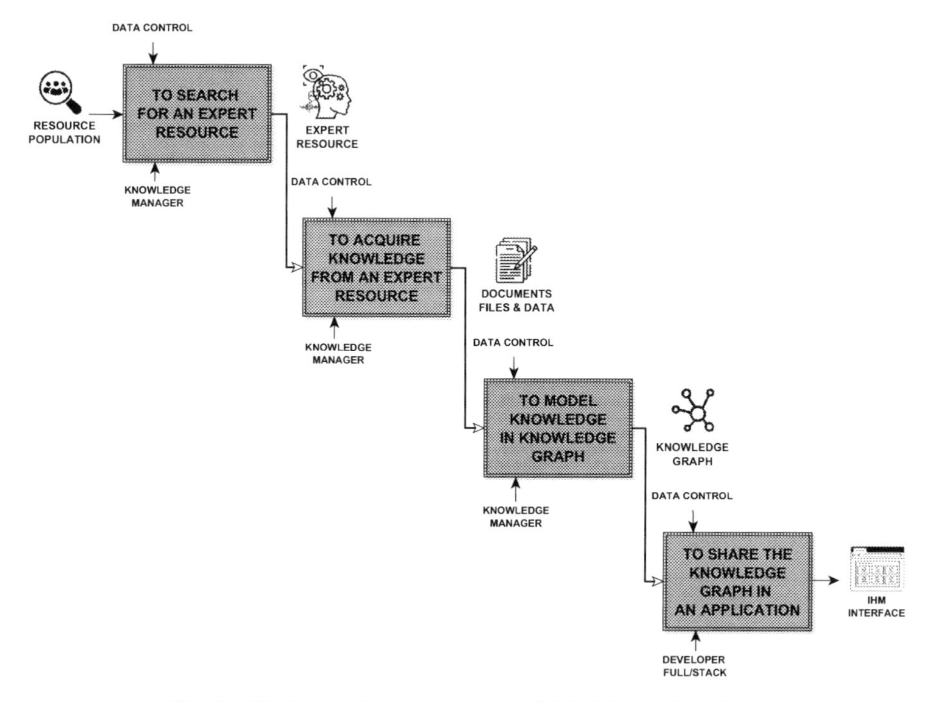

Fig. 3. FAME: The four phases of the KARMEN methodology.

3.2 [MATERIAL AND METHOD] Software Used and Method

The Table 1 shows all the software used in **KARMEN** methodology.

Table 1. Documents and software identified within the **KARMEN** methodology.

Software type	Software name	Description
Graph Data Platform	4.4.7	Graph Database
Graph Apps	Neo4j Bloom 2.6.1	Interactive exploration of graph data
Graph Apps	Neo4j - NeoDash 2.2.1	Dashboard Builder
CAD	CATIA V5 R 2013	Mechanical Design
Windows voice recorder	Tape recorder	Record the voice
Data science IDE	Jupyter Notebook/Python 3.9.7	Remote CATIA V5 with Python script
Py2neo library	Py2neo 2021.2.3	Remote Neo4J with Python script
Pycatia library	pycatia 0.5.7	To access the CATIA V5 Automation

3.3 [FIND] Expert Designer Identification for Knowledge Capture

This phase consists of finding an expert resource in a technical field. The **K**nowledge **M**anager searches and selects an expert in a database of internal or ex-ternal resources his or her organisation. This phase requires a current and updated database to identify the competent expert resource. The diagram, (Fig. 3), describes the scenario for selecting an expert resource. The **U**ser_**S**tory is as follows: As a **K**nowledge **M**anager, I want a search function to identify an expert resource and capture its knowledge.

3.4 [ACQUIRE] Information Gathering and Expert Knowledge

This is a key and important step in capturing the expert's knowledge in a technical field. This phase is called knowledge elicitation or knowledge capitalisation. It in-volves processes to capture and formalize expertise before its implementation in a system [22]. According to Bareiro et al., knowledge elicitation, is the process of obtaining knowledge from experts [23]. The **K**nowledge **M**anager conducts an interview with the expert to obtain information, knowledge and skill. The **K**nowledge **M**anager uses specific surveys. There are 4 types of surveys available to the **K**nowledge **M**anager:

- **ERS**- [Expert_**R**esource_**S**heet]
- **ICS**- [**I**dea and **C**oncept **S**heet]
- **PCS**- [**P**roduct **C**apitalization **S**heet]
- **PDS**- [**P**roduct **D**ata **S**heet]

The expert has to answer the questions of the various surveys. The **K**nowledge **M**anager writes the expert's answers on the sheets. The knowledge manager uses software to record voice and gaze simultaneously. The voice is recorded and converted to text. All the data and information collected in this way allow the expert's knowledge to be captured. The table (Table 2) describes the essential and necessary sheets required to conduct the interview. The **U**ser_**S**tory is as follows: As a knowledge manager, I want IT support and tools to capture the expert's knowledge in documents and files. Finally, the knowledge manager may ask the expert resource to design or redesign a part according to specific requirements. In this case, the expert will need to use a CAD software such as **CATIA V5** or 3**DEXPERIENCE**.

Table 2. Table of four sheets to capture the knowledge of expert.

File document	Description task
[Expert_**R**esource_**S**heet]	Capture the experience and skill using a scale of [0, 2, 4, 6, 8] on different items in the **AM** value chain
[**I**dea_and_**C**oncept_**S**heet]	Representing and illustrating the design concepts
[**P**roduct_**C**apitalization_**S**heet]	Product Capitalization Sheet
[**P**roduct_**D**ata_**S**heet]	Analyzing the Functions, Structure and Behaviors of the Product

3.5 [MODEL] Knowledge Graph Construction and Query Execution

This phases consists in gathering all the data collected in the "**Acquiring Knowledge**" phase (Fig. 3) to structure them in a **K**nowledge **G**raph. The data must be organized according to the **FBS** data model, **F**unction, **B**ehaviour, **S**tructure [5].

The **K**nowledge **M**anager has to collect and use all data and surveys to enrich the knowledge graph. The diagram (Fig. 3) describes the scenario for the knowledge manager to implement the **K**nowledge **G**raph. The **U**ser_**S**tory is as follows: As a **K**nowledge **M**anager, I want to retrieve the documents and survey to create a **K**nowledge **G**raph in the **Neo4J** application.

3.6 [EXCHANGE] Implementation of a Data Dashboard

This last step allows to exploit the **K**nowledge **G**raph and all the data collected in the previous steps of the **KARMEN** methodology. In this article, the **Neodash** software is used in the **Neo4j** environment. The **Neodash** tool enables the implementation of dashboards from knowledge graph data. The designer can then use this tool as a decision making tool. On this phase there are two user stories. The **U**ser_**S**tories are next: As a full/stack developer, I want to implement a **IHM** to facilitate novice designer tasks and foster knowledge sharing. As a novice designer I want a solution to assist me in my design choices and activities.

4 Results and Future Work

This section describes the experimental protocol in our study. It successively and exclusively presents the **ACQUIRE** and **MODEL** steps of **KARMEN** methodology. The **FIND** and **EXCHANGE** phases have not been performed.

4.1 [ACQUIRE] Information Gathering and Expert Knowledge

This phase consisted of interviewing the expert to gather information. Below are the details and description.

4.1.1 Participant

A specific expert in **AM** has been identified for the experimental protocol. This choice of expert aligns with the case study of redesigning a part for **A**dditive **M**anufacturing (**DFAM**).

4.1.2 Devices et Documents

A specific room was equipped with the necessary equipment to conduct the interview. The room was equipped with a laptop and the necessary software for knowledge capture. The three documents in table (Table 3) such as [Expert_**R**esource_**S**heet], [**I**dea and **C**oncept **S**heet], [**P**roduct **C**apitalization **S**heet] were used. The [**P**roduct **C**apitalization **S**heet] document was not used by the expert. Only the **K**nowledge **M**anager can use it to extract the structure of the **CATIA V5 CAD** model with a specific python script and generate the Structure part of the knowledge tree.

4.1.3 Acquisition Activities

The knowledge capture experience was conducted by me as a **K**nowledge **M**anager in a specific experimental room. The table below (Table 3) outlines all tasks of our experimental protocol with their execution times.

Table 3. Table of three sheets to capture the knowledge of expert.

File document	Nb questions	Interview time (mn)
[Expert_Resource_Sheet]	85	45
[Idea_and_Concept_Sheet]	1	5
[Product_Capitalization_Sheet]	5	10

4.2 [MODEL] Knowledge Graph Construction

The **MODEL** phase consisted of retrieving the documents and files collected in the **ACQUIRE** phase to complete and feed the **K**nowledge **G**raph.

4.2.1 Dataset and Query Execution

In order to prepare the data and feed the knowledge graph, the necessary documents were used and exploited: [**Expert_Resource_Sheet**], [**Idea and Concept Sheet**], [**Product_Capitalization_Sheet**], [**Product_Data_Sheet**]. These documents therefore allowed us to create the nodes and links in our **K**nowledge **G**raph. According to the FBS data model proposed by Gero et al. [5], a **K**nowledge **G**raph has been implemented.

There are also different data models inspired by the FBS model such as the **FBS-PPR** model. However, one has chosen to adapt the FBS model by adding the Resource node. According to Labrousse et al. [24] the **FBS-PPR** (**F**unction/**B**ehavior/**S**tructure-**P**rocess/**P**roduct/**R**esource) model consists of deploying the **FBS** model according to three views: the process view, the product view, and the resource view.

Here is the **K**nowledge **G**raph and its structure. There are ten nodes in the **K**nowledge **G**raph which are:

1) [**Product**] in grey colour is the central node of the **FBS-PPR** model.
2) [**Resource**] in orange colour represents the expert resource.
3) [**Process**] in red colour represents the manufacturing process (**AM**).
4) [**Technical Function**] in pink colour represents the technical function.
5) [**Physical Behaviour**] in green colour represents the expected physical behaviour of the product.
6) [**CAD_FILE**] in blue colour represents the product CAD file and links with other product (**CATPart** file). The "New Product" node represents the product redesigned (**DFAM**) by the expert resource.
7) [**Sheet**] in Purple colour represents and decomposes product geometry.
8) [**Field**] in Yellow colour represents the **AM** value chain.
9) [**Activity**] in Yellow colour represents the **AM** value chain.
10) [**Skill**] in Yellow colour represents the **AM** value chain.

The links between nodes are: **ADD_BODY;BEHAVIOUR_IS;FTn; FUNCTION_IS;HAS_ACTIVITY;HAS_BODY;HAS_REF;HAS_SHEET;HAS_SKILL; KNOWLEDGE_IN**(Value:[0,2,4,6,8]); **LINK_TO;PBn; PROCESS_IS; REDESIGN_BY; REDESIGN_IS; STRUCTURE_IS.**

The documents: [**Expert_Resource_Sheet**], [**Idea and Concept Sheet**], [**Product_Capitalization_Sheet**], [**Product_Data_Sheet**] are directly attached to their nodes: [**ERS**], [**ICS**], [**PCS**] and [**PDS**] (Fig. 4). When this **K**nowledge **G**raph is complete and fed with datasets including multiple expert resources and their sheets [**ERS**], [**ICS**], and [**PCS**], it will be possible to perform complex queries that will aid the designer. Novice designers will be able to search for parts similar to their design to identify design choices made by experts using powerful data analysis algorithms.

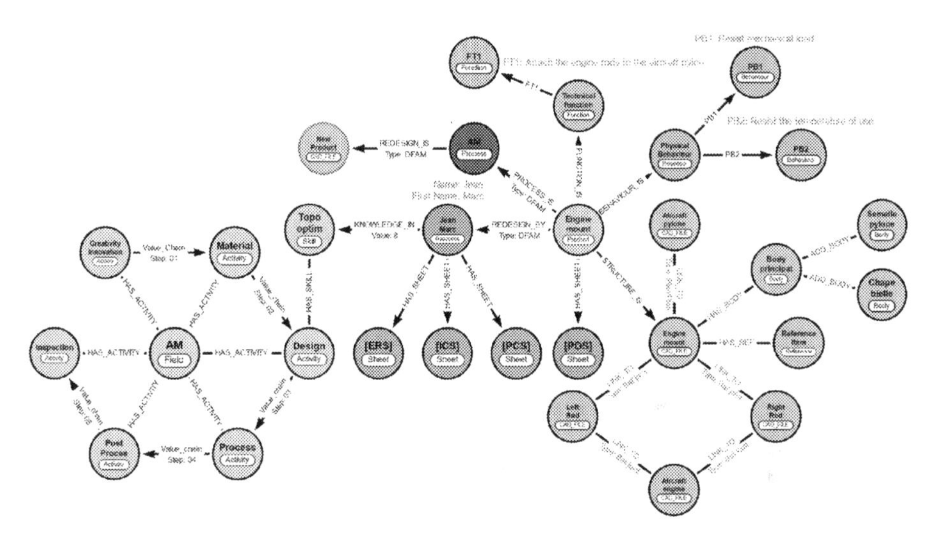

Fig. 4. Knowledge graph of the use case based on the FBS-PPR data model

4.2.2 Data Model and Graph Query

The proposed data model allowed to represent knowledge by structuring data with the data model approach based on the **FBS-PPR** model. The advantages of using a **K**nowledge **G**raph for design activities and knowledge transfer are:

- Clear and structural organization of information
- Easy navigation and information search
- Sharing of knowledge between employees
- Organizing specific mechanical design knowledge in a clear and structured format.
- Documenting efficient processes and practices for mechanical design, allowing for better understanding for new employees or future projects.

The **KARMEN KBE K**nowledge **G**raph allows to identify the skills of the expert resource within the entire Additive Manufacturing value chain. Thanks to **CYPHER** language queries, it is possible to write complex queries in a simple way. As an example, the query below allows to identify Jean's skills within the additive manufacturing value chain:

MATCH (r:Resource {Name: "Jean"})-[:KNOWLEDGE_IN]->(s:Skill) RETURN s)

Knowledge **G**raph makes it possible to capitalise on the design of the existing product and identify the mechanical links and interfaces with other products (For example: what is the structure of my tree in **CAD_FILE,** without opening my **CAD software**). Thanks to the **FBS-PPR** model and its ontology it is possible for a designer to search for information such as technical functions or specific behaviour on the product.

This allows her to make links and correlations with heterogeneous data [4].

Finally, the **K**nowledge **G**raph captures all knowledge and provides access to every step involved in the transformation from the initial product to the redesigned product.

5 Conclusion and Future Work

As a reminder, the research question was to answer the following: How can one capture the tacit knowledge of design experts within a **K**nowledge **G**raph?

The **KARMEN** methodology was proposed, consisting of four steps: **FAME** (**F**ind, **A**cquire, **M**odel, **E**xchange). This methodology was implemented within the graph-oriented database software **Neo4J**. The use of **Neo4J** offers a computer solution for implementing a structured **K**nowledge **G**raph with a precise data model in which we can perform powerful and complex queries to search for information. This data model is inspired by the **FBS-PPR** data model.

In future work, the proposed model can be enriched by adding other node data. Data such as audio and image were not exploited in this article. In the future, one could exploit artificial intelligence algorithms to extract knowledge and deduce relationships between data, such as inference. Finally, recent query tools such as **ChatGPT3** from **OpenAI** and its **API** will enable us to perform queries directly in natural language, assisting the designer and providing the right information to the right person at the right time.

In conclusion, it is important to acknowledge that the specific aspects related to capturing and representing tacit knowledge have not been addressed in this study. Future research could focus on integrating advanced techniques, such as the use of artificial intelligence algorithms and inference methods, to better handle tacit knowledge within the knowledge graph.

References

1. Furini, F., Rossoni, M., Colombo, G.: Knowledge based engineering and ontology engineering approaches for product development: methods and tools for design automation in industrial engineering. In: Systems, Design, and Complexity, Phoenix, vol. 11, p. V011T15A032 (2016). https://doi.org/10.1115/IMECE2016-67292
2. Hawisa, O.B.H., Tannock, J.: Knowledge management for manufacturing: the product and process database (2004). https://doi.org/10.1108/17410380410555826

3. Rocca, G.L.: Knowledge based engineering: between AI and CAD. Review of a language based technology to support engineering design. Adv. Eng. Inform. **26**(2), 159–179 (2012). https://doi.org/10.1016/j.aei.2012.02.002
4. Kim, L., Yahia, E., Segonds, F., Véron P., Mallet A.: i-Dataquest: a heterogeneous information retrieval tool using data graph for the manufacturing industry. Comput. Indust. **132**, 103527 (2021). https://doi.org/10.1016/j.compind.2021.103527
5. Gero, J.S., Kannengiesser, U.: A function–behavior–structure ontology of processes. AIEDAM **21**(4), 379–391 (2007). https://doi.org/10.1017/S0890060407000340
6. Miller, J.J.: Graph database applications and concepts with Neo4j (2013)
7. Goridkov, N., Rao, V., Cui, D., Grandi, D., Wang, Y., Goucher-Lambert, K.: Capturing designers' experiential knowledge in scalable representation systems: a case study of knowledge graphs for product teardowns. In: 34th International Conference on Design Theory and Methodology (DTM), St. Louis, vol. 6, p. V006T06A032 (2022). https://doi.org/10.1115/DETC2022-90697
8. Hoyle, D.: ISO 9000 Quality Systems Handbook, 4th edn. Butterworth-Heinemann, Oxford; Boston (2001)
9. Kiritsis, D., Bufardi, A., Xirouchakis, P.: Research issues on product lifecycle management and information tracking using smart embedded systems. Adv. Eng. Inf. **17** (3–4), 189–202 (2003). https://doi.org/10.1016/S1474-0346(04)00018-7
10. Terzi, S., Bouras, A., Dutta, D., Garetti, M., Kiritsis, D.: Product lifecycle management – from its history to its new role. IJPLM **4**(4), 360 (2010). https://doi.org/10.1504/IJPLM.2010.036489
11. Robinson, M.A.: An empirical analysis of engineers' information behaviors. J. Am. Soc. Inf. Sci. **61**(4), 640–658 (2010). https://doi.org/10.1002/asi.21290
12. Sandberg, M.: Knowledge Based Engineering - In Product Development, p. 16 (2003)
13. Verhagen, W.J.C., Bermell-Garcia, P., van Dijk, R.E.C., Curran, R.: A critical review of knowledge-based engineering: an identification of research challenges. Adv. Eng. Inform. **26**(1), 5–15 (2012). https://doi.org/10.1016/j.aei.2011.06.004
14. Stokes, M.: Managing Engineering Knowledge: MOKA: Methodology for Knowledge Based Engineering Applications. Professional Engineering Publication, London (2001)
15. Jayakiran, R., Sridhar, C.N.V., Pandu, R.: Knowledge Based Engineering: Notion, Approaches and Future Trends, p. 18. https://doi.org/10.5923/j.ajis.20150501.01
16. Camara, J.R., Véron, P., Yahia, E., Mallet, A., Deguilhem, B., Segonds, F.: Knowledge based engineering: systematic review and opportunities. Paper Presented at the CONFERE 2022, Bâle (2022)
17. Semantic Networks John F. Sowa. Semantic Networks. 1992. [En ligne]. Disponible sur: https://re-dock.org/wp-content/uploads/2012/12/semantic-net-sowa.pdf
18. Huet, A., Pinquie, R., Veron, P., Segonds, F., Fau, V.: Knowledge graph of design rules for a context-aware cognitive design assistant. In: Nyffenegger, F., Ríos, J., Rivest, L., Bouras, A. (eds.) Product Lifecycle Management Enabling Smart X, vol. 594, pp. 334–344. Springer, Cham (2020). https://doi.org/10.1007/978-3-030-62807-9_27
19. Presley, A., Liles, D.H.: The use of idef0 for the design and specification of methodologies (1995)
20. Lang, A., et al.: Augmented design with additive manufacturing methodology: tangible object-based method to enhance creativity in design for additive manufacturing. In: 3D Printing and Additive Manufacturing, vol. 8, no. 5, pp. 281–292 (2021). https://doi.org/10.1089/3dp.2020.0286
21. Lettori, J., Raffaeli, R., Peruzzini, M., Schmidt, J., Pellicciari, M.: Additive manufacturing adoption in product design: an overview from literature and industry. Procedia Manuf. **51**, 655–662 (2020). https://doi.org/10.1016/j.promfg.2020.10.092

22. Ammar-Khodja, S., Perry, N., Bernard, A.: Processing knowledge to support knowledge-based engineering systems specification. Concurr. Eng. **16**(1), 89–101 (2008). https://doi.org/10.1177/1063293X07084642
23. Barreiro, J., Martinez, S., Cuesta, E., Alvarez, B.: Conceptual principles and ontology for a KBE implementation in inspection planning. IJMMS **3**(5/6), 451 (2010). https://doi.org/10.1504/IJMMS.2010.036069
24. Labrousse, M., Perry, N., Bernard, A.: Modèle FBS-PPR: Des Objets d'entreprise a la Gestion Dynamique des Connaissances Industrielles, p. 19 (2010). https://doi.org/10.48550/arXiv.1011.6033

Author Index

Published by Springer Nature Switzerland AG 2024
C. Danjou et al. (Eds.): PLM 2023, IFIP AICT 701, pp. 323–325, 2024.
https://doi.org/10.1007/978-3-031-62578-7

Printed in the United States
by Baker & Taylor Publisher Services